AF540314

ELEMENTS OF HUMAN PALÆONTOLOGY

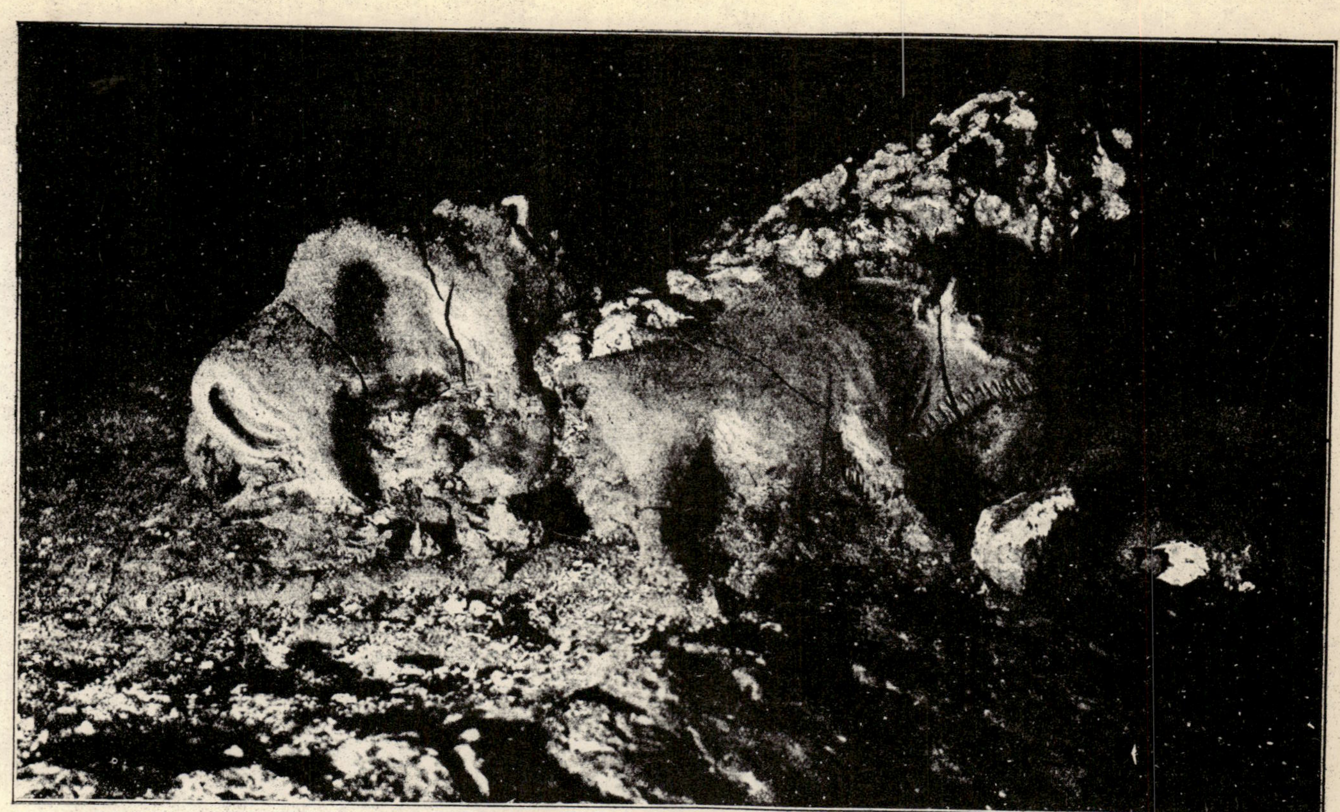

Bisons modelled in clay by Men of the Reindeer Age; in the Cave of Tuc d'Audoubert, Ariège, Southern France.
About one-ninth natural size. (Photo by M. Max Bégouen.)

Frontispiece (*see p.* 257).

ELEMENTS OF HUMAN PALAEONTOLOGY

BY

MARCELLIN BOULE

PROFESSOR IN THE MUSÉUM NATIONAL D'HISTOIRE NATURELLE, PARIS
DIRECTOR OF L'INSTITUT DE PALÉONTOLOGIE HUMAINE

TRANSLATED FROM THE FRENCH, WITH AN INTRODUCTION, BY

JESSIE ELLIOT RITCHIE

AND

JAMES RITCHIE, M.A., D.Sc., F.R.S.E.

Keeper of the Natural History Department, Royal Scottish Museum

WITH 250 ILLUSTRATIONS

1993

DISCOVERY PUBLISHING HOUSE

NEW DELHI - 110 002

ISBN 81-7141-207-6

Reprinted 1993

PUBLISHED BY :
DISCOVERY PUBLISHING HOUSE
4594/9, DARYA GANJ NEW DELHI - 110002

PRINTED AT :
EFFICIENT OFFSET PRINTERS
215, SHAHZADA BAGH
DELHI - 110035

TRANSLATORS' INTRODUCTION

France, by reason of her geographical position and her partial escape from the glaciers of the Great Ice Age, that swept the cumbered land surfaces of more northern countries, has treasured in her soil and caves many traces of the men of long past days. By a succession of fortunate chances, but still more by a series of deliberate and skilful excavations, a large number of the relics have been laid bare, and French scientists, by their description and interpretation of these, have taken a predominant share in piecing together the striking, though still incomplete, tale of the passage of the men of the Old Stone Age, an accomplishment for which the world of science must ever remain deeply in their debt.

It is fitting, therefore, that France should produce the most comprehensive account of the history of prehistoric men, and it is fortunate that the task should have fallen to the hands of so competent a scientist as Professor Marcellin Boule, whose wide knowledge of the various branches of prehistoric lore has stamped this work with a character of its own. For here are woven together strands gathered from the fields of zoology, palæontology and geology, of anthropology and archæology, to form a close fabric in which a pattern may faintly be discerned, the pattern of the slow evolution of mankind.

To English-speaking scientists Professor Boule and his works are well known, but a word may be said regarding them and the development of his scientific point of view by way of introduction to the general reader. Some forty years ago Boule began his scientific career under the aegis of Emile Cartailhac, whose outstanding contributions to the study of prehistoric Man placed him in the direct succession of the great French prehistorians of the first half of the nineteenth century. The influence of this great worker once and for all gave a bias to the mind of the young investigator; but already he had almost unconsciously been laying a broader basis for the study to which he was subsequently to devote himself. He has confessed that as a student he fell under the sway of a distinguished group of English scientists, learning geology from the works of Lyell, prehistory from those of Lubbock and John Evans, and gaining insight at once into comparative anatomy and method from the writings of Huxley.

Professor Marcellin Boule.

So much promise was shown by Boule's early work that he was appointed assistant to the great palæontologist Albert Gaudry, Professor of Palæontology in the National Natural History Museum of France, whose researches on fossil animals, particularly of Quaternary times, and whose palæontological generalizations in their turn contributed to the broad foundations upon which Boule was subsequently to build his story of Fossil Men. In 1903 our author was called upon to succeed Gaudry in the Chair of Palæontology, and in addition, on the creation, through the generosity of Albert I., Prince of Monaco, of the "Institut de Paléontologie humaine" in Paris in 1914, he was appointed Director of the first establishment devoted solely to the investigation of the history of prehistoric men.

His published researches during these years have covered a wide field, for Boule seems to be equally at home in the region of pure geology, as witnessed by his works on the central massif and the volcanoes of France, and on the geology of Velay, Auvergne, and Madagascar; in the field of palæontology, as shown by his researches in Madagascar, and his monographs on many Quaternary mammals in this country and in America, and on the Grimaldi caves; in general archæology, as the reports of his excavations of various prehistoric sites prove; and in human palæontology, as revealed by his masterly studies of the Fossil Man of La Chapelle-aux-Saints.

Outwith his official duties Boule has taken a prominent part in the advancement of the science of man's history. His editorship for some thirty years of the journal *L'Anthropologie* is reflected in the indispensable treasure-house which its pages have become for students of the subject the world over. The merits of his long life-work have been generally recognized by scientists in this country, where he has been elected a Foreign Member of the Geological Society of London, a Corresponding Member of the Zoological Society of London, an Honorary Foreign Fellow of the Royal Society of Edinburgh, and an Honorary Member of the Anthropological Institute of Great Britain, to which, on receiving the Huxley Medal, he delivered the Huxley Memorial Lecture in 1922.

How have these varied activities and achievements been reflected in Boule's scientific point of view, and what influence have they had upon his treatment of the story of prehistoric men? At the outset of his career his outlook was largely geological and palæontological; he was interested generally in the relations of the materials of the earth's crust, and in particular in their stratigraphical order, that is their time relations, which revealed the comings and goings of the animals of former ages. Thus he laid at the basis of his method of interpretation a sound knowledge of the geological successions in which remains of Fossil Men were likely to be found, and of the various faunas of extinct and migrated animals with which they were likely to be associated; and this is the first essential to a correct interpretation of the story of mankind. With this knowledge he was also able to picture with accuracy and in detail the various environments in which the different prehistoric men passed their existence; prehistoric topographies, climates, vegetations and animal life assumed a greater reality. His *Essai de Paléontologie stratigraphique de l'Homme*, published in 1888, was a notable contribution to the temporal development of Man.

But the story was yet far from complete. Boule's studies on the comparative morphology of animals readily led to a balanced appreciation of the structural differences of the various types of prehistoric man himself, and value was added to his comparisons by his knowledge of modern anthropology. A last link was required to complete the chain: discoveries of fossil bones, correctly interpreted, might indicate the physical evolution of Man, but they shed no light on his habits and manner of life. Boule, therefore, set himself to examine the evidences revealed by archæology; and in interpreting these in relation to his physical data he has been enabled to invest with a breath of life the dry bones of the geological succession.

In his *Fossil Men* he has focussed these many varied shafts of light upon the problem of Man's early history, so that from a mass of heterogeneous detail we see emerging a connected story of the development of Mankind upon the earth, a story still very imperfect, but apparently embodying the essentials of his physical and intellectual evolution. These remarks on Professor Boule's work may be concluded with a quotation from a letter by his old master, the late Professor Emile Cartailhac, to an American correspondent of the present writer, in which he refers to this volume as "un magistral ouvrage sur '*les hommes fossiles, éléments de paléontologie humaine*' très supérieur à tout ce qu'il y avait sur cette 'question suprême' dans l'ancien ou le nouveau monde."

While France has taken a first place in the advancement of the study of prehistoric man, it would be a mistake to suppose that Great Britain has lagged behind in making her contributions to the subject, and this Professor Boule freely acknowledges in the following pages.

It is true that remains of Palæolithic Man have been less frequently found here than on the continent, but indispensable additions have been made by the geological work of such as Lyell and Prestwich in former days, and in more recent times, particularly by the glacial geology of James Geikie and W. B. Wright and the glacial botany of Clement Reid; by the quaternary palæontology of such as Falconer, Boyd Dawkins, and Newton, and the pioneer explorations of Buckland in Kirkdale Quarry, and of M'Enery and Pengelly in Kent's Cavern, where, about the time of the researches of Tournal and Schmerling on the continent, M'Enery for the first time discovered dressed flints in intimate proximity with the bones of extinct animals, and declared them to be coeval—to the amusement of his contemporaries; by the archæological researches of John Evans, Lord Avebury, Macalister and many more; and by studies bearing more directly on prehistoric Man himself by such as Turner, Sollas, Smith Woodward, Keith and Elliot Smith. A noteworthy development of recent years has been the discovery and exploration in Scotland of various sites yielding clear traces of Azilian culture, and a continuation of this work in a land which so far has yielded no trace of Palæolithic Man promises to aid materially in bridging the gradually narrowing gulf which still separates the culture of the Old Stone Age from the dawn of Neolithic times.

Yet here we would echo for our own country Professor Boule's more general appeal for further investigation. We know that men of different palæolithic cultures inhabited Britain, but the connected history of these and of their immediate successors is still a play enacted upon a darkened stage. Fortunate chance discoveries, such as that which gave a first hint of the presence of Piltdown Man, may focus a brilliant light on an isolated spot of the darkened stage; and because of this each chance discovery should be reported at the finding to experts who may probe the possibilities to their utmost limit. But a general illumination of the prehistoric play, a consecutive story of the comings and goings of the forerunners of Modern Man in Britain, is likely to be achieved only by carefully planned and skilfully executed excavations, carried out with a view to deciphering the chronological succession of human settlements in sites inhabited by the peoples of former days.

This translation was already in page form when the revised

and enlarged second French edition made its appearance in the summer of the present year. While, therefore, it was impossible to amalgamate in the text all the many changes necessitated by the constant progress of discovery, the translation has been brought up to date by the inclusion, either in the text or in the Appendix, of all the fresh illustrations and textual additions and alterations of the French second edition. A few notes, indicated by "square brackets" [ ], have been added by the translators for the guidance of English readers. Throughout, the translators have had the inestimable assistance of the advice of Professor Boule, who read a proof of the work and to whom they wish here to express their gratitude.

J. R.

EDINBURGH, *October* 1923.

AUTHOR'S PREFACE TO FIRST EDITION

Barely a century and a half have elapsed since the prime question of the origin of Man was raised from the regions of dreams and fiction to the domain of science. Zoology and Comparative Anatomy first sought to solve it by methods soon found to be inadequate. Later, Palæontology became established as a separate science, having as its object the retracing of the history of beings of former ages, and since that time the highest hopes have been raised that this new science would reconstruct for us our own history.

During the second half of the nineteenth century, after the ever memorable discoveries of Boucher de Perthes in the ancient alluvials of the Somme, and of Edouard Lartet in the caves of Périgord, research in Human Palæontology has been consistently prosecuted. The advances made in the last twenty years are so striking that the "problem of Fossil Man" has become a universal subject of discussion among the general public, even more so perhaps than in professional scientific circles. From the very moment of its announcement each new discovery relating to our distant ancestors becomes the subject of numerous articles in the daily press or popular reviews; the substance of the articles could be easily criticized, but they bear clear witness to the lively and justifiable curiosity of the people. From all quarters comes a demand for a book which will give an account at once detailed and synthetic of our present-day knowledge of Fossil Men.

In France there are already certain valuable works of this nature, such as Hamy's *Précis de Paléontologie humaine*, G. de Mortillet's *Le Préhistorique*, E. Cartailhac's *La France préhistorique*, S. Reinach's *Catalogues du Musée de Saint-Germain*, Déchelette's *Manuel d'Archéologie préhistorique*, and the like. But these works are either too old, or else they present only the archæological aspect of Human Palæontology.

In foreign lands, where interest in all that touches upon prehistoric studies is no less keen than in France, there have appeared many popular works in which the items of text and illustration have been largely drawn from French publications. Some of these have been written by pure geologists, some by archæologists obviously unacquainted with biological problems, and others, by medical men or anthropologists, are too exclusively anatomical. Each has its merits, but none deals with the subject in all its fullness, that is to say in all its manifold aspects.

In writing in my turn this work on the *Elements of Human Palæontology*, I make no claim to more perfect achievement, but I have aimed at achieving something different. For close on forty years, from the time when I took up the science under the fond guidance of my very dear friend, Emile Cartailhac, I have never ceased to be interested in the natural history of Man. The variety of my researches, and the editing for twenty-five years of the journal *L'Anthropologie*, have perhaps placed me in a less unfavourable position than more strict specialists for gathering to a point the facts now garnered by the various branches of the science.

A few years ago, after having described at length the now famous skeleton from La Chapelle-aux-Saints, I sought to focus the present state of our knowledge of Human Palæontology. The copies of the memoir containing this attempt, of which too few were published, were soon exhausted; and besides it included a mass of technical evidence not easily grasped by readers insufficiently versed in the subject.

On several occasions I have discussed Fossil Men in the course of my Museum lectures; but such a method of diffusing knowledge is very limited, and my audiences and friends have urged me to publish these lectures in a form readily accessible to all naturalists and even to all educated persons who may wish to peer into the mysteries of the past. Hence the origin of this book, which attempts to sum up the main achievements of a science still young, in the establishment and development of which France has taken a preponderant part.

* * *

Fossil Man is revealed to us by two classes of relics, by two sorts of evidence. The first consists in the presence, in the heart of geological layers, of bones contemporary with these layers and more or less fossilized, like the bones of extinct animals with which they are often mingled. The second comprises various objects bearing traces of deliberate handiwork, that is to say, the products of human industry.

The latter evidences are much the more numerous, since they are most resistant to the various agents of destruction. They are to be met with in innumerable localities in almost all parts of the world. They provided the first proofs of the high geological antiquity of Man, and thereafter served as a basis for the establishment of a classification of prehistoric times. Above all, they reflect the intellectual and moral outlook of the oldest communities of Mankind.

The bone evidences, more friable and less easily preserved, are much more rare. Yet it is the scientific study of these alone that, bringing to our ken the chief anatomical characters of our most remote ancestors, is capable of throwing some light on their origin, on their zoological descent and their physical evolution, that is on the genealogical history of human beings or Hominians, the highest group of the Primates.

Although these two kinds of information both bear on Human Palæontology, although they should co-operate in establishing a History of Fossil Man, and are consequently inseparable, they belong to two somewhat different schools: for one is more zoological in nature, the other more archæological. With the second of these two points of view I shall deal much more briefly than with the first.

* * *

I have endeavoured to write this work with complete detachment of mind, confining myself exclusively to scientific ground. In my anxiety to avoid attributing conclusive value to any but positive facts, I have not hesitated to strew my text with more marks of interrogation than of affirmation; and this I have done in the belief that thus I may best advance science, and at the same time show most consideration for my readers.

Nevertheless, I am quite aware of my temerity. It is probable, indeed it is to be hoped, that almost as soon as my book is published, it may be already behind the times. Palæontological researches multiply on every side; new evidences will soon be brought to light, and we may expect them to be such that, if they do not controvert, they will at least modify our present theories, and present under unforeseen aspects the problems which engage our attention.

It cannot well be otherwise, considering the extreme poverty of our evidences. Our researches have as yet been directed upon but tiny portions of the habitable globe, and the European region, the only one which we are beginning to know, cannot be regarded as a centre of appearance or of dispersal. It is but an appanage of the Eurasiatic Continent, which through all time has been a great laboratory of life. In this appanage, this blind-alley, the history of the early human races cannot present an aspect of continuous and regular evolution: it is rather made up of intermittent contributions brought by successive waves of far-distant origin, from vast Asiatic and African territories of which our knowledge is still very scanty and vague.

The attempt I make is therefore only a provisional essay at focussing the subject. I am far from regarding it as a true *History* of Fossil Man: the writing of such a history will be possible only in the distant future. This work, which I have sought to render concrete by means of accurate and abundant illustrations, will none the less serve a useful purpose of its own. While satisfying so far as possible the justifiable curiosity of the general public as to the present aspect of the problem of our origin, it will, I hope, be of service to students and professional workers. It is with the second class of readers in view that I have given a large choice of bibliographical references, in order to aid more advanced study and to permit of references to the sources of information. I may be excused for particularly recommending the thirty volumes of *L'Anthropologie*.

To me it is a source of much gratification that this work should appear at the same time as the official inauguration of an establishment which forms the finest instrument for research till now available for prehistorians, the *Institut de Paléontologie humaine*, founded in Paris by a great patron of science, H.S.H. Prince Albert I. of Monaco.

MARCELLIN BOULE.

1921.

CONTENTS

CHAPTER VIII

THE MEN OF THE REINDEER AGE

CHAPTER IX

FROM FOSSIL MEN TO LIVING MEN

CHAPTER X

FOSSIL MAN BEYOND EUROPE

CHAPTER XI

GENERAL CONCLUSIONS

LIST OF ILLUSTRATIONS

CHAPTER I

HISTORICAL SUMMARY

CHAPTER II

TIME RELATIONS OF EARLY MAN

CHAPTER III

LIVING PRIMATES AND FOSSIL MONKEYS

CHAPTER IV

PITHECANTHROPUS—THE "APE-MAN"

CHAPTER V

THE PROBLEM OF TERTIARY MAN

CHAPTER VI

MEN OF THE CHELLEAN OR LOWER PLEISTOCENE AGE

CHAPTER VII

NEANDERTHAL MAN

CHAPTER VIII

THE MEN OF THE REINDEER AGE

CHAPTER IX

FROM FOSSIL MEN TO LIVING MEN

CHAPTER X

FOSSIL MEN BEYOND EUROPE

CHAPTER XI

GENERAL CONCLUSIONS

APPENDIX

ELEMENTS OF HUMAN PALÆONTOLOGY

CHAPTER I

HISTORICAL SUMMARY

By virtue of a supreme law of life, Mankind as a whole has had to pass through the phases of intellectual and physical evolution which to-day characterize the development of each individual in the human mass. In the beginning, the child is lulled by tales or songs of marvels: poetry is his first instructress. Later, his faculties of observation and reason awaken: truth compels him, and poetry is superseded by science.

So, regarding the "supreme question" of its origin, Mankind in its infancy had at first no source of information other than fairy tales, legends and stories of the miraculous. Then human intelligence developed; in certain bright spirits genius was made manifest; next calm observation, freed from all preconceptions, played its part; and finally, but only in later centuries when the reign of science began, there dawned some rays of Truth.

Our knowledge of Man's existence on the earth in prehistoric times is a conquest of modern science.

First Phase: from Ancient Times to the Renaissance. Simple imaginary Views.

Neither in ancient times nor in the Middle Ages does there seem to have been expressed any but imaginary conceptions of the origin of Mankind. In the Greek poets and philosophers of the pre-Christian era, vague references occur bearing on the low estate of the first Men. And so also in the Latin poets; everyone is familiar with the oft-quoted verses of Lucretius:—

Arma antiqua manus, ungues dentesque fuerunt,
Et lapides, et item sylvarum fragmina rami,
Et flammæ atque ignis postquam sunt cognita primum
Posterius ferri vis est, ærisque reperta.
Sed prior æris erat, quam ferri, cognitus usus.[1]

Similar views were expressed by Horace, Pliny, Strabo, Diodorus and others. Probably all these notions were purely intuitive, but it may be that they owed something to the persistence of ancient traditions. In any case they do not seem to have been based on true interpretations of relics of former

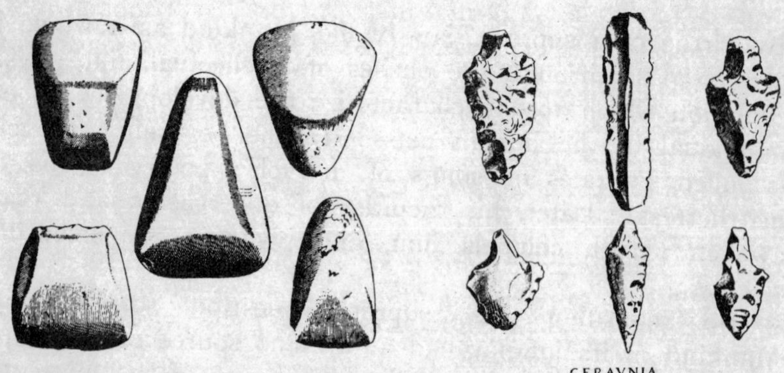

FIG. 1.—Thunder-bolts, figured by Mercati in his *Metallotheca*.

days, for even though stone axes and weapons, called *ceraunia* (Greek, χεραυνός = thunder), were already well known, their origin or real significance was unrecognized. They were regarded as produced or launched by lightning, and extraordinary powers were attributed to them (Fig. 1).

Such primitive ideas became widespread: with slight variations they have persisted to our own day in the popular superstitions of almost every country.[2]

[[1] The passage is thus rendered in English by Creech (1714):—
"And RAGE not furnish'd yet with SWORD nor DART;
With FISTS, or BOUGHS, or STONES the Warriours fought;
These were the only WEAPONS Nature taught:
But when FLAMES burnt the TREES, and scorch'd the Ground,
Then BRASS appeared, and IRON fit to wound.
BRASS first was us'd."]

[2] Cartailhac, É., *L'âge de pierre dans les souvenirs et superstitions populaires* (Paris, 1878).

And yet it is to a considered study of these ancient objects, it is to archæology, that we owe the first positive concrete facts of the great antiquity of Mankind.[1]

Second Phase: from the Sixteenth to the Eighteenth Century.

In regaining contact with Nature, lost since the time of the ancient Greeks, the scientific spirit reawoke with the Renaissance. Two great artists, Leonardo da Vinci and Bernard Palissy, propounded correct views regarding the nature of fossils. Yet although various authors, Agricola (1558), Gesner (1565) and others, described or figured polished stone axes and stone arrow-heads, they regarded them simply as curiosities. They still considered these objects, together with so many other "fossils," as sports of nature, of which they gave more or less quaint explanations.

FIG. 2.—Portrait of Mercati. Reduced from an engraving in the *Metallotheca*.

Mercati discovers the True Nature of the so-called "Thunder-bolts."

At the end of the sixteenth century, Michael Mercati, whose writings were not published till 1717, more than a century after his death (1593), discovered the true nature of the so-called thunder-bolts or *ceraunia*. "Most men," he says, "believe that *ceraunia* are produced by lightning. Those who study history consider that they have been broken off from very hard flints by a violent blow, in the days before iron was employed for the follies of war; for the earliest men had only splinters of flint for knives." And in this connection he quotes the verses of Lucretius.[2]

[1] See for the whole of the first part of this history: Hamy, E. T., *Précis de Paléontologie humaine* (Paris, 1870). Id., "Matériaux pour servir à l'Histoire de l'archéologie préhistorique" (*Revue archéologique*, 1906). Evans, Sir John, *Ancient Stone Implements*, 2nd ed. (London, 1897). Cartailhac, É., *La France préhistorique* (Paris, 1889). Reinach, S., *Description raisonée du Musée de Saint-Germain-en-Laye*. I. (Paris, 1889.) [Macalister, R. A. S., *A Text-book of European Archæology*, vol. i., *The Palæolithic Period* (Cambridge, 1921).]

[2] Mercati, M., *Metallotheca, opus posthumum*, Rome, 1717, p. 243. See on this subject, Vayson, "Les précurseurs de la préhistoire" (*L'Anthropologie*, xxxi., p. 357).

In 1636 Boetius de Boot, regardless of being "dubbed a fool," rejected the commonly accepted ideas; but he believed that he was dealing with implements of iron transformed into stone through process of time.

Various writers, however, Aldrovandus in 1648, Hassus in 1714, A. de Jussieu in 1723, Lafitau the Jesuit in 1724, and Mahudel in 1730, compared the old stone weapons of our countries with the stone weapons of living native tribes, notably the American Indians, and so initiated an excellent working method based on comparative ethnography, giving at the same time "the finishing blow to the erroneous beliefs regarding ceraunia."

Sequence of Prehistoric Ages.

In 1750, Eccard, after investigating old German burials, established a succession of different prehistoric ages; and in 1758, a learned magistrate, Goguet, published a remarkable work on the "Origin of Laws," in which he declared that a Stone Age had been followed by an Age of Copper and of Bronze, and then by an Iron Age. Later this classification was firmly established and developed by the Danish archæologists Thomsen and Worsaae.

Thus the hesitating science of the eighteenth century arrived at the same ideas as the ancient poets or philosophers; but these ideas were now based on the observation of material evidences. Nevertheless, although it was recognized that the historic civilizations had been preceded by uncivilized or crudely barbarous periods, the great antiquity of these primitive times was not suspected. The theories had first of all to be accommodated to the demands of biblical chronology. The new idea of Mankind beginning in a state of primitive destitution seemed incompatible with the idea of the physical and moral perfection of the terrestrial paradise. Hence arose these heated discussions, fierce battles of words, which to-day seem so naïve or ridiculous, especially when we consider, as M. Cartailhac has pointed out, that the most widely differing opinions regarding the date of the creation of Man did not diverge by more than 1500 years.

Buffon, who first suspected the immense duration of

geological time, although he sought to interpret the Scriptures "soundly," was familiar with the stones which "were believed to have fallen from the clouds and to have been formed by thunder, but which nevertheless are really the first relics of the art of Man in a state of nature." Yet, to his mind, the epoch of Man was only the seventh and last of his "Epochs of Nature," much later than the fifth epoch, characterized by the remains of the Elephant, the Rhinoceros and the Hippopotamus, which he found in the superficial soil.[1]

Third Phase: Man dates from a Geological Period prior to the Present Period.

With the nineteenth century Natural History sprang into sudden life and vigour. To the new sciences of Geology and Palæontology it fell to throw light on the great antiquity of Man. Up to that time it had been a question only of objects dating no further back than modern geological times, objects which to-day are known as *Neolithic*. But now attention had to be turned to stone implements much more ancient, found in the very heart of deposits which dated from a geological period preceding the *modern* period, and which were distinguished by the presence of remains of animals no longer existing to-day.

As early as 1715, Conyers, a pharmacist and antiquary of London, had found near that city, in the gravels of a former river and near the skeleton of an elephant, a flint worked after the manner now known as Acheulean. Bagford, a friend of Conyers, made the suggestion that the flint was a weapon used by a Briton to kill the elephant, brought over by the Romans in the reign of the Emperor Claudius!

John Frere.

In 1797, another Englishman, John Frere, made a similar discovery at Hoxne in Suffolk. He collected some dressed flints at a depth of four metres in a deposit containing bones of large extinct animals (Fig. 3). He was able to give his find a much more correct interpretation than Bagford, stating that it must certainly belong to a "very distant period, much more remote in time than the modern world." This observation, so full of judgment, almost of genius, passed unnoticed. It was brought to light again by

[1] Buffon, *Époques de la Nature* (Paris, 1778).

John Evans only after the memorable conflicts of Boucher de Perthes, of whom John Frere must be considered the forerunner.[1]

Nineteenth Century. Cuvier.

In 1823, a French geologist, Ami Boué, presented Cuvier with a human skeleton, exhumed near Lahr on the banks of the Rhine, from an ancient mud or loess containing also remains of extinct animals. This discovery was set aside by the famous palæontologist. "All the evidence leads us to believe," he said,[2] "that the human species did not exist at all in the countries where the fossil bones were found, at the period of the upheavals which buried them."

FIG. 3.—One of the flint weapons found by Frere at Hoxne, Suffolk. Half natural size. Facsimile of a figure published by Sir John Lubbock.

For these words the great naturalist has often been reproached; but it is easy to excuse him.[3] Cuvier had, as a matter of fact, examined all the evidences sent from various parts as remains of "antediluvian Man." Some were really human bones, such as those from Canstadt, from different German caves, from Lahr and from Guadeloupe; but no accurate observation, no decisive geological evidence justified the assertion of their high antiquity.

As for the other remains, Cuvier had recognized that the bones from Belgium were those of elephants; from Cerigo,

[1] Evans, Sir John, *loc. cit.*, p. 573. John Frere's account is to be found in *Archæologia*, vol. xiii., 1800, p. 204.

[2] "Discours sur les révolutions de la surface du globe" (in *Recherches sur les ossements fossiles*, 4th ed., vol. i., p. 217).

[3] Cartailhac, E., "Georges Cuvier et l'ancienneté de l'Homme" (*Matériaux pour l'Hist. nat. et primitive de l'Homme*, 1884, p. 27).

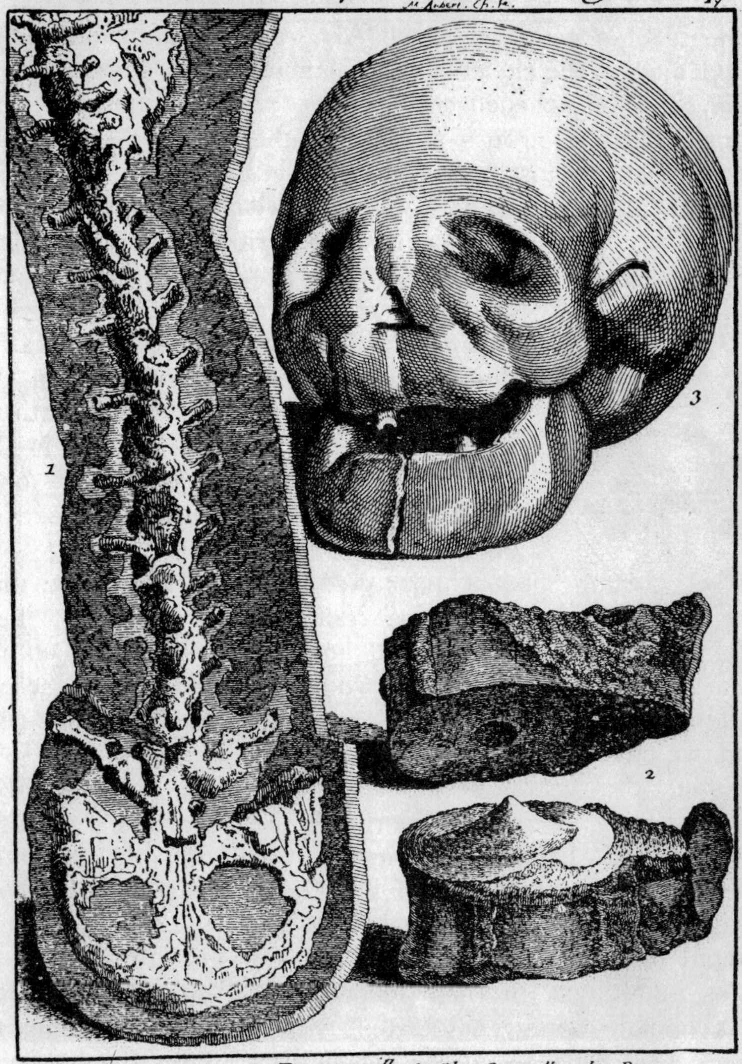

FIG. 4.—Remains mistaken, even in the eighteenth century, for those of fossil man. Facsimile of an engraving from *Oryctologie* by d'Argenville (1755).

1. Skeleton of a salamander found in a Miocene deposit in the neighbourhood of Constance and described by Scheuchzer as that of a "Human Witness of the Deluge" (*Homo diluvii testis*).

2. So-called "Petrified vertebræ from the back of a Man": in reality vertebræ of a fossil reptile, the Ichthyosaurus.

3. Modern head in a pathological condition (hyperostosis) which may still be seen in the Anthropological Gallery of the French National Museum of Natural History.

fragments of a cetacean; from Aix, remains of a chelonian; from Œningen, the skeleton of a salamander (the famous *Homo diluvii testis* of Scheuchzer) (Fig. 4). Such statements were well calculated to arouse scepticism, the more so as not the least trace of any fossil Ape had yet been discovered. Cuvier prudently added: "But I do not wish to conclude that Man did not exist at all before this period [that of the 'last upheavals of the globe']. He might have inhabited certain circumscribed regions whence he repeopled the earth after these terrible events; perhaps even the places he inhabited had been entirely swallowed up and his bones buried in the depths of the present seas, except for a small number of individuals who carried on the race."

FIG. 5.—Portrait of Tournal, from a photograph kindly lent by M. Cartailhac.

Cuvier died in 1832, just at the time when discoveries were imminent. "Perchance had he lived," wrote de Quatrefages, "he would have repeated the words he addressed one day to his fellow-worker Duméril: 'My dear friend, we have been mistaken.'"

Tournal and Schmerling.

About the year 1830, several naturalists in the Midi of France, Tournal in the Department of Aude, Emilien Dumas, de Christol and Marcel de Serres in the Departments of the Gard and of Hérault, continued in France such researches as Buckland had begun in England in 1820, by excavating the deposits accumulated in grottos and caves in their respective regions. There they found human bones, associated with numerous remains of animals belonging to species which had migrated or become extinct, bears, hyænas, reindeer, and others, the bones of which sometimes showed traces of cutting instruments. So clearly did Tournal recognize the importance of these observations that in 1829 he had no hesitation in writing: ". . . Geology, in supplementing our brief history, will at length awaken the pride of Man by revealing to him the antiquity of his race; for henceforth it lies in the power of geology alone to help us to some knowledge of the period when Man first made his appearance on the globe."[1] These words assuredly mark a very great step in advance.

Again, the Belgian author Schmerling published in 1833 an important work entitled *Recherches su les ossemen ts fossiles des cavernes de la province de Liége.* In this he not only demonstrated the co-existence of Man with the rhinoceros, bear, hyæna, and other animals, but further, he entitled his concluding chapter: "Relics worked by the hand of Man." These relics consisted of shaped bones, and in particular of an arrow-head and some flints. "Everything considered," he says, "it must be admitted that these flints have been cut by the hand of Man, and that they may have been used to make arrows or knives. . . . Even if we had not found human bones in circumstances strongly supporting the assumption that they belonged to the antediluvian period, proof would have been furnished by the worked bones and the shaped flints."

Some years later, in 1840, Godwin-Austen, continuing McEnery's studies of Kent's Cavern in England, arrived at the same conclusions.

Proof of the geological antiquity of Man was thus firmly established by these pioneers; but that is not to say that it was accepted by professional scientists, save perhaps by Constant Prévost.[2] To Boucher de Perthes belongs the merit of impressing it upon the learned world and of giving it common currency as well.

Boucher de Perthes.

Boucher de Perthes (1788-1868) was Controller of Customs at Abbeville.[3] He was a learned and prolific writer in diverse fields, a great lover of antiquities, "accustomed from childhood to listen to talk of fossils." Having devoted himself to the collecting of all sorts of ancient human remains, he had, towards the end of 1838, the good fortune to find "in diluvial beds" the "first diluvial axes," which he submitted to his fellow-members of the "Société d'émulation d'Abbeville." In 1846 he published the first

[1] *Annales des sciences naturelles*, vol. xviii., 1829, p. 258.
[2] Gosselet, J., *Constant Prévost*, Lille, 1896, p. 165.
[3] See Ledieu, A., *Boucher de Perthes, sa vie, ses œuvres* (Abbeville, 1885).

volume of his *Antiquités celtiques et antédiluviennes*, entitled *De l'industrie primitive ou des arts à leur origine*. In this work, Boucher de Perthes declared that the ancient alluvial soils, *diluvial* as he called them, in the suburbs of Abbeville, contained many stones worked by "*antediluvian*" Man buried at various depths along with bones of large animals belonging to extinct species. "In spite of all their imperfection," he says, "these rude stones prove the existence of Man as surely as a whole Louvre would have done."

This assertion, although founded on minute observations and on excellent evidence, at first met with the utmost disfavour. "Contradictions, jeers, scorn, were unsparingly heaped upon the author," wrote M. de Saulcy. He was regarded as a dreamer, as a kind of visionary, and the scientific world, priding itself on its detachment, allowed him to talk without further concerning itself with facts which, he maintained, he had forcefully introduced into the domain of practical science.[1]

FIG. 6.—Portrait of Boucher de Perthes, from a lithograph.

Far from being discouraged, Boucher de Perthes continued, with fine perseverance and good nature, to combat this systematic and often sarcastic opposition. Soon two camps were formed in the learned world. The first included several naturalists of independent spirit, among them A. Brongniart and Constant Prévost, who, while maintaining a certain caution, supported Boucher de Perthes. In the second and by far the largest camp, that of the extremists, with Elie de Beaumont at their head were to be found the more

[1] See Meunier, Victor, *Les Ancêtres d'Adam*, Thieullen Ed. (Paris, 1900). This failure was probably due in part to the fact that Boucher de Perthes associated with true primitive instruments, as if they were of the same significance, other stone *figures* or *symbolic* stones which were only "sports of Nature," and which are now recognized as of no account. But how was it possible at that time to separate the tares from the wheat?

academic scientists, the disciples and successors of Cuvier, who, while repudiating any suggestion of prejudice, nevertheless exaggerated their master's scruples. "Before the intervention of English geologists and archæologists had deprived this great question, raised and solved by a Frenchman, of its wholly French bearing, for so long the entire French Academy followed the lead of its Permanent Secretary, like a flock of sheep on the heels of the shepherd."[1]

In 1854, Dr Rigollot of Amiens, having found in the sand-pits at Saint-Acheul "axes" similar to those from the gravels of Abbeville, was the first to associate himself wholeheartedly with the views of Boucher de Perthes, which till then he had strenuously opposed. Further, a distinguished naturalist of the Midi, Dr Noulet, brought forward favourable evidence when he announced the occurrence at Clermont, near Toulouse, of an "alluvial deposit containing remains of extinct animals, mingled with stones shaped by human hands."

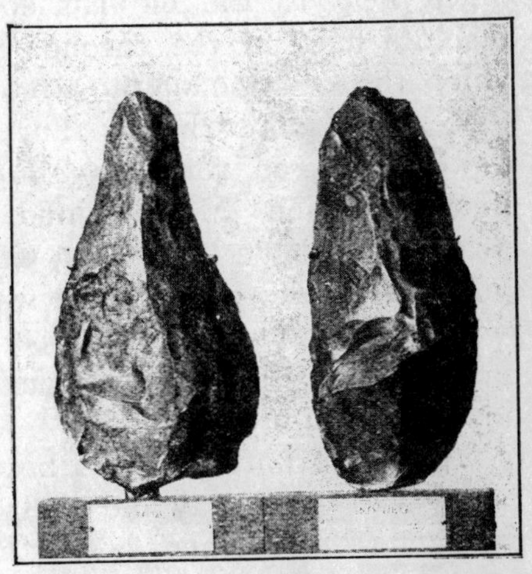

Fig. 7.—Two of the flints obtained by A. Gaudry from the gravels at Saint-Acheul. About one-third natural size. Now in the Palæontological Gallery, French National Museum of Natural History.

In 1859, after repeated study of the facts on the spot, several distinguished English scientists, the palæontologist Falconer, the stratigrapher Prestwich, the archæologist John Evans, the anatomist Flower, and the famous geologist Lyell, who soon afterwards published his celebrated work *The Antiquity of Man proved by Geology*, all clearly and decidedly declared their adherence to the theory.[2]

In the same year Albert Gaudry, a palæontologist then at

[1] Meunier, V., *loc. cit.*, p. ix.

[2] For an account of this intervention, see Falconer, H., *Palæontological Memoirs*, vol. ii., p. 596; Prestwich, "On the Occurrence of Flint Implements, associated with the Remains of Extinct Mammalia" (*Proc. Roy. Soc.*, 1859).

the outset of a brilliant career, went to Amiens to study the deposits and to carry out excavations. Having made up his mind never to leave his workmen, he himself succeeded in extracting, along with teeth of a large ox, "nine axes" from the "diluvium" at a depth of 4½ metres, and at a level from which, a short distance off, had been obtained bones of the Rhinoceros, Elephant and Hippopotamus[1] (Fig. 7). Gaudry's evidence made a deep impression upon the minds of several independent scholars, but opposition continued in the Institute, which held to the old conception of the deluge, and had absolute faith in the chronology of the Bible, according to which the creation of the world dated no further back than 4000 years before Christ. This opposition was carried to such a point that, on 18th May 863, a geologist in the highest official position, Member and Permanent Secretary of the Academy, Elie de Beaumont, went so far as to say: "I do not believe that the human race was contemporary with *Elephas primigenius*. M. Cuvier's theory is born of genius; it is still undemolished."[2] He even wondered if the dressed flints were not of Roman origin. . . .[3]

Academic immortality is but a senile illusion. Permanent Secretaries pass away and their names fall into oblivion; but the name of Boucher de Perthes will shine forever in the firmament of Science.

A very great advance in Science had been made by the discovery that beyond the limits of History stretched a vast Prehistory, which is finally lost in the obscurity of geological

[1] Gaudry, A., "Contemporanéité de l'espèce humaine et de diverses espèces animales aujourd'hui éteintes" (*Comptes rendus de l'Académie des Sciences*, 3rd October 1859).

[2] *Comptes rendus de l'Académie des Sciences*, 18th May 1863.

[3] The persistence of this injurious influence, which continues even to our own day in a more or less feeble or unconscious form, is shown by the following facts: At the death of Boucher de Perthes, his works were withdrawn from sale by decision of his family and sold for waste paper. Some years afterwards, Victor Meunier wrote his book, *Les Ancêtres d'Adam, Histoire de l'Homme fossile*. The book was printed in 1875, but was never published. It gave an account of the "martyrdom" of Boucher de Perthes, and the publisher, afraid of incurring the displeasure of the Academy, suppressed the whole issue. In 1900 the firm of Fischbacher published a new edition edited by A. Thieullen, a warm admirer of Boucher de Perthes. It is a work of great interest.

time. Henceforth the origin of Man became a problem for palæontology, on a par with the problems of the origins of the animals. The impulse was given; everywhere zealous workers devoted themselves to investigations, with good results. A new science, that of "Human Palæontology,"[1] was on the point of being definitely established.

Edouard Lartet.

Edouard Lartet, who was born and died in Gers (1801-1871), was the chief founder of this new science. At first a lawyer by profession, he awoke to his true calling on seeing a molar tooth of a Mastodon, found by a peasant in his village. Deeply interested, he read Cuvier's works, studied osteology, and devoted himself to the investigation of fossil bone-remains, which abounded in the ground about his family estate. From 1836 onwards, he explored and made famous the rich beds of Sansan, which date from Mid Tertiary times. There he discovered, among other strange forms entirely new to science, remains of an anthropoid ape, an ancestor of the modern Gibbons, and this he named *Pliopithecus*.

Fig. 8.—Edouard Lartet, from his only portrait, for which we are indebted to his son, Louis Lartet.

P. Fischer, author of one of the biographies of E. Lartet, points out the importance of this discovery from the point of view of the question of fossil Man: "Cuvier, in an enlightened and needful criticism of the so-called bone-remains of man and of contemporary monkeys of extinct species, exposed their lack of authenticity. He accordingly inferred that monkey and man were late in appearing. 'What astonishes me,' said he, 'is that, amongst all these mammals, the majority of which have at the present day congeners in warm regions, there is not a single Quadrumana; and also that there has been found not a single bone, not a single

[1] The expression is due to Serres—"Notes sur la Paléontologie humaine" (*C. R. Ac. Sci.*, xxxvii., 1853, p. 518).

tooth of a Monkey, even of any extinct species. Neither is there any Man: all the bones of our species which have been collected along with those I have referred to were present by accident.'"

"In thus associating the date of Man's appearance with that of monkeys," Fischer continues, "Cuvier prepared the way for the great reception accorded to the discovery of the Sansan Ape, and it could be foreseen that the discovery of a fossil Ape would be followed by that of fossil Man."[1]

The insight of Etienne Geoffroy Saint-Hilaire did not err. Cuvier's distinguished adversary had pointed out "the important bearing on natural philosophy" of Lartet's discovery, destined "to inaugurate a new era of knowledge relating to human life." But he added, "the time for philosophical research is not yet."

Even in 1845, Lartet boldly admitted the possibility of Tertiary Man. "This corner of ground," he said, speaking of Sansan, "once supported a population of mammals of much higher degree than those here to-day. . . . Here are represented various degrees in the scale of animal life, up to and including the ape. A higher type, that of the human kind, has not been found here, but we must not hastily conclude from its absence in these ancient deposits that it did not exist. . . ." These were prophetic words. It seems as if Lartet had "a presentiment of the important part he was to play later, in the scientific discussion regarding the co-existence of Man with the large Quaternary mammals."

About the year 1850, E. Lartet went to Paris to continue his researches. He settled near the Museum, the scientific treasures of which attracted him, and where he found none but friends. In 1856, he described the jaw of a new anthropoid ape, *Dryopithecus*. Three years later he published a comprehensive monograph on the fossil Proboscidians. But his writings on the animals of former times constantly led him back to the great problem of fossil Man. With great sympathy and interest he followed the efforts of Boucher de Perthes.

[1] Fischer, P., "Note sur les travaux scientifiques d'Edouard Lartet" (*Bull. de la Soc. géolog. de France*, 2nd Ser., xxix., p. 246).

On the 19th of March 1860, E. Lartet sent to the Académie des Sciences a note on the occurrence of Man in Western Europe in geological times, entitled, "Sur l'ancienneté géologique de l'espèce humaine dans l'Europe occidentale." The Académie has been accused of refusing to print this memoir, and the fact is that only the title appears on p. 599 of Volume L. of the *Comptes rendus*. For the text, reference must be made to the *Archives des Sciences de la Bibliothèque universelle de Genève*, or to the *Quarterly Journal of the Geological Society of London*, which received it with enthusiasm.[1]

Now this memoir was of prime importance. Along with a description of the celebrated cave of Aurignac, which the author had just explored, it contained certain suggestions of the greatest significance, which were renewed and developed the following year (1861) in the *Annales des Sciences naturelles* under the title: "New researches on the co-existence of Man and of large fossil Mammals regarded as characteristic of the last geological period."[2]

It would seem that even from the time of his first purely geological writings, E. Lartet had been an opponent of the cataclysmic theory of the world's development. It required a great deal of independence and true courage to challenge a theory held by the scientific pundits. This courage he showed, a fact which sufficiently explains the hostile attitude of Elie de Beaumont.

In 1858, in his note "On the Ancient Migrations of Mammals of the Present Period,"[3] he had already assailed the idea of deluges or other catastrophes. "The day is perhaps not far distant," he said, "when the erasure of the word *cataclysm* from the vocabulary of practical geology will be proposed." Or again: "It is an abuse of the technical

1 "It was too soon to announce these truths to the Académie des Sciences; it did not understand that, in refusing to publish the forecast of E. Lartet, it was placing itself in the backwash of geological and anthropological progress, and that a day would come when it would be a cause for deep regret to find in a foreign publication seven pages so creditable to French science, rejected by the Institute of France."—E. Cartailhac, *in litt.*

2 "Nouvelles recherches sur la co-existence de l'Homme et des grands Mammifères fossiles réputés caractéristiques de la dernière époque géologique."

3 "Sur les migrations anciennes des Mammifères de l'époque actuelle."

language of science to use such high-sounding expressions as *upheavals of the globe, cataclysms, universal disturbances, general catastrophes,* and so on, for they immediately give an exaggerated significance to phenomena geographically very limited. . . . The great harmony of physical and organic evolution on the surface of the globe has in no case been affected. Aristotle perfectly understood these alternating movements of the earth, which have at different times changed the relations of continents and seas; he knew equally well how to reduce to its proper regional proportions the Deucalian Deluge, exaggerated and embellished by poetic fiction. Apparently this great naturalist also had to combat the fantastic ideas of the cataclysmic philosophers of his time, and the severe reproach he flung at them, might just as well, after 2000 years, be applied to certain of our geologists or palæontologists of the present day: 'It is absurd, on account of small and transitory changes, to invoke the upheaval of the whole universe.'"[1]

The memoir contains another new and suggestive idea. The history of Man, like that of animals or like any geological history, is indeed a continuous story, and demands a chronological method. "Were it possible to establish that the disappearance of the animal species characteristic of the last geological period was successive and not simultaneous, a means would be discovered of establishing, at one and the same time, the relative chronology of the unstratified fossil deposits and their time relations with these diluvial beds whose geognostic bearings are well defined." Accordingly, Lartet proposed a "palæontological chronology," which for the first time allowed a classification to be made of the beds in which traces of fossil man had been found up to that time. "Thus, in the period of Primitive Man there would be the Age of the Great Cave Bear, the Age of the Elephant and of the Rhinoceros, the Age of the Reindeer and the Age of the Aurochs; much after the manner recently adopted by archæologists in their divisions of Stone Age, Bronze Age, and Iron Age."

This classification could not be perfect, but its actual

[1] "Ridiculum enim est, propter parvas et momentaneas permutationes, movere ipsum totum" (γελοιδον γαρ, etc., Aristotle, *Meteorol.*, 1, I, c. 2).

existence was of great value, in that it asserted the geological nature of the problem of Man's existence, showed how the history of our ancestors must be sought for in bygone ages, and fixed some mile-stones on the long journey. So a broad path was thrown open to investigators. In his eulogy of Lartet, Hamy has well said: "To the doctrine of the antiquity of Mankind, Aurignac converted a number of disciples, who were the more valuable in that they translated enthusiasm into productive activity."

Soon after, in 1864, E. Lartet discovered the famous engraved mammoth from La Madeleine, where, in delightful fashion, one of our distant forebears had himself inscribed

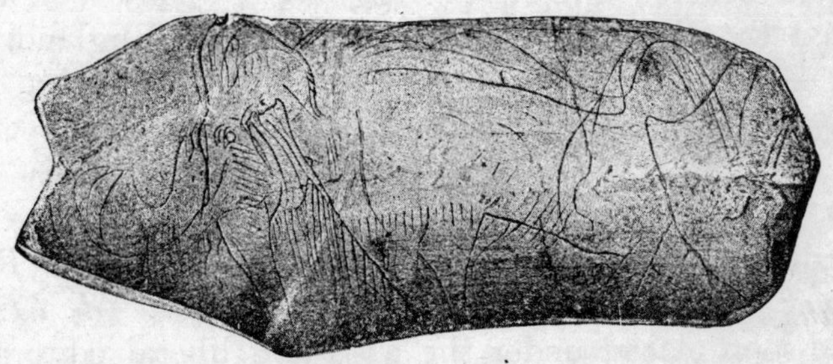

FIG. 9.—Fragment of ivory from La Madeleine, bearing an engraving of the Mammoth. One-third natural size. From the plate by E. Lartet. The original belongs to the French National Museum of Natural History.

decisive proof of his geological antiquity (Fig. 9). Along with Christy, an Englishman, he undertook the investigation of the deposits of the Vézère valley, the fame of which is now world-wide. Thus he succeeded in revealing the astonishing artistic culture of the men of the Reindeer Age. The work in which so many fine discoveries were to have been described and expounded has unfortunately never been completed.[1]

In 1869 Lartet was chosen to succeed d'Archiac in the Chair of Palæontology in the French National Museum of Natural History. He was then sixty-eight years of age, and he died some months later, without having delivered his first lecture.

[1] Lartet, E., and Christy, H., *Reliquiæ aquitanicæ:* being contributions to the archæology and palæontology of Périgord (Paris, 1866-1875, 1 vol. in 4to, with 102 plates).

If I have spoken at length of Edouard Lartet, it is, first, from admiration for so independent and disinterested a man of science; secondly, to show the outstanding part which, through him, France played in the creation of the science of Human Palæontology; and, finally, because the achievement of our illustrious countryman has not always been sufficiently understood. To the public at large it is unknown, and the scientist has not appraised it at its true value. And yet the passing of the years only adds to the fame of Edouard Lartet.

Successors of Edouard Lartet.

Lartet's example was followed in France by numerous scholars and investigators, P. Gervais, de Vibraye, A. Milne-Edwards, Louis Lartet, Piette and others; whilst, in Belgium, Dupont took up and completed the work of Schmerling; and in England, where a good fight had also been waged, Lubbock, John Evans, and Boyd Dawkins published very valuable works on Prehistory.[1]

In 1864, in order that the progress of the science might be recorded, Gabriel de Mortillet founded a special Review, *Matériaux pour l'histoire naturelle et primitive de l'Homme*, which he soon placed under the able and liberal editorship of Emile Cartailhac. Keeping the archæological standpoint especially in view, G. de Mortillet revised Lartet's classifications. With a lucidity that appealed to the comprehension of every investigator, he grouped systematically the innumerable facts of a science whose birth he had seen, and to the development of which he had largely contributed.

It was not long before prehistorians began to hold international congresses, where results in one country were compared with those in others, where general questions were discussed, and where interdependent labours were planned, for discoveries had meantime spread to every continent. So, step by step, we reach the present day, when researches in prehistoric archæology have become the fashion, when everyone

[1] Lubbock, John, *Prehistoric Times* (London, 1867, 7th ed., 1913); French translation by Barbier, under the title *L'Homme avant l'Histoire* (Paris, 1867, 2nd ed. 1871). Evans, John, *Ancient Stone Implements* (London, 1872, 2nd ed., 1897). Dawkins, W. Boyd, *Cave Hunting* (London, 1874); *Early Man in Britain* (London, 1880).

grubs in the most ancient of our archives, too often, alas, with an utterly inadequate scientific training.

* * * *

Thus arose the science of Prehistory or Prehistoric Archæology, founded on facts which, although supplied by all kinds of material things, nevertheless throw a tolerably clear light on the intellectual and moral character of the Men regarding whom History is silent.

Bones of Fossil Man.

In the meantime what progress had been made in research regarding the physical and zoological characters of Man himself? What steps had marked the progress of Human Palæontology in the strict sense of the words, the sense in which they are mainly used in this work?

After the discovery by Ami Boué, in 1823, of a human skeleton in the loess of the Rhine Valley, a discovery the significance of which Cuvier, as we have seen, utterly repudiated, there followed a barren period. Every find of human bones was now regarded *a priori* with suspicion. But when the great antiquity of Man was demonstrated by means of dressed flints and proved by geology, discoveries of human bones seemed more natural: they increased in number.[1] No fewer than eighty have been recorded from the beginning of the nineteenth century to our own day. Palæontology would thus seem to have been provided with material sufficient to enable it to attain to great results and to frame important conclusions.

Review of the Discoveries.

Unfortunately these discoveries are far from being of equal value because of uncertainty regarding the age or even the authenticity of many of them. It is very easy to fall into error in dealing with such material. In many a place the earth is but human dust. Nothing alas! is more common in superficial soils than the skeletons of our fellows. Of course the physical characters of the bones vary according to the date of their burial; and the burials of historic times present features which would hardly deceive a practised eye. In the case of prehistoric

[1] See Quatrefages, A. de, and Hamy, E. T., *Crania ethnica: Les Crânes des races humaines* (Paris, 1882). Première partie. *Races humaines fossiles.*

burials or of bone-remains of the Quaternary Period, one important character must be taken into account, that of fossilization, by which is meant the physical and chemical transformation of a bone, which having lost its organic substance has become pervaded by mineral matter and so more dense. But this character is not sufficient; the degree of fossilization may vary according to certain conditions of the environment, independent of age. Appeal must then be made to the conditions of the soil deposit, to geological and palæontological criteria. When a discovery is made, however, a competent observer is rarely on the spot, ready to make the necessary investigation. At the present day, now that the attention of an enlightened public has been directed to such events and their importance is understood, the assistance of professional scientists is usually invited; and several recent discoveries have also been made following upon systematic excavation conducted by experts. Formerly this was not the case, for then the geology and the palæontology of the Quaternary formations had barely been outlined. Many human skulls and skeletons, carelessly exhumed without scientific investigation, have been placed in museums, where anthropologists study them without sufficiently enquiring into the record of remains the origin and exact bearings of which cannot now be accurately determined.

As the question of age is a factor of prime importance in palæontology, scientific accuracy demands a courageous elimination of all those osteological evidences the high antiquity of which is not assured. After close scrutiny of all the discoveries recorded up to the present day, I retain for consideration in this book only those whose authenticity and age are beyond dispute. Here it is better to err through excess rather than through lack of prudence.

The story of these discoveries, which form the foundations of Human Palæontology, will be given in the following chapters in order of the age of the relics disclosed, that is, in their respective chronological order. For the present, I wish only to indicate briefly the main steps of progress up to the present day.

The Neanderthal Discovery. The first and one of the most important stages centred in the discovery, in 1856, of the famous brain-pan or cranium at Neanderthal in Rhenish Prussia. This object was examined in succeeding years by various naturalists. With its considerable dimensions, its receding forehead, its enormous orbital ridges and its flattened brain-box, the skull presented an extraordinary appearance (Fig. 10). Schaaffhausen in Germany, and Huxley in England, declared it "the most bestial of all known human skulls," and emphasized its simian or monkey-like characters.

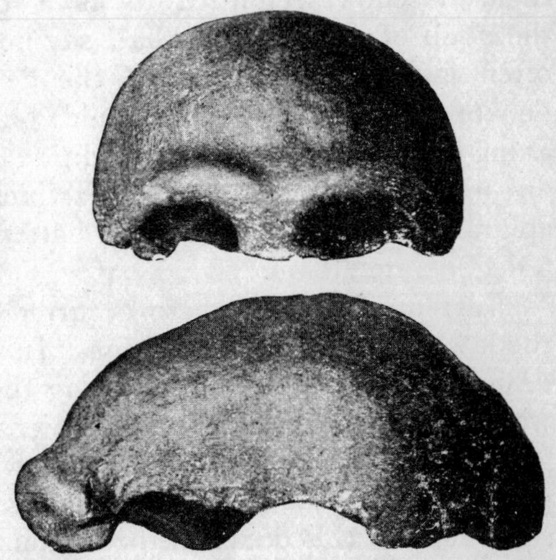

FIG. 10.—The Neanderthal Skull, seen full face and in profile. One-third natural size. Photographs of a cast.

This happened at a time when the scientific world was in a state of effervescence. Evolutionist ideas had begun to spread. Lamarck, who, long before Darwin, had not hesitated to attack the formidable problem of the origin of Man, and who conceived it as occurring through the modification of a Quadrumane, had been forgotten before he had even been understood or appreciated. But now Darwin published *The Origin of Species*, Boucher de Perthes began to gain ground, and Albert Gaudry made public the results of his first researches on the transformations of fossil mammals; Broca founded the Société d'Anthropologie de Paris, and Huxley wrote his celebrated memoir on the *Evidence as to Man's Place in Nature* (1863), which was followed soon after by Carl Vogt's excellent *Vorlesungen über den Menschen* (1863).

The Neanderthal skull, by reason of characters obviously of low type, and a conformation resembling that of the skulls of certain large Apes, supported the evolutionist theory; in the eyes of philosophic naturalists it appeared to be a sort of primitive form lessening the depth of the gulf which now separates the Apes from Men.

But this interpretation was not to the liking of anti-evolutionists of the old school. The scientific value of the skull was disputed and denied. As it had been found by workmen, geologists and palæontologists took exception to the obscurity of its origin. Eminent anthropologists, among them Virchow, regarded it as a pathological specimen or the skull of an idiot. I shall say nothing of the zealous and often foolish intervention of the defenders of religion, in a debate to which religion could only contribute arguments animated by sentiment, by tradition or by prejudice. It was an intervention of this kind which provoked the famous epigram of Huxley, that it was better to be a perfect Ape than a degenerate Adam.

Just at this time there occurred the notorious episode of the jawbone of Moulin-Quignon. In 1863, Boucher de Perthes, desirous at all costs of discovering the fossil bones of the Man who had dressed the flints of Amiens and of Abbeville, found a human jawbone in conditions which stirred up lengthy polemics and caused floods of ink to flow. It would indeed seem as if on this occasion the famous and worthy archæologist had been the victim of a fraud. The English scientists who had so emphatically supported his views regarding the dressed flints, refused to believe in the authenticity of the jawbone; and one of them, John Evans, pronounced upon it a *Requiescat in pace*, of which the echoes have not yet died away. This, clearly, was not calculated to add to the credit of the new theory.

La Naulette.

But in 1865, Ed. Dupont, in the course of scientific explorations organized by the Belgian Government in the caves of that country, found a human lower jaw in one of the excavations on the left bank of the Lesse, the *Naulette pit* (Fig. 11). The circumstances of its deposit left no loophole for criticism. Now this jaw, taken from a deep bed, where it lay along with bones of the Mammoth, Rhinoceros, Reindeer, etc., differed from the jaws of all modern

Men in one important character which struck the observer at first glance, the absence of a chin. Here again was the stamp of the ape, associated none the less with other characters which were purely human. One was tempted to associate the jaw from La Naulette with the Neanderthal skull, as belonging to a similar lowly type.

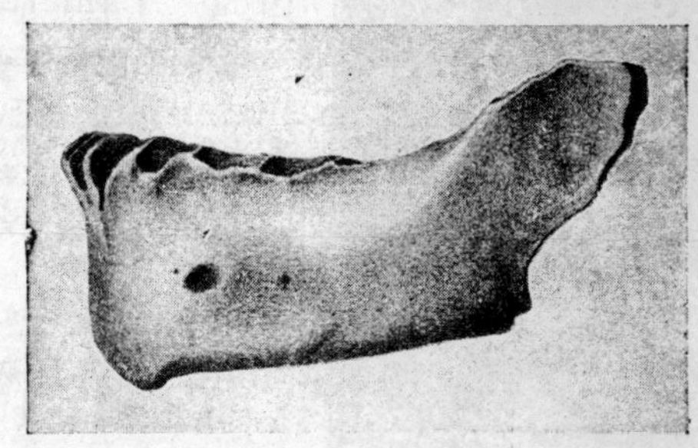

FIG. 11.—The jawbone from La Naulette. Three-quarters natural size. After de Quatrefages and Hamy.

Cro-Magnon.

In 1868, Louis Lartet, following with distinction in his father's footsteps, described the rock-shelter of Cro-Magnon on the banks of La Vézère, in the Dordogne, from which several human skeletons had already been obtained. On this occasion the skeletons presented all the features of modern Man; so much so indeed, that their great antiquity was not acknowledged by most anthropologists, who could not bring themselves to abandon their preconceived notions and to throw so far back into the past the physical type of *Homo sapiens*. So it was also with the skeleton found in 1872 by M. Rivière in one of the caves of Grimaldi (Fig. 12). The "Mentone Man," closely resembling the Cro-Magnon type, was considered to be Neolithic. The geological bearings were, however, perfectly definite.

On the other hand, far too much importance was laid on some skeletons obtained, about the same time, from more or less ancient and more or less disturbed river deposits of the Seine, at Clichy, Grenelle, and elsewhere.

In 1870 Hamy[1] published a summary of the state of the science at this time, in a book which may still be consulted with profit. In the following year, Darwin,[2] tackling the great problem of the descent of Man, published a work in which palæontological facts do not and could not as yet play any

[1] Hamy, E. T., *Précis de Paléontologie humaine* (Paris, 1870)

[2] Darwin, C., *The Descent of Man* (London, 1871).

but a secondary part, but in which the famous naturalist expounded in all its bearings the theory of the animal origin of Man, formerly precisely stated by the great Lamarck.[1] To this theory the German naturalist Haeckel had just given his strong support in his *Generelle Morphologie der Organismen* (Berlin, 1866).[2]

FIG. 12.—"Mentone Man," found by M. Rivière. Anthropological Gallery of the French National Museum of Natural History.

About the same time, Broca[3] published some excellent studies on the comparative morphology of Apes and Man, and thus placed his great craniological knowledge at the service of human palæontology. During the years 1873 to 1882, de Quatrefages and Hamy contributed to this branch of science a great work,[4] in which descriptions of the principal cranial types of modern Man were preceded by long systematic discussions on all the fossil or pseudo-fossil evidences then known.

The Spy Men.

The year 1887 was marked by an interesting discovery of two human skeletons in a cave at Spy in the province of Namur. This was an event of considerable scientific importance, fortunate in two respects: first, in that the Quaternary Age of the deposit, investigated by geologists, was

[1] Lamarck, *Philosophie Zoologique*, 1809, i., p. 337.

[2] See also Haeckel, E., *Histoire de la Création* (Fr. trans., Paris, 1874). *Anthropogénie ou histoire de l'évolution humaine* (Fr. trans., Paris, 1877). *État actuel de nos connaissances sur l'origine de l'Homme* (Paris, 1900). [English editions of these works appeared as follows: *The History of Creation* (1st Eng. ed., London, 1875; 3rd, 1883). *The Evolution of Man* (Eng. ed., London, 1879). *Our Present Knowledge of the Descent of Man* (1898).]

[3] Broca, P., "L'ordre des Primates" (*Bult. de la Soc. d'Anthrop. de Paris*, 2nd Series, vol. iv., 1869).

[4] Quatrefages, A. de, and Hamy, E. T., *Crania ethnica*.

not open to question; secondly and especially, because the Spy skulls resembled in every way the Neanderthal skull. The hypothesis of the pathological nature of the latter was definitely destroyed by the fine report of Fraipont and Lohest, which helped to confirm the opinion of those who believed in the actual existence of an ancient human type very different from, and of lower nature than modern types.

FIG. 13.—E. T. Hamy, from a photograph.

Pithecanthropus.

This opinion was notably strengthened some years later, in 1894, by the work of Dubois on the remains of *Pithecanthropus*, discovered in Java in 1891. In a succeeding chapter we shall discuss this famous fossil at some length; it is sufficient at present to state the indisputable fact, that the skull-cap of *Pithecanthropus* really embodies a morphological type ideally intermediate between the skulls of anthropoid apes, such as the Chimpanzee or the Gibbon, and a human skull.

These fine discoveries instigated others. A positive fever took hold of investigators; and excavations carried out in almost every part yielded many evidences, but of very unequal value.

Later Discoveries.

Amongst the most important of the later discoveries, first in order of time must be mentioned that at Krapina in Croatia, which brought to light many human remains of Neanderthal type.

Next come the results of the important explorations undertaken by the Prince of Monaco, Albert I., in the Grimaldi Caves. Several human skeletons were exhumed in the Grotte des Enfants: some belonged to the Cro-Magnon type, the Palæolithic Age of which was here definitely established; while another, the most ancient, revealed to Professor Verneau the existence of a different type, of negroid character, the "Grimaldi type."

The Mauer Jawbone.

In 1907, a new fact of prime importance was brought forward. Up to that date, the Man of the oldest dressed flints was known only by the products of his handiwork—no authentic relic of his skeleton had been obtained. Then Schœtensack described a jawbone found in the ancient gravels of Mauer near Heidelberg. And this jawbone, very much older than those from La Naulette, from Spy, or from Krapina, presented a still more primitive appearance.

La Chapelle-aux-Saints.

By systematic excavations carried out in France, the Abbés Bouyssonie and Bardon, Capitan and Peyrony, and Henri Martin, discovered in human settlements, deep in the caves or shelters of La Chapelle-aux-Saints in the Department of Corrèze, of La Ferrassie in Dordogne, and of La Quina in Charente, several skeletons and portions of skeletons of men of Neanderthal type.

Human palæontology has thus been furnished with records of exceptional value, which have enabled us to gain a fuller knowledge of this ancient type than we possess of many modern savages.[1] This *Homo neanderthalensis* will be the subject of one of the longer chapters of this volume.

Piltdown.

After a considerable period of relative inactivity in the sphere of human palæontology, England, which can claim a most honourable part in the foundation and development of the science, was seized with new enthusiasm for it. In addition to some recent discoveries, the importance of which was overestimated, such as that of the Ipswich skeleton, considered to belong to a period more remote than the Quaternary, but in reality barely prehistoric, there occurred the Piltdown find, studied from the anatomical point of view by the palæontologist Smith Woodward. Although its particular and general significance are still disputed, it is certainly a most important discovery, as we shall see later.

Fossil Man beyond Europe.

So far, I have spoken only of Europe, and solely of Western Europe, a very small part of the globe, yet the rest is almost unknown from the point

[1] Boule, M., "L'Homme fossile de La Chapelle-aux-Saints" (*Annales de Paléontologie*, 1911-1913).

of view which interests us here. Researches carried out in the two Americas, especially noteworthy being those of Ameghino in South America, have not yet produced any conclusive discovery. Asia, the outstanding importance of which will one day become apparent, has yielded no results to speak of, except of an archæological nature. Quite recent discoveries at Boskop in South Africa,[1] as well as at Talgai in Australia (see the last chapter of this book), show that, whenever investigations are undertaken with sufficient resources in these different parts of the globe, great results will be forthcoming.

Rather than seek to revive the importance of the odds and ends of old bones, collected in past times without sufficient scientific guarantees, upon which many authors still dwell with too great satisfaction, we must now devote ourselves to making new explorations, with all the care, the precaution, and the scientific method now incumbent upon palæontologists and anthropologists having a true understanding of the problems to be solved.

The results obtained during the last few years, following the method I advocate, at Grimaldi, at Mauer, at La Ferrassie, at La Quina and at Piltdown, are most encouraging.

[1] [And at Broken Hill Mine, Rhodesia.]

CHAPTER II

TIME RELATIONS OF EARLY MAN

1. *Relative Chronology*

CHRONOLOGY is the basis of all genealogy, and it is therefore important, before entering upon the history of Fossil Man, to see how the evidences relating to our distant ancestors can be arranged according to the order of their relative antiquity. For this, some idea of geology is indispensable.[1]

The Geological Ages.

The history of the earth falls into several Ages :—

(1) The Archæan.
(2) The Primary.
(3) The Secondary.
(4) The Tertiary.
(5) The Quaternary.

The Archæan Age, by far the longest, corresponds to the formations deposited in the primitive seas of the globe. Under the influence of central incandescent masses and of mechanical processes resulting from their contraction, these deposits have been subjected to innumerable transformations, which have caused the fossils they must once have contained to disappear almost entirely.

[1] Boule, M., *Géologie* (6th ed., Paris, 1919); *Conférences de Paléontologie* (2nd ed., Paris, 1910). These two works are very elementary. For greater detail see the *Abrégé* and particularly the *Traité de Géologie* of Albert de Lapparent. The admirable works of Albert Gaudry may also be read with much pleasure and profit—*Enchaînments du Monde animal* and *Essais de Paléontologie philosophique.*

[English readers may consult the articles dealing with Stratigraphical Geology in any standard Encyclopædia, or the corresponding sections in text-books, such as Lapworth's *Intermediate Text-book of Geology*, Lake and Rastall's *Text-book of Geology*, or A. Geikie's more detailed *Text-book of Geology.*]

The succeeding Ages are characterized by the development of certain groups of organisms, and they are distinguished from one another owing to great changes in the physical world, such as upheavals of mountain ranges, alterations in the distribution of lands and seas, and so on.

The Primary or Palæozoic Age begins with the formation of the first deposits in which fossils are well preserved. At that time living things belonged to the lower groups or types of the organic world: the animals comprised only Invertebrates and, just towards the end of the Age, some Fishes and certain primitive four-footed creatures.

The Secondary or Mesozoic Age is marked by an advance in the animate world. It is the Age of Reptiles, which were then much more numerous, more varied, and more powerful than at the present time.

The Tertiary, or Cainozoic, shows fresh progress. The great Reptiles have disappeared, and their place has been taken by the more highly-organized animals, Birds and especially Mammals.

Finally, the Quaternary Age, which still lasts, is characterized by the reign of the most highly perfected of mammals, Man.

The appearance and the development of the great zoological groups have thus followed in order of degree. Invertebrates appeared before Vertebrates, and amongst the latter, the lowest, the Fishes, came first, then the Batrachians or Amphibians, then the Reptiles, the Birds and the Mammals.

The diagrams (Figs. 14 and 15) show that the importance of each Age is in close relation to the depth of the corresponding formations, and therefore in close relation to its duration.[1]

One is struck by the disproportion between these Ages. The following figures, although only very approximate, will give some idea of it. If the figure 1 be taken to represent the thickness of the Quaternary formations, or the duration

[1] The figures placed in the third column (duration in years) are purely hypothetical. Only their relative values need be considered. From the absolute point of view their proportion alone need be borne in mind. Further, we probably ought to compute by hundreds, rather than by tens, of millions of years.

of the corresponding Age, the duration of the Tertiary Age must be represented by 20, of the Secondary Age by 30, and of the Primary Age by 150. And it is probable that the

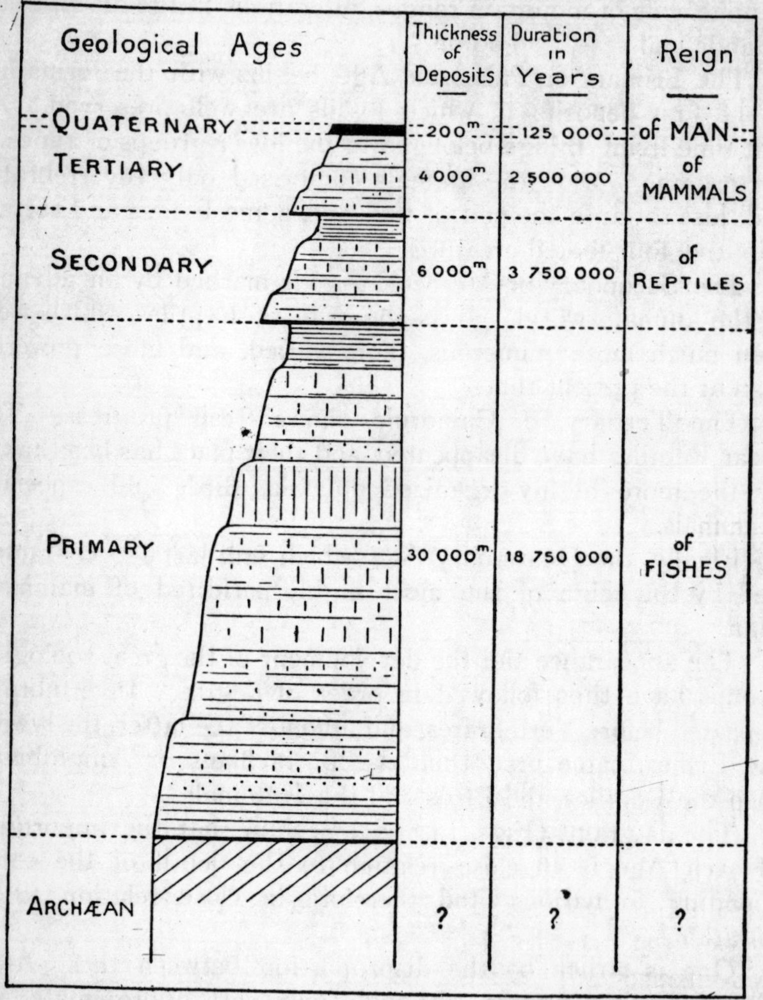

FIG. 14.—Diagram of the general succession of the Geological Ages, of their relative duration (estimated according to the thickness of the corresponding formations), and of their principal palæontological characters.

Archæan Age corresponds to a lapse of time equal to or greater than that which separates us from the beginning of the Primary Age!

The Tertiary Age and its Mammals.

Man, a mammal, could not appear before the Tertiary Age, for the mammals of the Secondary Age were small and very primitive animals. The more highly-developed types of mammals of the Tertiary Age are not all of the same antiquity. We must,

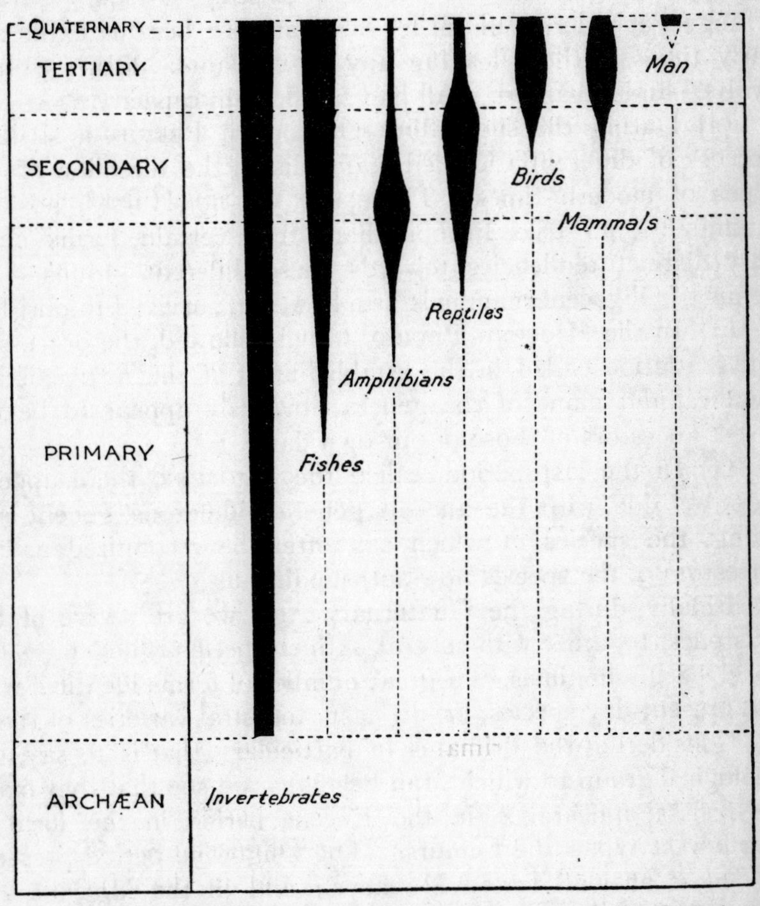

FIG. 15.—Diagram showing the development and distribution in time of the Invertebrates as a whole, of the various classes of Vertebrates, and of Man.

therefore, examine the divisions of the Tertiary Age, as well as the relative development of mammals in general, and more particularly of the most highly organized group, the Primates, of which Man is the most perfect type.

Geologists divide the Tertiary Age into four great Periods—

(1) As regards mammals, the EOCENE PERIOD, the most ancient, possessed only Marsupials (*Didelphia*), or else primitive placental mammals of synthetic types, presenting such a mixture of characters that it is often impossible to classify them in any of the now existing Orders. Certain strange Pachyderms of rather large stature became extinct in early times, without leaving any descendants. Other groups evolved and advanced. All had small brain capacity.

(2) During the succeeding OLIGOCENE PERIOD, a striking process of differentiation led mammals in the direction of the types of modern times. There are yet no Horses, no true Ruminants, no true Proboscidians, but certain forms show very distinct tendencies towards these different groups. No genus of Oligocene mammals has, however, survived to our day.

(3) In the MIOCENE PERIOD, which followed, the process of differentiation ended in the establishment of the now existing families, and many of the genera obviously appear to be the direct ancestors of those of our own day.

(4) In the last period, called the PLIOCENE, there appear, side by side with the ancient genera, numerous genera still living, the species of which can often be recognized as the ancestors of the species now surrounding us.

Finally, during the Quaternary Age, we are aware of the existence, together with several extinct species which have not survived to our times, of a great number of forms identical with the present-day species, or, at least, ancestral varieties of them.

Considering the Primates in particular—that is to say, the zoological group to which Man belongs—we see that they make their first appearance in the Eocene period in the form of the lowest type—the Lemurs. The Oligocene period presents the most ancient Tailed Monkeys; and in the Miocene and Pliocene periods appear numerous Anthropoid Apes.[1] So far as present knowledge goes, fossil Man is only known in the Quaternary Age. Here, also, we thus have a continuous progression: Man, the most highly perfected of the Primates, is also the last-comer.

[1] For a more detailed account, see Chap. III.

The Quaternary Formations.

No very great difficulty would be likely to arise in indicating the chronological order of discoveries made in Tertiary deposits, for their succession is well established. The task is more difficult and much more delicate in relation to Quaternary formations, which are rich in human remains. It demands more precision, and I must therefore deal with this subject in some detail.

The classification of Quaternary times may be based—

(1) On the study of the formations and of their relative positions (Stratigraphy).

(2) On the study of fossils (Palæontology).

(3) On the study of human relics (Prehistoric Archæology).

The Quaternary deposits are numerous and of varied character. A complete acquaintance with them is indispensable to the study of human palæontology.[1] They may be classified as follows :—

I. Deposits of marine origin.

II. Sedimentary deposits of continental origin. { Glacial formations. Alluvial deposits of plains and valleys. Cave deposits.

III. Deposits of volcanic origin.

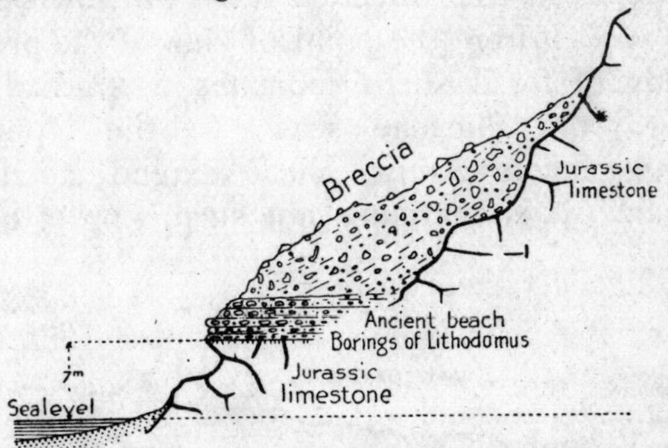

FIG. 16.—Geological section of the shore of the Mediterranean at the headland of Grimaldi, near Mentone. An ancient sea-beach, containing shells of warm seas, is almost 7 metres higher than the level of the present sea. It bears breccia or fallen debris, also Pleistocene, but of a later period.

[1] For an account of our present knowledge of the Quaternary formations, see E. Haug's *Traité de Géologie*, vol. ii. (Paris, 1911). [English readers should consult A. Geikie's *Text-book of Geology*, vol. ii. (London)].

Marine Deposits.

Marine deposits are of relatively little importance from our point of view. At the end of the Tertiary Age the Pliocene seas already possessed approximately the configuration of the seas of the present day. In the Quaternary Age, the changes were relatively insignificant. Along certain coasts, however, there are to be seen, at altitudes generally not exceeding a few metres, old littoral strands or ancient beaches which indicate slight oscillations of the sea-level (Fig. 16). The shells found on these raised beaches are identical with those of modern seas; certain slight differences in geographical distribution correspond to climatic differences, but they are scarcely noticeable (Fig. 17).

FIG. 17.—Shells from Quaternary marine deposits in France.

To the left a *Strombus* of warm seas (*Strombus bubonius*). To the right a species from cold seas (*Mya truncata*). One-third natural size.

It is very difficult, if not impossible, to establish a chronology of the Quaternary Age from these marine deposits alone. Attempts in this direction have till now been far from satisfactory, at least from the point of view of the present study.

Deposits of Glacial Origin.

The study of fossil shells indicates a gradual cooling of the Pliocene seas. In the Upper Pliocene layers Arctic shells extend as far as the Mediterranean. It is therefore not surprising to observe that

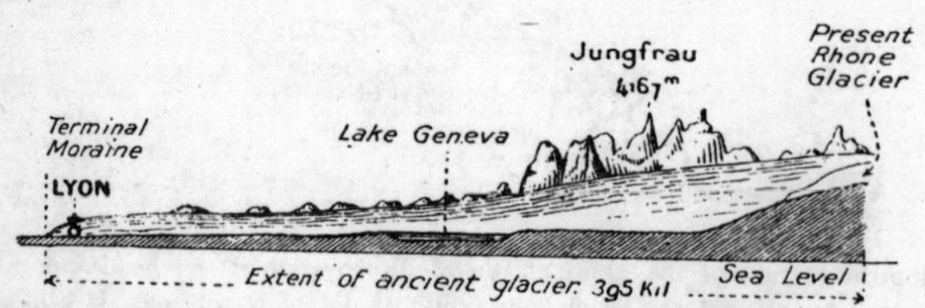

FIG. 18.—Sketch showing the extent of the Quaternary glacier of the Rhone Valley contrasted with the present-day glacier.

towards the end of the Tertiary era—that is to say, at a date when the Pyrenees were still young, when the Alps had just effected their greatest thrust, and when the volcanic eruptions of Central France were nearing completion—all these mountains were covered by glaciers greater by far than the greatest modern glaciers of the Alps.

Under the influence of causes still not fully understood,

FIG. 19.— Map of Europe in the height of the Ice Age.

these glaciers underwent an extraordinary development at the commencement of Quaternary times.[1] From the summit they reached the plains, where they extended vast distances from their starting-points, dispersing or accumulating all along their tracks enormous "erratics" and immense moraine deposits.

[1] This development is not peculiar to the later geological periods; it followed all the great upheavals of mountain chains, at all periods and in every country. We now know that vast glacial fields existed during the Eocene, the Permian, the Silurian, and even the Archæan Ages.

In the northern countries of Europe, conditions were even more severe. On the heights of Scandinavia there originated rivers of ice, which flowed in all directions and joined in one continuous sheet, similar to the *Inlandsis* of modern Greenland, but even much vaster. On one side, these masses of ice, overflowing the North Sea, came into contact with the glaciers which covered a part of the British Isles. On the other side, they spread over Denmark, Holland, Germany and Russia, the soil of which is still covered with moraines and erratic blocks of Scandinavian origin (Fig. 19).

Geologists have come to recognize several stages in the advance of the glaciers—*glacial* phases—separated from each other by stages of recession of the glaciers, or *interglacial* phases. The number of these periods varies according to different authors, or according to the countries. J. Geikie[1] defines six for the British Isles. Penck and Bruckner[2] have described four in the Alps. In France there have long been distinguished, in most of the mountain regions, traces of three principal glacial phases, the oldest dating from Pliocene times, while the two others are Quaternary.

The moraine formations, as one would naturally expect, contain neither fossils nor human remains; they cannot then be of direct interest in the study of palæontology, but they contribute to it valuable data regarding the climatic conditions of Quaternary times, and sometimes help us to establish useful stratigraphic relations.[3]

Ancient Alluvial Deposits.

Away from the bed of a stream or tidal river and up the sides of the valley are to be found, above the flood-plain, areas of varying size arranged in terraces covered with gravel or rolled pebbles (Fig. 20). These

[1] Geikie, James, *The Great Ice Age*, 3rd ed. (London, 1894). *The Antiquity of Man in Europe* (Edinburgh, 1914).

[2] Penck, A., and Bruckner, E., *Die Alpen im Eiszeitalter* (Leipzig, 1909).

[3] A German school, too much in vogue in France, seeks to make glacial formations predominant in the classification of Quaternary times. But specialists in the different countries have not yet reached agreement; there is every shade of opinion, from that admitting only one single glacial phase, to that asserting at least half a dozen. Apart from other considerations, this diversity of opinion must make us cautious and inclined to doubt the value of a system of chronology based solely on glacial phenomena.

alluvial deposits have been laid down in Quaternary times by floods, due either to an extraordinary increase in rainfall,

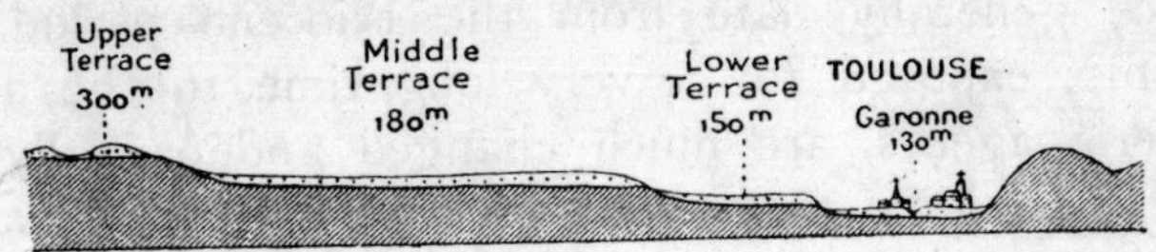

FIG. 20.—Outline across the valley of the Garonne at Toulouse.

or to the rapid melting of the glaciers. As each valley shows formations disposed in terraces at several levels, it may be concluded that the hollowing out of the valley was not

FIG. 21.—An old Sand-quarry at Chelles (Seine-et-Marne).

There may clearly be distinguished two sets of alluvial deposits occupying the bottom of the Marne Valley and resting uncomformably the one upon the other (along the line x x). The lower set is of Chellean age. The upper set, cutting into the former, is of more recent age, Acheulean or Moustierian. Phot. D'ACY.

continuous, but was subject to periods of cessation, followed by new activity; here is a phenomenon of periodicity parallel to, and often correlated with that of the glacial phases.[1]

[1] These features can and must often be explained also by changes in the general slope of water-courses in consequence of movements of the earth's crust, or to displacements in the line of the sea-shore.

In general, *these alluvials are the more ancient the higher the level they occupy.* Those which cover the high plateaux in France generally date from the Pliocene period. Their components, exposed for a very long time to the action of atmospheric agents, are much changed and partially decomposed. The middle terraces of the rivers of France and their large tributaries belong to the beginning of Quaternary times, and the lower terraces to more recent periods. By this definitive altitudinal arrangement, there may be established the relative age of the relics yielded by these alluvials, such as the bones of animals, the bones of man, or the products of man's industry.[1]

A serious source of error arises, however, from the alterations undergone by these gravels, which are constantly undermined, displaced or swept away by the streams themselves. Often, indeed, in ancient alluvials at a definite level, the presence of elements derived from all the older levels may be observed. An imported fossil must be distinguished from a fossil contemporary with the layer in which it is found. It must not be forgotten that *the true age of an alluvial deposit is that of its most recent fossils.*

Muds. Peat-Bogs. Calcareous Tufas.

In many countries, particularly the north of France, Belgium and Germany, the ancient alluvials are covered with *silts* (or *loess*), due either to the transport of dust by the wind, or to the violent rush of rain down the slopes, but most often to both causes together. These muds contain shells of land molluscs (Fig. 22). They are of the utmost importance in our study, because of their wealth of human fossil remains of all kinds. We cannot apply to them the rule laid down above with regard to alluvials, and take their altitude as a basis for establishing the relative ages

FIG. 22.—Shells of land Molluscs from loess. Left, *Helix hispida.* Right, *Pupa muscorum.*

[1] It must be noted that the rule is not absolute. Certain valleys, like that of the Marne at Chelles (Fig. 21), may have been completely hollowed out even below their modern channel since the Lower Pleistocene. New alluvials, due to change in the general gradient of the streams, may then have filled up the bottom of the valley, superimposing themselves upon the older alluvials, and occupying a higher altitudinal level than they did.

of the layers. The muds form a kind of cloak cast over the sides of the valleys and their alluvial formations (Fig. 23); these may belong to different periods, but, owing to their continuity, *the muds may actually be contemporaneous on the plateaux and in*

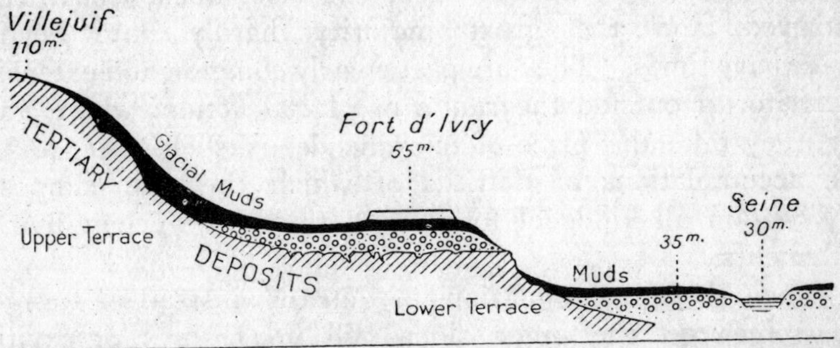

FIG. 23.—Geological section near Paris, to show the disposition of the Pleistocene muds, extending from the summit of the plateaux to the bottom of the valley, whilst the ancient alluvials are clearly deposited in terraces.

the bottoms of the valleys. Ignorance of this fact has led many prehistorians, on occasion, into unfortunate mistakes.

I shall merely mention the *peat-deposits*, which generally date, in France at any rate, from a relatively recent period (Neolithic), and which consequently do not possess the same interest for us.

Calcareous tufas, deposited by springs charged with carbonate of lime, encrusting springs, often contain imprints

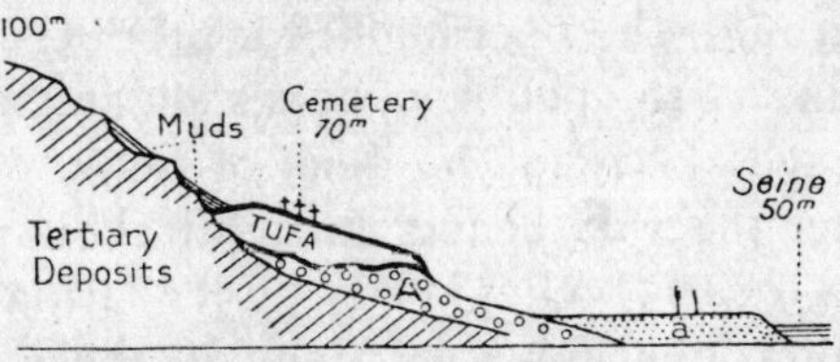

FIG. 24.—Geological section of Pleistocene tufas at La Celle-sous-Moret (Seine-et-Marne). The tufas, with warm flora, rest on Pleistocene alluvials, overlaid at one point by muds containing a cold fauna. A, Pleistocene alluvials. *a*, Recent alluvials.

of plants, remains of invertebrates, and bones of mammals, which give us valuable information relating to climatic conditions. Certain of these tufas have even yielded human relics, such as La Celle-sous-Moret, near Paris (Fig. 24), Cannstadt, in Germany, and others.

The debris on the slopes is due to the disintegration of the underlying rocks of a region. These accumulations may belong to any age, for the influence of the atmospheric agents to which they are due has been exerted at all times, but the most ancient have been destroyed, and the great majority hardly date beyond Quaternary times. They are particularly common and extensive in territories outside the range of glacial action, where, in a way, they take the place of moraine deposits (Fig. 16, p. 33). The accumulations of detritus of which they are composed may contain all kinds of fossils and traces of ancient human settlements.

Screes and Debris.

When these accumulations of detritus, instead of forming talus-screes on the outer slopes, fill up fissures or cavities hollowed out in the solid rock by natural agencies, agglomerates containing angular fragments are formed, known by geologists as *breccias*. They are often composed of bones of animals, and in such a case are termed *bone-breccias*. If the fissures or cavities thus filled attain any considerable extent or depth, they afford examples of every gradation up to grottos and bone-caves.

Grottos and Bone-Caves.

In regions formed of calcareous rocks riddled with subterranean cavities, the play of atmospheric erosion and the progressive hollowing of the valleys has established communications between these caves and the outer air. Through the openings thus produced, fierce torrents have carried in earth, pebbles, bones of animals, and so on. Further, the caves became the dens of lions, of bears, and of hyænas; thither the wild beasts brought their prey, and there left their own dead bodies (Fig. 25). Finally, they often served as habitations for the Men of these times. From our point of view, two kinds of subterranean caves may be distinguished: first, caves serving as dens, the deposits of which contain mainly animal remains, and second, the habitations of cave-dwellers. The latter did not penetrate so far into the interior of the earth's crust, and sometimes they only occupy the entrance of the deeper caves. Often, indeed, they are only simple shelters under overhanging rocks (Fig. 26). In such cases the cave deposits are formed chiefly by human agency, cinders from hearths and kitchen debris, in which are mingled products of man's industries and sometimes even the bones of the ancient inhabitants themselves.

All these remains of Quaternary times have been gradually buried under depths of rubble, so that the deposits in the interior of grottos or caverns now form, for the naturalist, archives as valuable as they are time-honoured. The more such deposits lend themselves to detailed and stratigraphic

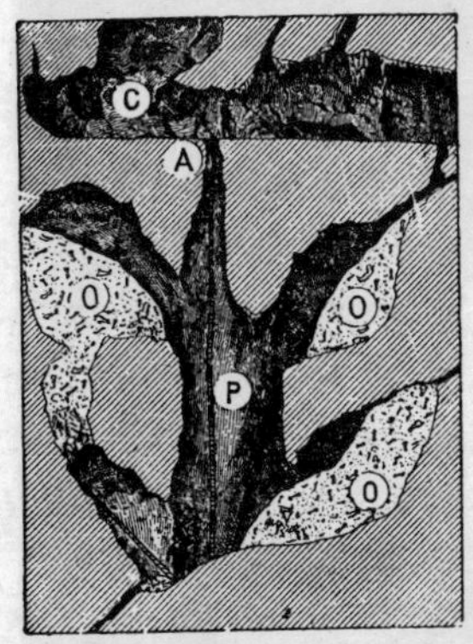

FIG. 25.—Section of a Bone-pit in the Gargas Cave (Haute-Pyrenees).

C, bottom of cave. A, entrance to pit. P, O, clay, yielding skeletons of bear, hyæna and wolf, now in Palæontological Gallery of National Museum of Natural History, Paris.

FIG. 26. — Rock-shelters at Bruniquel (Tarn-et-Garonne), inhabited during the Reindeer Age.

analysis, the more exact is the information they yield regarding the ancient stages of the human race in our country; and in this connection may be mentioned the Cave of Mas d'Azil, in the Pyrenees, and the Grimaldi Caves, near Mentone (Fig. 27).

Volcanic Formations.

Volcanic countries are very often rich in fossils. The dense streams of lava, ejected at burning heat, do not of course contain them, but the lava-flows are generally poured into the depths of the valleys, and cover the contemporary alluvials. These alluvials and all their palæontological content have thus been, in a sense, sealed up and preserved from all outside interference, and are most valuable for chronological study.

Volcanic discharges, cinders and lapilli, may bury the dead bodies of animals ; they are often borne along and redistributed by water, in association with all kinds of debris scattered over the surface of the soil, and with it are deposited further on as stratified layers of volcanic tuffs. Pithecanthropus and the

FIG. 27.—The Grimaldi Caves.
Photograph taken by Davanne about 1870, before the construction of the railway.

Denise skull from near Le Puy, were found in layers of this nature.

Classification of Quaternary Times. The Stratigraphical Method.

Even the detailed study of these different geological formations would not in itself be sufficient to enable us to establish a correct chronology or classification of Quaternary times. The *stratigraphical method*, which is, in general, the most certain and the most precise, is here more difficult of application than in the preceding geological ages. The Quaternary deposits, indeed, varied as they may be, are isolated or placed side by side, rather than superimposed. Marine formations, so important in the consideration of the preceding ages, no longer cover great expanses, and it is difficult to correlate them with the land formations. In short, we have only local and limited successions, each representing only a fraction of the duration of Quaternary times.

Quaternary Fossils, and the Method of Palæontology.

The *palæontological method* also has its weaknesses. There are no longer marine fossils so widely distributed as were the Graptolites of Primary times, the Ammonites of Secondary, and the Nummulites of Tertiary times. And, moreover, the marine invertebrates of the ancient Quaternary raised beaches cannot afford much help, for all the Quaternary invertebrates present a degree of specific development closely resembling their modern state.

The Quaternary plants were identical with the plants of to-day. They were too greatly influenced by topographical conditions to permit of their being used successfully to elucidate the problems of chronology. But they furnish valuable information regarding climate. For it has been found that, in one and the same region, certain layers contain species of warm climates, whilst others yield species of cold climates. It is permissible to suppose that the latter date from a glacial phase, while the warm flora corresponds to an interglacial phase.

Extinct or Emigrated Animals.

The study of fossil vertebrates, particularly of mammals, is of great assistance. The Quaternary species are usually so closely akin to the modern species that it is often difficult to distinguish them. But there are also some extinct species: the Ancient Elephant (*Elephas antiquus*), the Mammoth (*Elephas primigenius*), Merck's Rhinoceros (*Rhinoceros mercki*), the Woolly Rhinoceros (*Rhinoceros tichorhinus*), the Sabre-toothed Tiger (*Machairodus*), the Cave Bear (*Ursus spelæus*), and others. These animals are capable of furnishing excellent chronological data, for their extinction was not simultaneous. Thus of the three species of fossil elephants in our country, easily distinguished by their molar teeth (Fig. 28), *Elephas meridionalis* became extinct before *Elephas antiquus*, and the latter before the Mammoth. In the same way the Woolly

Rhinoceros persisted longer in our country than Merck's Rhinoceros (Fig. 29).

Side by side with these forms which are now extinct, there is a fairly large number of species still alive, but no longer inhabiting our part of the globe; they are species which have *migrated*, and these are equally useful for study. At the present day the mean temperature of our regions is less high than it was during the warmest period of Quaternary times, and it is less low than during the coldest period. So that, following Lartet's example, the Quaternary mammals may be divided into two groups: a *warm climate group*, which migrated to the south; and a *cold climate group*, which migrated to the north. The principal animals of the southern group are the Hippopotamus, which could live only in a country where the rivers do not freeze; the Cave Lion, which differed from the modern Lion only in its slightly larger build; the Cave Hyæna, which exactly resembled the Spotted Hyæna of Central Africa, although it was somewhat stronger in build; the Porcupine, and the Barbary Ape. . . .

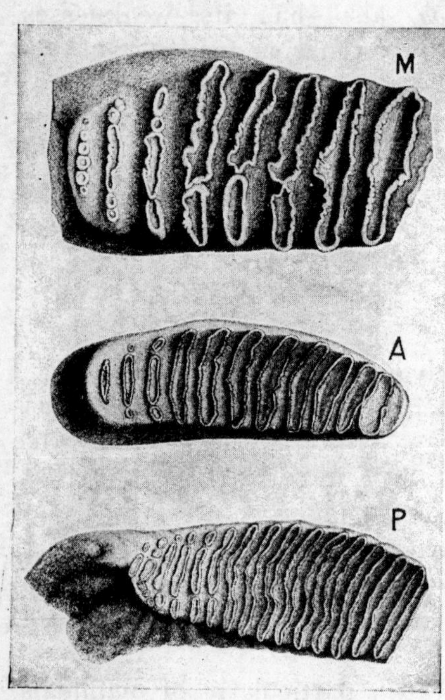

FIG. 28.—Molar Teeth of the three principal species of fossil elephants found in our country.

M, Southern Elephant (*Elephas meridionalis*); the tooth is very broad, the enamel plates are thick and widely separated.

A, Ancient Elephant (*Elephas antiquus*); the tooth is narrow, the enamel plates are less thick but again widely separated.

P, Mammoth (*Elephas primigenius*); the tooth is broad, the enamel plates are much more numerous, more compressed, and less thick.

Among the animals of the northern group, which now live only in polar regions, must be mentioned the Blue Fox, the Glutton, certain rodents such as the Lemming, and the Musk-ox.

The Reindeer deserves special mention. This ruminant, which cannot now live in a wild state in a degree of latitude lower than 60° N., was remarkably widespread throughout the whole of France at a particular epoch of Quaternary time,

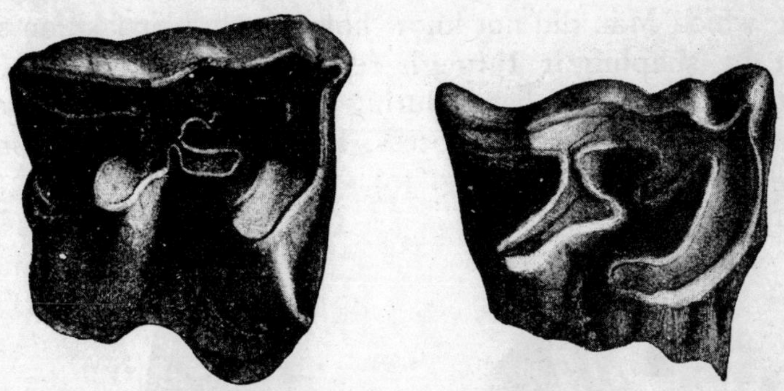

FIG. 29.—Upper Molars of *Rhinoceros mercki* (left), and of *R. tichorhinus* (right). Three-quarters natural size.

called on that account the Reindeer Age. At the same time, the Ibex and the Chamois, now confined to the high peaks of the Alps and of the Pyrenees, frequented the lowest plains of France.

Excellent fossil remains of all these animals have been obtained, but they must be made use of with judgment. The geographical distribution of animals had already become so complex during Quaternary times, that each region might have, and indeed did have, its own peculiar fauna; and this renders the synchronizing of deposits far from each other a task of much delicacy. Finally, we know that changes of climate have led to migrations of faunas, so that a species need not belong to the same age in two different localities.

Other species, such as the Horse, the Pig, the Ox, etc., are found almost everywhere, at all levels; we must therefore regard their presence as of no definite significance.

The Archæological Method.

For the establishment of the chronology of Quaternary times, a third method is available, the most ancient of all, the *archæological method.* It is based on the study of human industry, which has developed and changed, becoming by degrees more and more perfect; the least destructible of its products, implements of stone, are everywhere widespread.

Thanks to them, it was early recognized, as we have seen, that an age of stone preceded the age of metals. And in the Stone Age an older or *Palæolithic* phase can be distinguished, during which Man did not know how to work or fashion stone, except by shaping it through repeated blows; and a more recent or *Neolithic* phase, during which Man had learnt to fashion stone by another process, by wearing it down through friction and polishing it (Fig. 30).

FIG. 30.—A Palæolithic *dressed* axe, and a Neolithic *polished* axe. One-third natural size.

Further, in each of these phases a succession of periods has been determined, characterized by different forms of objects or implements. Archæologists have thus arrived at results of great value to collectors, and at the same time of great interest to science.[1] But classifications of this kind can have only local significance, for, even at the present day, very different industries may be contemporaneous in different regions.

Each method has thus its good qualities and its defects, and each is insufficient in itself. Certain classifications, established by pure geologists like James Geikie and Penck, incur the grave reproach of not taking sufficient account of palæontological

[1] Mortillet, G. de, *Le Préhistorique*, 1st ed. (Paris, 1883). Mortillet, G. and A., *Musée préhistorique* (Paris, 1881).

and archæological facts, of being based entirely upon a single physical phenomenon—the glacial phenomenon. Classifications based solely on fossils, like those of Lartet and of Boyd Dawkins, can serve only for the major divisions. In appealing to archæological facts exclusively, prehistorians have come to make a much greater number of stages in the series of prehistoric times and to establish a more detailed chronology. Unfortunately such classifications, initiated or perfected by Lartet, Gabriel de Mortillet, Piette, Cartailhac, Breuil, and others, can only be applied *a priori* to restricted areas of our continent alone. It is not possible to establish with certainty synchronisms or time-correlations at great distances, for the human factor is essentially mobile, changing and varied in its manifestations. Archæological aspects tend to be local, rather than to show sequences of ethnographical facts having uniform, general and contemporaneous significance. The idea of *facies* should play as important a part in prehistory as in geology.

Proposed General Classification of Quaternary Times.

In 1888 I proposed a classification of Quaternary times, based on the use and correlation of the three kinds of data.[1]

Such discoveries as have since been made in all parts do not seem to necessitate any alteration in the main lines of this first attempt, which I here reproduce, simplified and modified in some points of secondary importance.

First, I refer to geological and palæontological facts. They have a more general significance and bearing than ethnographical facts, because they are independent of human action, and because they dominate it. It therefore falls to geologists and palæontologists to establish the great divisions of Quaternary times, since these are of geological significance; and it is within the limits fixed by them that prehistorians may, in their turn, make subdivisions of an archæological and of a more localized nature.

A first problem is the delimitation of Tertiary and Quaternary times, a problem the more difficult to solve

[1] Boule, M., "Essai de Paléontologie stratigraphique de l'Homme" (*Revue d'Anthropologie*, 1888 and 1889).

because no great physical or biological fact can be appealed to in order to establish a division between them. In reality the Tertiary era still exists in every respect. It would certainly be best to suppress the term "Quaternary" in our classifications, and to include it in the Tertiary under another name, *Pleistocene*, as a period analogous to the Pliocene, Miocene, and other periods which preceded it.

Apart from this—the most rational view of the matter, which, however, does not seem likely to be readily adopted—we may either consider the Quaternary era as starting with the age of *Elephas meridionalis*, as I formerly thought, or immediately after this epoch, which would then remain in the Pliocene, and this is my present opinion. In any case, the matter is of little importance from our present point of view. Problems of grouping are somewhat arbitrary, for they represent only mental conceptions; what must be established is the succession of events and the synchronism of the phenomena.[1]

I leave, then, in the Pliocene, the fauna which includes *Elephas meridionalis*, and which is found in geological formations indicating a topography, and sometimes even a geography, somewhat different from those of Quaternary times. It appears now to be established that this period was marked by a great glacial extension (*Gunzian* or *Mindelian* of Penck's system) of which the moraines, few in number and badly preserved, appear to merge with the alluvials now covering the plateaux between the modern valleys. The principal fossils characteristic of these alluvials of the plateaux and of other contemporary formations are: the Southern Elephant (*Elephas meridionalis*), the Etruscan Rhinoceros (*Rhinoceros etruscus*), a Horse (*Equus stenonis*), the Great Hippopotamus (*Hippopotamus major*), and others.

The deposits of Saint-Prest in Eure-et-Loir, of Solilhac in

[1] M. Haug, in his *Traité de Géologie*, wished to extend the range of the Quaternary era, making it begin much sooner and attributing to it half of the Pliocene. This innovation, unjustifiable from a scientific point of view, has besides the serious drawback of creating inextricable confusion in the minds of many naturalists interested in various ways in Quaternary times, for it diverts the significance of expressions currently employed by all these naturalists, and results in a regular terminological dispute.

TABLE OF QUATERNARY TIMES.

Geological Divisions.	Geological Phenomena and Formations.	Palæontological Characters.	Archæological Divisions.	Fossil Men.
QUATERNARY — HOLOCENE or RECENT	Recent alluvials Peat mosses Climate almost as to-day	Species now found in the same country. Domestic animals	METALS { Iron, Bronze, Copper NEOLITHIC Transition Period—AZILIAN	MODERN MAN **HOMO SAPIENS**
QUATERNARY — PLEISTOCENE — Upper	Upper cave deposits. Upper part of the loess Climate cold and dry; regime of steppes and tundras **Post-glacial phase**	Steppe Fauna REINDEER Age — Tundra Fauna	PALÆOLITHIC — Upper { MAGDALENIAN SOLUTREAN	Fossil HOMO SAPIENS { **Race of CHANCELADE** **Race of CRO-MAGNON**
QUATERNARY — PLEISTOCENE — Middle	Maximum cave deposits. Loess. Alluvials only at low levels Moraines of the last **great glacial phase** Climate cold and moist	MAMMOTH Age — Mammoth—*Elephas primigenius* Woolly Rhinoceros—*Rhinoceros tichorhinus*, etc.	PALÆOLITHIC — Upper { AURIGNACIAN	Fossil HOMO SAPIENS { **Race of GRIMALDI**
QUATERNARY — PLEISTOCENE — Lower	Ancient cave deposits Alluvials of mid and lower terraces. Calcareous tufas **Great interglacial phase** Climate mild Moraines of the **penultimate great glacial phase**	HIPPOPOTAMUS Age — Hippopotamus—*H. amphibius* Ancient Elephant—*Elephas antiquus* Merck's Rhinoceros—*R. mercki*	PALÆOLITHIC — Lower { MOUSTIERIAN ACHEULEAN CHELLEAN	NEANDERTHAL MAN **HOMO NEANDERTHALENSIS** PILTDOWN MAN **HOMO DAWSONI** HEIDELBERG MAN **HOMO HEIDELBERGENSIS**
TERTIARY — PLIOCENE — Upper	Plateaux alluvials **Great interglacial phase** **Great glacial phase**	Age of the SOUTHERN ELEPHANT — Southern Elephant—*Elephas meridionalis* Rhinoceros—*R. etruscus* Horse—*Equus stenonis*		

Haute-Loire, of the Forest-bed in England, of Süssenborn in Thuringia, etc., may be reckoned to be on the boundary between the Pliocene and the Pleistocene, that is to say, as forming the transition between the Tertiary and the Quaternary eras. They seem to represent a great interglacial period, the *Norfolkian* of Geikie.

The Quaternary era comprises the *Pleistocene* period and the *Holocene* or Modern period.[1] The Pleistocene of geologists corresponds exactly to the *Palæolithic* of archæologists, the Holocene to the *Neolithic* and the ages of the *metals*.

Geological Divisions.		Archæological Divisions.	
Quaternary Era.	Holocene Period.	Period of the Metals.	Iron Age. Bronze Age. Copper Age.
		Neolithic Period.	
	Pleistocene Period.	Palæolithic Period.	

Divisions of the Pleistocene Period. Lower Pleistocene.

The Pleistocene period may be divided into three ages corresponding to as many stages. The following are their main characteristics, to which, however, I shall return at greater length when dealing with the fossil men of each age.

With the Lower Pleistocene is associated, in the first place, a fresh and apparently the most important of the extensions of the ice-sheet, the moraines of which extend furthest from the mountain centres (Penck's *Rissian* phase).

Following this glacial phase there came, in the second place, a great interglacial phase, distinguished by a fauna with species indicating a warm climate: the Ancient Elephant (*Elephas antiquus*), Merck's Rhinoceros (*Rhinoceros mercki*), the Hippopotamus (Fig. 31), the Sabre-toothed Tiger (*Machairodus*), monkeys, and so on. The remains of these animals are found sometimes in the oldest alluvials of the recent series of freshwater deposits, sometimes in the deepest or the lowest deposits of caves.

[1] The term "Quaternary" is often used as synonymous with "Pleistocene." This is erroneous, since it mistakes the part for the whole; but the term, nevertheless, has passed into current use.

It is also in the deposits of this period that the oldest stone implements universally recognized as authentic—the amygdaloid flints—are found. From the archæological point of view these represent the *Chellean*, a name derived from the locality of Chelles, in the Department of the Seine-et-Marne, and which I consider excellent, at least in our present state of knowledge. Up to recent years, no bone-remains of Chellean Man had been recorded. It will be necessary to consider two finds recently made in formations which, according to all the evidences, must have dated from the Lower Pleistocene. These consist of a few fragments of skull and jawbones, to which have been given the names *Homo heidelbergensis* and *Eoanthropus dawsoni*.

FIG. 31.—Skull of fossil Hippopotamus, about one-tenth natural size. Palæontological Gallery of French National Museum of Natural History, Paris.

Mid Pleistocene.

The Mid Pleistocene is characterized, from the geological point of view, by a return of those great glacial developments which have left the best-preserved moraines, situated, as a rule, within the limit of the moraines of the preceding glacial extension (*Wurmian* phase of Penck). It is also characterized by the formation of alluvial sheets, sometimes superimposed upon those of the Lower Pleistocene, but most often situated below the level of the latter, and forming terraces *lower* in relation to the middle terraces. This is the period of the principal deposits of cave debris, and of the formation of great masses of mud or *loess* in France, for both these deposits have a similar origin. It is also the period of the last hollowing out or filling up of the valleys, of the penultimate sculpturing of the surface configuration of the present day.

The fauna is very different from that of the Lower Pleistocene. The Hippopotamus no longer inhabits our countries. The Ancient Elephant is replaced by the Mammoth (Fig. 32). Merck's Rhinoceros persists a little longer, and was succeeded by the Woolly Rhinoceros which, as we know, was adapted, like the Mammoth, to the rigors of a glacial climate. In France the Mid Pleistocene saw the reign of the great Carnivores: the Bear, the Hyæna and the Cave Lion (Fig. 33). It corresponds fairly exactly to the archæological period called *Moustierian*, after the Cave of Moustier in the Dordogne.[1] Between the Chellean and the Moustierian is interposed the period known as Acheulean, after Saint-Acheul, near Amiens.

FIG. 32. — Skeleton of Mammoth from Beresowka (N. E. Siberia). In Petrograd Museum. Height 3·25 metres.

In the cave-deposits of the Mid Pleistocene were found many human bones, sometimes almost complete skeletons, which revealed a prehistoric type very different from modern man, namely Neanderthal Man (*Homo neanderthalensis*), which we shall study at length, from French material of exceptional importance.

Upper Pleistocene.

The Upper Pleistocene has no clearly defined characters from the geological or stratigraphical points of view. The deposits corresponding to it present a topography and hydrography very much resembling those of the present day, but, especially towards the end of the period, they were contemporaneous with a dryer climate. In the caves the deposits are principally formed by relics of Man, cinders from hearths and kitchen refuse.

In spite of some forward movements of the ice, gradually decreasing in importance, but considered by certain authors

[1] Generally written "Mousterian." It seems to me more correct to write Moustierian.

as glacial phases, this period marks the general retreat of the great glaciers of the preceding period; it is, therefore, often distinguished as *Post-glacial*.

The fauna is still essentially that of the Mid Pleistocene. The large species are the same (Mammoth, Woolly Rhinoceros, and so on), and in this respect the Mid and Upper Pleistocene may be regarded as forming one geological unit, as distinct from the Lower Pleistocene. But the stratigraphy of the caves generally allows of the recognition of

FIG. 33.—Exhibit of Quaternary Carnivores in the Palæontological Gallery of the French National Museum of Natural History, Paris.

From left to right—Skeleton of Cave Hyæna; three skeletons of Cave Bears; two skeletons of Cave Lions; in the right foreground Cave Wolf.

several successive layers enclosing slightly different faunas. It is impossible, however, always to affirm that these differences are not due to the interference of man. His intervention must certainly have brought to the settlements, to the exclusion of many others, the bones of certain animals which perhaps could not have come there by the simple play of natural causes. The Reindeer is accompanied first by a very cold fauna, known as the fauna of the *tundras*, because it contains the principal members of the fauna of the frozen and desolate countries which bear this name to-day, such as the Musk-ox, Glutton, Blue Fox, Lemming, Snowy Owl, and others, whose remains are found in association in certain layers of the Arctic peat-bogs. Then comes a less cold fauna, indicating a climate and habitat similar to those of the Russian and Siberian steppes, and including the Saiga Antelope, Prairie Squirrels, Jerboas,[1] and others, whilst our low plains are invaded by the Chamois, the Ibex, the Marmot, and, everywhere abundant, these large ruminants, the Bison and Great Extinct Ox. But the Reindeer (Fig. 34) is always the dominant species through-

FIG. 34.—Modern Reindeer.

out the whole period, which has long been known as the *Reindeer Age*.

Archæology has been of special service in enabling a very clear line to be drawn between the two great periods of Mid and Upper Pleistocene. The distinction between the Mammoth Age and the Reindeer Age, first made by Lartet, is justified much more by archæology than by palæontology. It crops up again in the almost parallel terms of Moustierian and Magdalenian, invented by G. de Mortillet. Modern prehistorians are agreed in dividing the Palæolithic into the *Lower Palæolithic*, comprising the Chellean and the Moustierian

[1] See various works of Nehring, A., notably *Über Tundren und Steppen der Jetzt-und-Vorzeit* (Berlin, 1890).

periods, and the *Upper Palæolithic* or Reindeer Age, thus affirming the archæological and anthropological individuality of two periods difficult to distinguish by the resources of geology and palæontology alone.[1]

Ethnography goes even further, and allows of the introduction into the Reindeer Age, or Upper Palæolithic, of several divisions of real importance, at least for Western Europe. The names given to these are: *Aurignacian* (from Aurignac in the Haute-Garonne, where there is a grotto explored by Lartet), *Solutrean* (from Solutré, Saône-et-Loire), and *Magdalenian* (from La Madeleine, in the Dordogne).[2]

The fossil men of these periods are fairly numerous and very different from Neanderthal Man. They have already the characters of Modern Man (*Homo sapiens*), from whom they cannot be distinguished except on the ground of race or of ancestral variety.

Certain deposits, of which the grotto of Mas d'Azil may be taken as typical, correspond to a period called *Azilian*, which exhibits, as we shall see later, all the characters of a transition from Pleistocene to Holocene.

2. *Absolute Chronology.*

Relative and Absolute Chronology.

Relative chronology, which we have just been studying, is distinct from absolute chronology. The latter must furnish dates in units of time—years, centuries or millenia. History has its absolute chronology, which, it may be said, is the more determinate or precise as it deals with events more or less close to us. On every hand there arises the demand for an absolute chronology for prehistory. The first question asked by the

[1] The geological expressions, Upper, Middle, and Lower Pleistocene, should not be absolutely confounded with the archæological expressions, Chellean, Moustierian, and so on. The latter can only have a local or regional significance, while the former should be, in principle at least, of universal application. It will be useful to employ, as the case demands, one or other of these two nomenclatures, in order to avoid confusions which prehistorians often make.

[2] Breuil, H., "Les subdivisions du Paléolithique supérieur et leur signification" (*Congrès international d'Anthropologie*, Geneva meeting, 1912).

public at large, in connection with any prehistoric fact, has to do with its date.

Now we must honestly admit that, in the present state of the science, it is almost impossible in most cases to answer such a question.[1] The temerity with which certain writers, without sufficient reservation, present figures which are all the more fallacious because their very detail suggests precision, makes an unfortunate impression on the minds of scientists aware of our ignorance. I would not for a moment suggest that the problem may or must remain unsolved, for no one can fix a limit to the progress of the human mind.[2] But I hold that the first duty of the scientist is to respect the confidence of those who come to him in good faith, and to set before them only such facts as are sufficiently well established.

In order to satisfy as far as possible the curiosity of my readers, which is after all quite legitimate, I shall briefly summarize the results of the leading attempts at establishing a definite chronology, and shall draw from these results such conclusions as seem to me most reasonable.

There is no need to dwell here upon the estimates of the ancient writers, of biblical legends, of the Chaldean, Egyptian or Chinese traditions and chronicles, and others, which in this matter cannot be considered reliable.

Different Methods of Time Estimation.

Since the great antiquity of Man was scientifically established, men of science have constantly striven to translate it into figures. They have conceived all kinds of time-measuring methods, based on astronomy, on geology, and even on biology.

In principle, astronomical methods ought to be most

[1] Thanks to certain historical dates furnished by Chaldea and Egypt, and to archæological comparisons established step by step, satisfactory results have been obtained for the Ages of Metals in the West. (See Chap. IX). But these ages only represent the end of our prehistoric era. What I am discussing here relates to an infinitely more remote past.

[2] "When science undertook, scarcely more than half a century ago, to measure the mass and the dimensions of atoms, its first estimates varied, in the ratio of 1 to 20, to such an extent that the problem seemed to certain minds insoluble and even non-existent. It is, however, solved to-day." (Houllevigue, "Causerie scientifique," in the *Temps*, 16th Dec. 1911).

satisfactory, since astronomy has given us our units of time. The difficulty is to discover a geological event which can be correlated at once with the antiquity of Man and with a measurable astronomical phenomenon. It was believed to have been found in the development of the glaciers, which were attributed either to variations in the eccentricity of the earth's orbit (that is to say, to differences in the distance of the earth from the sun), or to variations in the obliquity of the earth's axis (precession of the equinoxes). As astronomy can calculate the dates of such periodic variations, it should be enough to choose from among the dates those which correspond to the variations most favourable for the establishment of a glacial regime, in order to know the absolute age of the Glacial Period and, in consequence, the age of the men contemporary with this period.

Thus, according to the English astronomer Croll, the Glacial Period would have begun with the last great cycle of eccentricity in the earth's orbit (corresponding to the maximum distance of the earth from the sun, that is, to a maximum of coldness) 240,000 years ago; further, it would have lasted 160,000 years and would have come to an end 80,000 years ago. According to Lyell, the Great Glacial Period should have coincided with an older period of great eccentricity, and this would make it date back to 800,000 years ago.[1]

This theory has been attacked alike by astronomers, physicists and geologists. It has not been proved that the Glacial Period or periods were related to the astronomical phenomena in question; and we shall see that there are excellent reasons to-day for believing that the end of the Glacial period did not date so far back as was believed by Croll and Lyell.

It will be enough merely to mention a few attempts at time-estimation based on palæontology and the phenomena of organic evolution. It was believed that a numerical appraisement could be made of the periods corresponding to changes undergone by the fauna of a country through extinction,

[1] Lyell, C., *Principes de Géologie*, Fr. ed., Paris, 1873, t. 1, ch. xiii., p. 351. [*Principles of Geology*, 10th Eng ed., London, 1867, vol. i., ch. xiii., p. 291 *et seq.*]

transformation or the appearance of species. The hints thus obtained are altogether vague and indefinite; they have, however, the merit of impressing on the mind the idea of the immensity of even the most recent geological times, when we learn, for example, that the changes which have taken place in the animate world since the end of the Palæolithic period are imperceptible or insignificant.

The geological methods of estimating time are more numerous, and almost all depend upon much the same principle: A geological phenomenon is selected, the effects of which can be measured throughout a unity of time—that is, the rapidity of its progress in the present period, or throughout a lapse of time determined by historical data, can be gauged. From this result, by simple rule of three, the time may be calculated necessary to produce effects conforming to the prehistoric period the date of which we desire to discover. If we know, for example, the absolute value of the time required for the deposition of a certain thickness of alluvium, it will be easy to estimate the time required for the deposition of a greater mass of alluvium, the formation of which corresponds to a definite geological or archæological period.

Unfortunately, conditions are far from being so simple as this. The first difficulty lies in really knowing what we wish to measure, especially in the case of events the beginning and the end of which it is impossible to determine with exactitude. The expressions we must necessarily employ, such as the "beginning" or "end" of the Glacial Period, the "date of Man's appearance," and those serving to designate the divisions we make in the succession of the ages and of the events correlated with them, are perforce very vague; they correspond to no absolute reality, for such abrupt divisions exist only in our minds.

Further, the duration of the same event may have varied in different countries. It is clear, for example, that the Glacial Period came to an end much sooner in France than in Scandinavia. Another, and still more serious objection, is that the geological methods of time-estimation depend on an assumption of the permanence and unbroken regularity of natural processes; and this is extremely improbable. In any event, in the majority of cases it is impossible to make allowance for the counteraction of destructive forces, which may have worked in inverse ratio to the constructive forces, weakening, diminishing and sometimes annihilating the effects of the latter.

But whether these criticisms be justified or not, and their significance is evident, investigators, at any rate, have turned their attention to all kinds of natural accumulations. They have examined, in their turn, the peat-mosses, alluvials, glacial moraines, cones of mountain-torrent detritus, lake-deposits, deltas, marine ooze, calcareous concretions, alteration in rocks by atmospheric agents, the hollowing out of gorges, the backward cutting of waterfalls, the rate of progress or of retreat of glaciers, oscillations in the level of the sea, etc. I have noted more than forty attempts of this kind, and others must certainly have escaped me.

The Results, and a Discussion of them.

On the following page will be found in tabular form a number of estimates of the duration (1) of the Quaternary era; (2) of the Glacial period; (3) of Post-Glacial times. The figures denote the number of years.

The numbers in the first two columns vary, in round numbers, from 10,000 to 1,000,000, that is to say, in a ratio of 1 to 100. The difference, therefore, is enormous; partly because many of the estimates are largely sentimental, the lowest having certainly been influenced by the philosophic or religious views of their authors. But, for the most part, such a discrepancy witnesses to the inaccuracy or inadequacy of the methods employed; and amongst them all it is very difficult to choose one that is better than the rest.

The discrepancy of the figures in the third column, relating to the duration of Post-Glacial times, is much less great. Ignoring certain evaluations clearly erroneous, the 80,000 years of the astronomer Croll and the 100,000 years of the geophysicist Forel, we find that the difference is only between 4,000 and 30,000, that is, in the ratio of about 1 to 7. It is still far too great, but it shows some improvement.

SUMMARY OF TIME-ESTIMATIONS IN YEARS.

Author.	Quaternary Times.	Glacial Period.	Post-Glacial Period.
ARCELIN and FERRY, French geologists	10,000	...	4 to 5,000
HOLST, Swedish geologist	30,000	17,000	5 to 6,900
PRESTWICH, English geologist	...	25,000	8 to 10,000
WARREN UPHAM, American geologist	100 to 150,000	20 to 30,000	6 to 10,000
BECKER, American geologist	...	50,000	...
HEIM, Swiss geologist	...	100,000	16,000
G. DE MORTILLET, French prehistorian	230 to 240,000	100,000	16,000
RUTOT, Belgian geologist and prehistorian	...	140,000	7 to 8,000
CROLL, English astronomer	240,000	160,000	80,000
A. M. HANSEN, Norwegian geologist	...	130 to 190,000	7 to 9,000
J. LUBBOCK, English prehistorian	...	200,000	...
WALCOTT, American geologist	400,000	...	...
SOLLAS, English geologist	...	400,000	17,000
OSBORN, American palæontologist	500,000	500,000	25,000
J. GEIKIE, Scottish geologist	620,000 at least	...	...
DANA, American geologist	...	720,000	...
LYELL, English geologist	...	800,000	...
PENCK, German geographer	500,000 to 1,000,000	500,000 to 1,000,000	20,000
L. PILGRIM, German geographer	1,620,000	1,290,000	...
R. S. WOODWARD, GILBERT, RUSSEL, WINCHELL, E. ANDREWS, EMERSON, American geologists	...	...	7 to 10,000
HICKS, American geologist	...	...	15,000
SARAUW, Danish prehistorian	...	...	10 to 25,000
DE GEER, Swedish geologist	...	...	12,000
GÔSSE, Swiss prehistorian	...	...	18,280
NUESCH, " "	...	...	24 to 29,000
FOREL, " "	...	...	100,000

An analysis, into which I cannot enter here, of the processes of calculation which give the lowest estimates, shows that these processes, for example those of Arcelin and Ferry, are marred by error. If, then, we take as the extreme limits the figures 8,000 and 24,000, the ratio of difference is reduced to 1 to 3. Here we approach an agreement sufficiently close to give us confidence. Further, if we note that the majority of the estimates, in spite of the fact that they are based on very different phenomena, fluctuate between 8,000 and 15,000, that is to say, in ratio of 1 to 2, we are led to place a certain amount of trust in dates which now belong clearly to the same scale of magnitude.

These figures would certainly have been still closer to each other had they really been applied to the same duration of time; but this is not the case, for the meaning of the expression *Post-Glacial times* varies remarkably, according to the authors who use it and the countries in which it is used.

If, in this case, agreement is much less imperfect than in that of the duration of the Glacial Period or of the sum total of Quaternary times, it is, first, because the space of time concerned is much shorter; secondly, because the phenomena that must be studied are much nearer to us, so that their effects are better preserved and more exactly apprehended; and finally, because these phenomena themselves more closely resemble the happenings of modern times, so that the latter afford more equivalent terms for comparison.

Duration of Post-Glacial Times.

One of the most ingenious and most satisfactory methods, too lengthy for description here, is that by which the Swede, de Geer,[1] has been able to measure the rate of retreat of the last great Scandinavian glacier. If we compare the data published by this geologist with the results obtained by Sarauw, Kjellmark and other European scientists, on the one hand, and by various American geologists on the other, we reach a very satisfactory degree of agreement. I consider that we may reckon at 10,000 years ago, the extreme end of the Glacial Period in France, the departure of the Reindeer, the beginning of the superficial peat-bog formations, and the first evidences of the Neolithic civilisations.

The round number I suggest shows that I have only an approximation in view; and I consider it is rather moderate than excessive. It is the only suggestion I am able to make.

The Immense Duration of Quaternary Times.

I have not the assurance to compute the duration of the older archæological or geological periods, being convinced, unfortunately, that the estimates put forward do not, so far as is known, rest on secure foundations; but I hasten to add that everything tends to give the impression that an immense lapse of time separates our period from that of our first forefathers.

[1] Geer, G. de, "A Geochronology of the last 12,000 years" (*Congrès géolog. intern. de Stockholm*, 1910, p. 241.

Let us leave on one side the archæological aspect of the question, in order to consider only physical and biological facts. Since the end of the Pleistocene times of the geologists, or the Reindeer Age of the archæologists, that is to say, since about 10,000 years ago, the geological changes undergone by our country are inappreciable. This will serve as a starting-point.

The deposits corresponding to the period immediately preceding, that is to the Reindeer Age, are already of remarkable thickness. This is not a matter solely of human contributions, which may have accumulated with great rapidity. I have specially in view these sterile layers due to the activity of atmospheric agents, which often alternate with the hearths, and correspond to long periods of disuse of the caves and shelters. I believe that we can, without fear of error, attribute to this period a duration equal to, if not greater than, that which separates us from it.

Now these deposits of the Reindeer Age, in spite of their relative importance, which impresses us in excavations, in reality play only an insignificant part in the topography of our country. The bones they contain are barely fossilized; nearly all of them belong to animals or to men in every way similar to their modern representatives. But when we go back as far as the Moustierian period, we observe in other directions certain important changes. Here everything bears witness to a different topography, brought about by physical forces of which impressive traces are everywhere visible: the demolition of mountain regions, the accumulation of moraine deposits over thousands of square miles, the last stages of the sculpturing of the valleys, and the formation of the lower alluvial terraces; enormous deposits of silt over the land surfaces, and of clays containing bones in subterranean caves; variations in the shore-line; repeated volcanic manifestations, and so on.

These physical phenomena are accompanied by appreciable changes in the fauna, particularly by the disappearance of several species of large mammals, whose bone-remains are more fossilized than those of the Reindeer Age.

Man, at least in our region, belongs to a species different from Modern Man (*Homo sapiens*). Who can hope to have any accurate notion of the duration of this period? And yet, can we refuse to accord to it an impressive number of millenia?

We are still, however, very far from the time at which Man's earlier presence in our country has been indubitably proved.

This Man, belonging to very old Palæolithic times, lived in a physical and biological environment altogether different from that of his successor. The topography and even the geography of the Acheulean and Chellean periods show further changes of very considerable extent and duration. The seas and the continents had not yet assumed their present configuration; the British Isles were attached to the continent; Europe and Africa were connected by land-bridges. The sculpturing of our modern valleys, which was particularly the work of the latter part of the Pliocene, had not yet been completed. That was the period of the middle terraces of the Seine, the Somme, the Garonne, etc. Hundreds of craters, now breached and broken, still illumined the Central Massif of France. The geological formations of the period consist of deposits the antiquity of which is still clearly evident in the alteration undergone by their contents.

As different are the scenes revealed by organic nature. Europe was peopled by Asiatic or African mammals, of which many have long been extinct. Others have changed, and their evolution can only have occurred with very great slowness. The human industries of these remote times show everywhere the same obvious characteristics. The few bone-remains of the makers of the primitive implements reveal types very different from those of to-day. Everything now fades away in the mists of antiquity. The only chronological impression we can gather from these pictures is of an immense duration of time; the giddiness of immeasurable space begins to overwhelm us.

And so, if none of the figures suggested for the duration of Quaternary time and for the antiquity of Chellean Man can satisfy our desire for precision, none can surprise, still less can any astound us.

Even now, we are only at the dawn of Quaternary times. If, pursuing our giddy course, we penetrate into the Tertiary era, in our search for the real beginnings of Mankind, we must needs count no longer by hundreds, but by thousands of millenia!

CHAPTER III

LIVING PRIMATES AND FOSSIL MONKEYS

BEFORE entering into the heart of our subject, we must make a fresh preliminary study. We cannot well understand and appreciate the morphological characters of our distant ancestors, until we have gained some notion of the general zoological bearings of the Primates, and reviewed the discoveries relating to fossil monkeys.

Living Primates.

The term *Primates*, "the first" of the mammals, was created by Linnæus, who used also the designation *Anthropomorphs*.

The Primates include animals of somewhat different appearance, but distinguished from all other mammals by a combination of the following characters: a large brain-box, containing a highly-developed brain; fore-limbs adapted for grasping, and terminating in hands with flat nails; teeth adapted for a mixed diet; two mammary glands on the breast.

Classification of the Primates.

As this definition applies to Man as well as to all the other Primates, the classification of the group has been the subject of special discussion. Some naturalists, as Isidore Geoffroy Saint-Hilaire and de Quatrefages,[1] wished to set up a separate kingdom for our species, because of the superiority of its intelligence and of its "religiosity." But in the classification of animals we do not take as a basis their intellectual characteristics. Why change the method when we come to Man? Linnæus and

[1] Quatrefages, A. de, *L'espèce humaine* (Paris, 1876); *Introduction à l'étude des races humaines* (1887).

Lamarck clearly comprehended this; and Darwin observed that much greater differences are to be found in the class of insects, for example between an ant and a parasitic insect, than between man and monkeys; and he added with reason: "If Man had not been his own classifier, he would never have thought of founding a separate order for his own reception."[1]

Buffon classified Man as *Bimana*, Monkeys as *Quadrumana*; and Cuvier revived these terms.[2] In the genus *Homo*, Linnæus included Man and the large monkeys without tails—the Anthropoid Apes; here he had *Homo sapiens*, with his varieties *ferus*, *europæus*, *asiaticus*, and the rest, but here also he placed *Homo sylvestris*, or Orang, and *Homo troglodytes*, or Chimpanzee.

Cuvier completely separated Man (Bimana) from Monkeys (Quadrumana), and among the latter he distinguished a lower group under the name of *Lemurs*. To-day we may still accept the following general classification[3]:—

PRIMATES .	Bimana	Arms shorter than legs. Very large brain.	MEN or HOMINIANS. Very highly developed Primates.
	Quadrumana.	Orbits completely closed; arms longer than legs.	MONKEYS or SIMIANS. Primates definite and diversified.
		Orbits incompletely closed; arms shorter than legs.	LEMURS. Primitive Primates still little differentiated.

Lemurs.

The Lemurs, also called Prosimians, Pseudo Monkeys, or Fox-faced Monkeys, have neither the vivacity nor the intelligence of the true monkeys. The face is hairy instead of smooth. The skull is easily

[1] Darwin, Charles, *The Descent of Man*, London, 1871, vol. i., p. 191.

[2] For this they have been criticized, and not unreasonably, for *anatomically* the lower extremities of monkeys are *feet*, just like the lower extremities of Man. But Cuvier knew this as well as Huxley or Broca. It is therefore rather in a physiological sense that he uses the terms *Bimana* and *Quadrumana*. This reservation made, we may, I think, continue to use the terms, especially in a work such as this.

[3] For the history of the classification of the Primates, see Topinard, P., *L'Homme dans la Nature* (Paris, 1891); Gregory, W. K., *The Orders of Mammals* (New York, 1910).

FIG. 35.—A Representative of the Lemurs and of each great group of Monkeys.

Above, left—A Lemur, the Pied Lemur (*Lemur varius*), after A. Grandidier.

Above, right—A Flat-Nosed or New World Monkey, a Marmoset (*Hapale jacchus*).

Below, left—A Flat-Nosed or Cebian Monkey from South America (*Cebus capucinus*), after Elliot.

Below, right—An African Long-Nosed Monkey (*Cercopithecus brazzai*), from a specimen in the Natural History Museum of France.

Centre—An Anthropoid Ape, the Chimpanzee (*Troglodytes niger*), from a specimen in the collections of the same Museum.

distinguished from a monkey's skull: the occipital foramen is placed behind as in other mammals; the orbits are widely separated, directed laterally, and they communicate with the temporal fossa; the lachrymal foramen opens in front of the orbits. The cerebral hemispheres of the Lemurs do not completely cover the cerebellum; their structure is relatively simple. The Lemurs again have their lower incisors set in front and projecting forwards; they have long legs and short arms. They are still very primitive quadrumana and for the most part lead an arboreal existence.

Formerly widely distributed in America and in Europe, they are now found only in Africa, in Indo-Malay, but chiefly in Madagascar, where nine-tenths of the species occur, and where the group was formerly represented by strange giant forms, such as *Megaladapis*.

Monkeys or Simians.

The true monkeys comprise a graduated series ranging from low and insignificant forms, like the Marmosets, to powerful forms and those most akin to Man, the Anthropoid Apes; but they have a certain number of anatomical characters in common. The brain-box is highly developed in relation to the face, which is fore-shortened. The cerebrum, overlapping the cerebellum and covering it, is divided by fissures into lobes; it shows well-developed convolutions. The occipital foramen lies *beneath* the skull. The orbits, directed forwards, lie in front, are closed by a bony septum, and no longer communicate with the temporal fossa; the lachrymal canal opens into the orbits. The teeth are arranged in a continuous series; the lower incisors are upright like their neighbours, not inclined. The fore-limbs are longer than the hind-limbs. The forearm, gifted with the power of movements both of pronation and supination, has become a perfect instrument for grasping.

Monkeys have expressive faces; their intelligence is superior to that of other mammals. They have strong passions; it has been truly said that they reflect Mankind in its lowest qualities. They inhabit warm or temperate climates, living in trees or among rocks, and feeding on fruit, buds, eggs, and insects.

We may classify the monkeys as follows :[1]—

MONKEYS OR SIMIAN APES.

32 teeth (2 premolars). Narrow nasal septum (nostrils close together) CATARRHINIANS. Long-nosed, or Old World Monkeys FIG. 36.	No tail	ANTHROPOMORPHIC GROUP (Anthropoid Apes). Chimpanzee Gorilla Orang Gibbon
	Tail more or less long	CYNOMORPHIC GROUP (Dog-faced Monkeys). Baboons (*Cynocephalus*) Macaques Cercopithecus Sacred Indian Ape (*Semnopithecus*) and its allies
Large nasal septum PLATYRRHINIANS. Flat-nosed or New World Monkeys FIG. 37	36 teeth (3 premolars)	CEBIAN GROUP. Capuchins Sapajous or Weepers Ateles (Spider Monkeys) Howling Monkeys Sakis
	32 teeth. Digits with claws	ARCTOPITHECUS GROUP, or HAPALIDES. Marmosets Midas Monkey

[1] This classification and nomenclature are essentially French. In other countries, the grouping and particularly the terms employed are rather different; and the latter may cause some confusion which it is well to point out. Thus the general group of Monkeys or Simians is sometimes called the *Anthropoidea*, a term long in use in France, concurrently with that of "Anthropomorphes," to describe the higher monkeys—those most nearly related to ourselves. The Cynomorphic Apes of French authors constitute the *Cercopithecidæ* of British authors, and the Anthropomorphic Apes of the former are called *Simiidæ* in Britain, a name too closely resembling the expression Simian Apes, used by French writers to describe monkeys as a group.

General works to read or consult regarding the monkeys are: Gervais, P., *Histoire naturelle des Mammifères* (Paris, 1854) (old but still good). Flower, W. H., and Lydekker, R., *An Introduction to the Study of Mammals* (London, 1891). Elliot, D. G., *A Review of the Primates* (New York, 1912), (a large and superb monograph).

The HAPALIDÆ or MARMOSETS (genus *Hapale* and *Midas*) are the smallest of the monkeys. In certain characters they are related to the Lemurs, for their digits are not opposable. They have only 32 teeth, as in the large monkeys and Man, but they have 3 premolars and 2 molars, in place of 2 premolars and 3 molars.

The CEBIANS, like the Marmosets confined to South America, are inferior in organization and intelligence to the Old World monkeys. Their nostrils are widely separated, their long tails are generally prehensile. The thumb is barely opposable. They have 36 teeth (3 premolars and 3 molars on each side, above and below).

The CATARRHINE or LONG-NOSED MONKEYS of the Old World, with nostrils close together, have the same dental formula as Man (2 premolars and 3 molars). The CYNOMORPHS or DOG-FACED MONKEYS, of which numerous species are widely distributed in Asia and in Africa, are the most common representatives of the Primates in our menageries. Their tails, sometimes rudimentary, as in the Barbary Ape, are never prehensile.

Manlike Apes (Anthropomorphs or Anthropoids).

The ANTHROPOID or MAN-LIKE APES (*Anthropomorphs*) are of still greater interest to us as being the higher monkeys most nearly related to ourselves in their whole organization.[1]

The Gibbons (genus *Hylobates*), of south-eastern Asia and the Malay Archipelago (Sumatra), are the smallest of the anthropoids, the largest species, the Siamang, being scarcely a metre high. They show certain traces of low degree, suggestive of the tailed monkeys. Their fore-limbs are very long, but they generally walk erect, and in this, as in certain skull characters, they resemble Man.

The Orang-Outang or Pongo (*Simia satyrus*), of Sumatra and Borneo, is much larger and stronger. Its legs are very short; its brain-box is round (brachycephalic) and high, and bears sagittal and occipital ridges. The males are armed with strong canine teeth.

[1] See Hartmann, R., *Les Singes anthropoides* (Paris, 1886). Various papers by Owen, Deniker, Selenka and others.

The Gorillas (genus *Gorilla*) inhabit the Gaboon region of western equatorial Africa. Largest of the apes, they are stronger and more robust than Man. Their skull is elongated (dolichocephalic) and, in old males, bears great bony crests, strong supra-orbital ridges and powerful canine teeth. The ears are small.

The Chimpanzees (genus *Troglodytes* or *Anthropithecus* or *Pan*), probably represented by several species, are also African, inhabiting the equatorial forests of the west. In several respects they resemble Man more closely than do the other anthropoid apes. Although the male still possesses stronger canine teeth than the female, sexual dimorphism is nevertheless less marked than in the Gorilla. The skull is also elongated, but it lacks parietal and occipital crests, although the supra-orbital arches are prominent. Such morphological characters, together with the playfulness and intelligence of these apes, add a peculiar interest to the Chimpanzees.

Man and his Group (Hominians).

HUMAN BEINGS (or HOMINIANS) comprise one family only, reduced in our day to the single genus *Homo*, a genus the divisions of which have created more disagreement among naturalists than any other in all the class of mammals.

The divergences of the classifications of Modern Man are truly extraordinary: ranging from the monogenistic schemes of Linnæus, Blumenbach, and Cuvier, with their four or five varieties or races of *Homo sapiens*, and incidentally that of Isidore Geoffroy Saint-Hilaire, with four principal types and twelve secondary races, to the modern schemes, such as that of de Quatrefages, with his five stems and eighteen branches, themselves dividing into numerous smaller branches; or that of Deniker, who recognizes twenty-nine races and sub-races; or the polygenistic scheme of Sergi, which includes three genera, eleven species and forty-one varieties or sub-varieties; not to mention the composite scheme of Giuffrida-Ruggeri, etc.[1]

[1] Quatrefages, A. de, *L'espèce humaine: Introduction à l'étude des races humaines.* Deniker, *Races et peuples de la Terre* (Paris, 1900). Sergi, G., *Le Origini umane* (Turin, 1912). Giuffrida-Ruggeri, *L'Uomo attuale* (Milan, Rome, Naples, 1913).

The discussions between "monogenists" and "polygenists," resulting in floods of ink, have never reached any settlement of a question which has been described, and with some reason, as "immaterial."

Anthropologists have now returned to the ideas of the older naturalists, who divided Modern Man into three main groups: WHITE, YELLOW (including Red-Skins), and BLACK.

According to the monogenists, these three divisions represent only races of the single species *Homo sapiens*. According to the polygenists, they are separate species. (1) *Homo albus* or *Homo caucasicus* (an old and quite inappropriate term), or *Homo indo-europæus*; (2) *Homo flavus* or *H. mongolicus*; (3) *Homo niger* or *H. æthiopicus.*

The sub-divisions of these three great groups are usually based much more upon community of language, religion, and morals, than on physical characteristics. They are most often merely ethnographical, a result due to the important part played by migration. Yet even if the hypothesis of the existence of several types, zoologically quite distinct in origin, be admitted, it is clear that innumerable minglings and crossings, repeated during thousands and thousands of years in complex ways, must have markedly reduced the primitive physical differences, and disguised them under a cloak of more uniform characters.

Anatomical Differences between Man and Monkeys.

The anatomical differences separating Man, or the various types of Man, from the highest monkeys, the Anthropoid Apes, are numerous but of different values. These differences we shall have frequent occasion to cite, and we must, therefore, now enumerate the most important.[1]

The most significant concerns the great development of the cerebral portion of the skull or brain-box, in which the brain is lodged, and the associated reduction in size of the facial part of the skull (Fig. 38). This relative development of the

[1] For greater detail, see Huxley, *Man's Place in Nature.* Vogt, Carl, *Leçons sur l'Homme* (Fr. trans., Paris, 1865; 2nd ed., 1878). [*Lectures on Man; his place in Creation, etc.*, London, 1864]. Broca, P., *L'Ordre des Primates* (Paris, 1869). Topinard, P., *L'Homme dans la Nature* (Paris, 1891). A good summary is to be found in Vianna de Lima, A., *L'Homme selon le transformisme* (Paris, 1888).

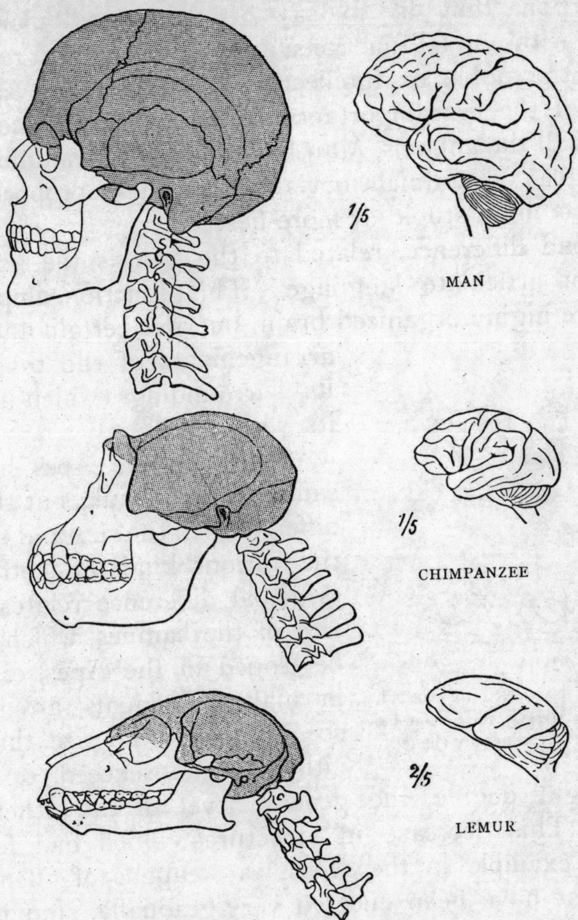

FIG. 38.—Comparative Morphology of the Skull, the Brain, and the Cervical Vertebræ of a Lemur, an Ape, and Modern Man.

In the Pied Lemur (*Lemur varius*), the cerebral portion of the skull (shaded in figure) is small in relation to the facial part. The brain is not large; the cerebrum still lies in front of the cerebellum. The foramen magnum is situated at the hinder portion of the skull; the direction of the vertebral column is obviously in line with the axis of the brain.

In the Ape (Chimpanzee), the cerebral portion of the skull is larger and the facial portion reduced. The brain is larger; the cerebrum has become complicated, and overlaps the cerebellum. The foramen magnum has moved nearer the base of the skull. The vertebral column is more oblique in relation to the axis of the brain; the cervical vertebræ have long spinous processes which stand at right angles to the bodies of the vertebræ.

In Modern Man, the cerebral portion of the skull greatly exceeds the facial portion, here reduced to a minimum. The very large cerebrum, highly convoluted, overlaps and extends beyond the cerebellum. The foramen magnum occupies a still more forward position. The vertebral column forms a right angle with the base of the skull; the spinous processes of the cervical vertebræ are inserted obliquely on the vertebral bodies and are directed downwards.

facial portion, that is to say of the mandibular region, connected with animal functions, and of the cerebral region, set apart for the noble or intellectual functions, is a zoological character of the first importance, as Cuvier long since pointed out: "Of all the animals, Man has the largest cranium and the smallest face; as animals deviate from these proportions, so they become more stupid or more fierce."

A second difference, related to the first, is the possession by Man of articulate language. This function implies not only a more highly organized brain, but also certain anatomical arrangements of the tongue and its surroundings which facilitate its working.

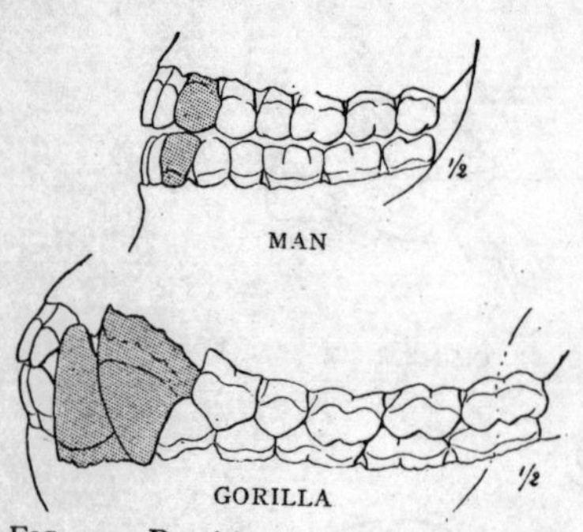

FIG. 39.—Dentitions, viewed from the side, of a Man and an Ape (Gorilla). One-half natural size. The canines are shaded.

Man and the Apes have the same dental formula, but there are differences in the size and shape of the various kinds of teeth. The principal difference relates to the size of the canines, which, highly developed in the Apes, especially in male individuals, are in Man so diminished in size that they either do not exceed, or exceed only in slight degree, the general level of the other teeth (Fig. 39). This decrease in structures which had formerly served, for example in the Apes, as weapons of offence and defence, must have been effected very gradually, step by step with the attainment of the upright position, of the setting free of the fore-limbs, and of the correlated development of the brain.

The perfectly erect attitude is characteristic of Man alone. The Anthropoid Apes, even the Gibbon, possess it only in an imperfect degree; and this incapacity is accounted for by anatomical differences. In Man, the vertebral column is inserted at the base of the skull, so that, in his normal upright position, the skull is almost naturally balanced on the first vertebræ; and thus the occipital foramen (foramen magnum) lies in a horizontal plane and is placed beneath the skull. In the Lemurs, as in almost all four-footed mammals, the axis of the skull and of the vertebral column lie in line with each other, and the occipital foramen, placed almost in a vertical plane, occupies the hinder portion of the skull. In the true Monkeys the occipital foramen has begun to move towards the base of the skull; and in the Anthropoids it approaches, without actually attaining, the human position (Fig. 38).

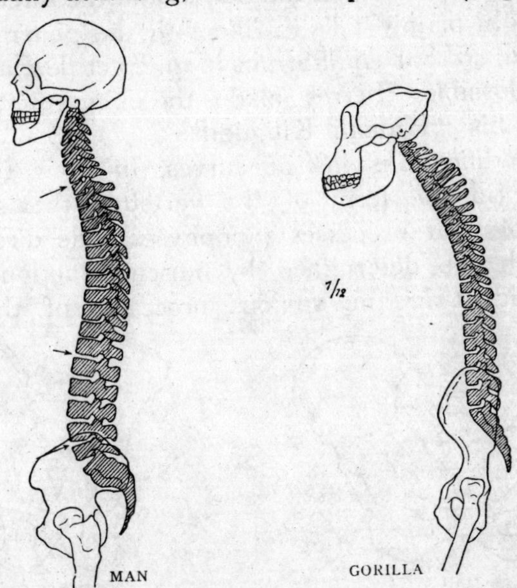

FIG. 40.—Skull, Vertebral Column, and Pelvis of Man and of Gorilla.

In the Gorilla, the vertebral column throughout its whole length only shows one single curve, concave in front; its general direction is not vertical. The skull, with heavy face, falls forward, and has to be held back by powerful muscles and ligaments at the nape of the neck. The pelvis is narrow.

In Man, the vertebral column shows several curves, alternately convex and concave in front (the arrows indicate the points where these curves change), in such a way that the whole forms a balanced system on which the head rests naturally in equilibrium, and which extends past the large pelvis in a likewise vertical direction towards the lower limbs.

Certain differences in the vertebral column are also entailed by the more or less erect attitude (Fig. 40). In the Monkeys, whose body is always inclined forward, the column shows only two curves, a dorsal and a sacral, both concave in front. In the unborn or in the new-born human child, only these two curves occur. At a later stage childish endeavours to stand and walk erect induce a profound modification in the form of the spinal column, which before long shows four curves: a cervical curve, concave behind; a dorsal curve, concave in front; a lumbar curve, concave behind; and a sacral curve, concave in front. As these four curves follow each other alternately, in first one and then the other direction, the general trend of the column is vertical. In this manner, the weight of head and trunk bears principally upon the pelvis, and the direction of the resultant weight plainly falls in line with the general direction of the column, so that equilibrium in an erect posture becomes as easy as possible. Serres made these curves the special attributes of his "Human Kingdom."

These modifications in the curves, in their turn, entail modifications in the form of the vertebræ themselves, and particularly of their processes (apophyses), the direction and size of which are determined by muscular action. This is especially evident in the spinous processes of the cervical

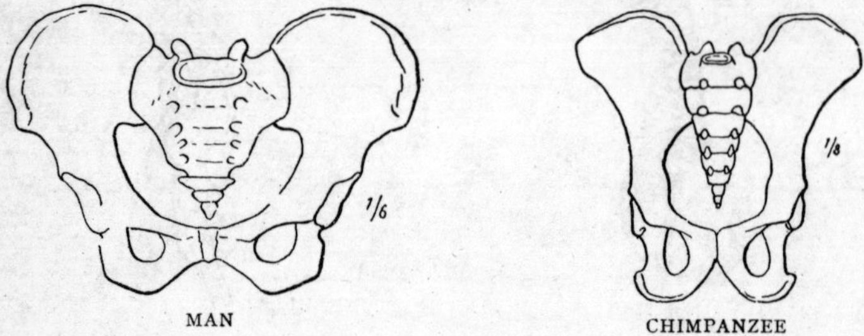

FIG. 41.—Comparative structures of the pelvis of Man and of an Ape (Chimpanzee).

vertebræ. In consequence of this easy balance in Man which I have just described, the actions of the extensor muscles of the nape of the neck, of the spinal muscles, and of the cervical ligament may be, and indeed are, much less powerful in him than in quadrupeds, whose head and trunk tilt forward. In Man and in the quadrupeds, then, the spinous processes differ both in their development and in their direction (Figs. 38 and 40). And in these characters, Monkeys, and especially the Anthropoid Apes, are intermediate between the exclusively quadruped mammals and Man, the most perfect of the bipeds. Finally, the human races are not all exactly similar to each other, for the lower races, as we shall see later, still preserve certain characteristics of the stage now represented by the Anthropoid Apes.

Further, the form of the pelvis also is correlated with the vertical position and erect walking. In Man, the haunch bones (ilia) are enlarged, widened, and spread out in the form of a bowl or basin, so offering easy support to the abdominal viscera. In Monkeys, the pelvis does not bear all the weight of the intestines, and the much narrower haunch bones are placed almost parallel to the sacrum, thus more closely resembling those of quadrupeds (Fig. 41).

Another difference which Owen, the great English anatomist, considered to be fundamental, is the formation of the foot (Fig. 42). In the Anthropoid Apes, the first toe, widely

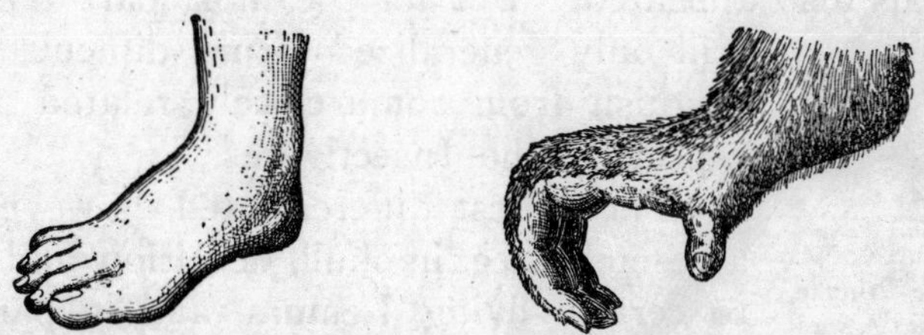

FIG. 42.—Foot of Man and of a Gorilla.

separated and much shorter than the others, is readily opposable to the latter and plays the part of a thumb. Thus the foot, physiologically speaking, has become a hand (whence the term *Quadrumana*). In Man, the great toe is the largest of the digits and is closely pressed against the others, to which it cannot be opposed. Here we have a true foot, that is to say, an organ for support; and this exclusive adaptation to a particular function is reflected in all the bones of the human foot, for their structures differ from those of the corresponding bones of the hind hand, or shall we say the prehensile foot of Anthropoid Apes.

In spite of the existence of some transitional grades of structure, observed in certain Monkeys and Man, all these characters are practically so definite that there can be absolutely no confusion between the most highly-developed of living Monkeys and the lowest of living Men.

It must therefore be acknowledged that zoology, the teaching of which I have just summarized, defines with tolerable clearness the boundaries of Man's place in Nature.

Can Palæontology supply any supplementary hints regarding the relationships between Monkeys and Man? To answer this question we must first review the discoveries bearing on fossil Monkeys, and this I shall now do briefly, in chronological order.

Fossil Monkeys.

The Eocene Primates.

The precursors of the great group of Primates appeared in North America towards the base of the Eocene, almost at the beginning of the Tertiary era, when the archaic representatives of other orders of mammals also appeared. But for the most part they were still only generalized forms, difficult to distinguish from some closely related groups, especially the Insectivores.

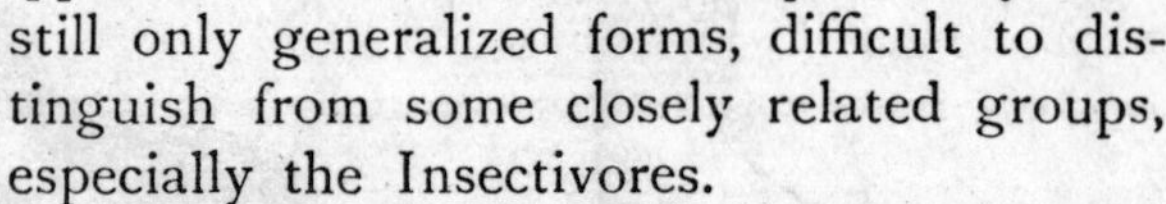

FIG. 43.—Skull of a small Eocene Primate, *Anaptomorphus homunculus.* Natural size (after Osborn).

Those best differentiated, however, bear a resemblance, in skull, dentition and brain, to certain living Lemurs, and in particular to the curious Tarsier of the Malay Archipelago. The very interesting *Anaptomorphus* (Fig. 43) may be taken as an example of this highly evolved series. The famous palæontologist, Cope, was inclined to regard this fossil as the common ancestor of Monkeys and Man, because of the rounded form of its skull, and because, of all the Eocene mammals, it seemed to possess comparatively the largest brain. Perchance in the older deposits there may yet be discovered forms more primitive still.

In the Mid Eocene of the United States, the increase in numbers and differentiation of these early Lemurs becomes marked. The Anaptomorphus group, fruit-eaters, with rounded skulls, is represented by several genera. Other forms, such as *Pelycodus*, and *Notharctus*, omnivorous in diet and possessing elongated skulls, show characters indicating evolutionary tendencies towards modern types.[1]

As there has not yet been observed the least trace of a well-defined Primate in the older deposits of other parts of the

[1] Osborn, H. F., "American Eocene Primates" (*Bull. American Mus.*, xvi., 1902). Matthew, W. D., *ibid.*, xxxiv., 1915. Gregory, W. K., On the structure and relations of *Notharctus* (*Mem. Amer. Mus.*, n.s., ii., 1920).

world, we must suppose, pending further information, that the group originated in North America, or rather on a northern Americano-European continent, whose former existence is revealed by geology. Moreover, as the types which I have just described disappeared from America with the Upper Eocene, there must have been migrations in various directions.

The oldest remains of authentic Old World Lemurs[1] have been found in the Mid Eocene deposits of France and Switzerland. Their remains become abundant in the Upper Eocene and Lower Oligocene of Europe (Fig. 44). The rich

FIG. 44.—Skull of a Lemur from the phosphorites at Quercy (*Adapis magnus*). Three-fifths natural size. Palæontological Gallery, French National Museum of Natural History.

layers of phosphorites at Quercy have yielded, in a fine state of preservation, complete skulls of animals such as *Necrolemur*, *Pronycticebus*, and *Adapis*.[2]

Afterwards, having migrated to Asia, Africa, and especially Madagascar, these various types have persisted in these regions to the present day and have become differentiated: they divided

[1] *Plesiadapis*, from the lowest Eocene at Cernay, near Rheims, at first believed to be a Lemur, is also related to the Insectivores. *Protoadapis*, from a level above the Lower Eocene, at first taken for a Lemur, was subsequently regarded as an Insectivore. To-day the tendency is to replace it among the Lemurs. Such examples show the difficulties which arise in classifying types of mammals with primitive and generalized characters.

[2] See especially Filhol, H., *Recherches sur les Phosphorites du Quercy* (Paris, 1877).

into numerous branches, some of which thereafter developed into giant and specialized forms, such as the *Megaladapis*.[1]

The most ancient forms of the Primates likewise spread from North America to South America. It is probable that, by the loss of a premolar, and the increase of their brain-box at the expense of their face, they became transformed into true Monkeys, the ancestors of the modern Flat-nosed Monkeys (Platyrrhinians).

In an Oligocene or Miocene deposit in South America, known as the *Santa-Cruzian*, there have actually been found remains of old-time Sapajous, to which the late Argentine palæontologist, Ameghino, attached the greatest importance, considering them "as the ancestors of all the monkeys of the Old and of the New Continent," and even of Man. He gave them very expressive names: *Homunculus patagonicus*, *Anthropops perfectus*, and so on.[2] They are certainly interesting relics; but they are very fragmentary, and quite insufficient to support Ameghino's far-reaching conclusions. For the present they may simply be regarded as ancestral forms of the modern Cebians or Flat-nosed (Platyrrhinian) Monkeys, a group the evolution and differentiation of which has taken place and is still in process in that very country.

Although fossil monkeys are absolutely unknown in North America, Ameghino nevertheless announced the existence of higher monkeys in South America, on account of certain anatomical remains which have been the subject of much discussion. The late palæontologist has described an atlas vertebra and a femur from Monte-Hermoso, which he considers a Miocene locality, but which the majority of geologists and palæontologists who have studied the sedimentary deposits of South America regard as scarcely even Pliocene. He sought to prove that these bones belonged to a creature which united his *Homunculus* to the genus *Homo*, and which he named *Tetraprothomo*, attributing to it a stature of 1 to 1.10 metres.

1 Grandidier, G., "Recherches sur les Lémuriens disparus" (*Nouvelles Archives du Muséum*, 1905).

2 Ameghino, F., "Les formations sédimentaires du Crétacé supérieur et du Tertiaire de Patagonie" (*Anales del Museo nacional de Buenos Aires*, xv., 1906).

He was thus led to conceive a complete new human phylogeny: *Tetraprothomo* was said to have been followed by *Triprothomo*, this by *Diprothomo*, which was supposed to have given rise to *Prothomo*, from which in its turn would have arisen the genus *Homo*. We shall see later that these stages were purely figments of the imagination.[1]

Oligocene Monkeys.

To return to the Old World. Until recently it was impossible to affirm that the Catarrhine or Long-nosed Monkeys were very old in the geological sense of the word. Dog-faced Monkeys (*Cynomorphs*) and Anthropoid Apes seemed to appear simultaneously in the Miocene, and the first fossil anthropoids were even older than the known fossils of tailed monkeys. Such a state of affairs justified the expectation of important discoveries in the older deposits. About ten years ago one such discovery was made.

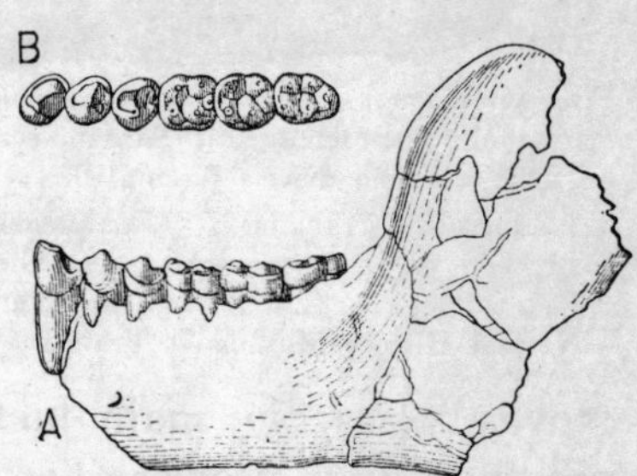

FIG. 45.—Lower Jawbone of *Propliopithecus hæckeli*, after Schlosser.

A, Side view, natural size.
B, Teeth seen from above.
C, The same teeth enlarged two times.

The Eocene and Oligocene deposits at Fayum in Egypt are famous for their palæontological treasures, and particularly for the remains of primitive Proboscidians, which form the subject of important works by Andrews. In 1910 Schlosser announced the discovery of remains of several Primates in the Oligocene deposits at Fayum, and he has since published a detailed memoir on these curious fossils.[2]

Of the three species described, all of small size, as is usual in archaic forms, two bear some resemblance to the Eocene Lemurs of the United States, and some to the early Cebian Monkeys of South America. They are evidence of an evolu-

[1] For further details and bibliography, see that part of the last chapter of this book dealing with "Fossil Man in South America."

[2] Schlosser, M., "Beiträge zur Kenntnis des Oligozänen Landsäugetiere aus dem Fayum" (*Beiträge zur Paläontologie und Geologie Oesterreich-Ungarns*, xxiv., 1911). Résumé with figures in *L'Anthropologie*, xxiii., p. 417.

tionary stage from which may be derived, according to Schlosser, all the higher Primates, Man as well as Apes.

The third species has been given the very suggestive name of *Propliopithecus hæckeli* (Fig. 45). This monkey is represented by two lower jawbones, and Schlosser considered it to be very closely related to the Gibbons (*Pliopithecus*) from the Mid Miocene deposits at Sansan and other localities. This form may possibly be a descendant from a species of the Anaptomorphus group, which had passed perhaps through a Cebian stage, and given rise directly to *Pliopithecus*, from which both the Anthropoids and Man may have been derived.

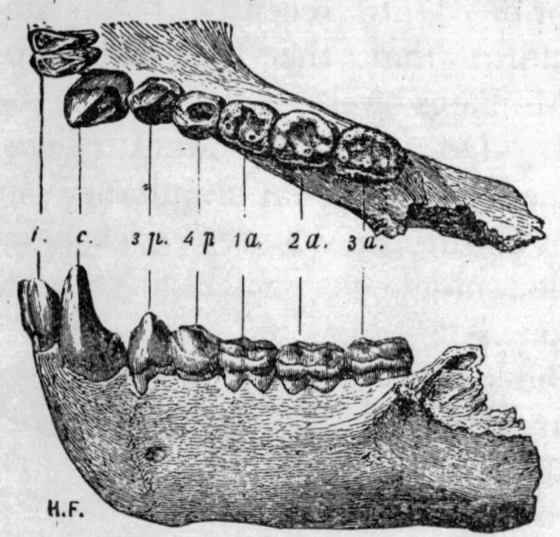

FIG. 46.—Lower Jaw of *Pliopithecus antiquus* from Miocene deposits at Sansan, seen in profile and from above. Natural size.

i, incisors; *c*, canines; 3*p*, 4*p*, premolars; 1*a*, 2*a*, 3*a*, molars. After A. Gaudry. Palæontological Gallery of the National Natural History Museum of France.

Certain of Schlosser's views on the phylogeny of monkeys and of Man have been criticized as being based on too incomplete evidence, and consequently as being very rash. Nevertheless one main fact is brought out in his work: the presence, in a Lower Tertiary, or at any rate an Oligocene deposit, of several generalized forms of true monkeys, and particularly of one form which might pass for a primitive type of Anthropoid Ape.

Miocene Monkeys. Pliopithecus.

The Miocene deposits of several countries have yielded fairly numerous remains of Monkeys which show most distinct affinities with living forms.

Pliopithecus antiquus (Fig. 46), discovered by Edouard Lartet in 1837 at Sansan (see Chap. I.), and since found in certain contemporaneous deposits in several European countries, was related to the Gibbons,[1] and we have just seen

[1] Lartet, E., *Comptes rendus de l'Académie des Sciences* (16th January and 17th April 1837).

that Schlosser regarded it as a direct descendant of the Egyptian *Propliopithecus*.

Oreopithecus bambolii, from Monte Bamboli, in Tuscany, first described by Gervais,[1] and later by Ristori, bears some resemblance to the Dog-faced Monkeys (*Cynomorphs*), the Guenons (*Cercopithecus*), and the Anthropoid Apes (Fig. 47).

Dryopithecus. A geological horizon in South France, very like that at Sansan, the calcareous sand of Saint-Gaudens, has yielded several remains of a large Monkey. A jawbone was first described in 1856 by Edouard Lartet

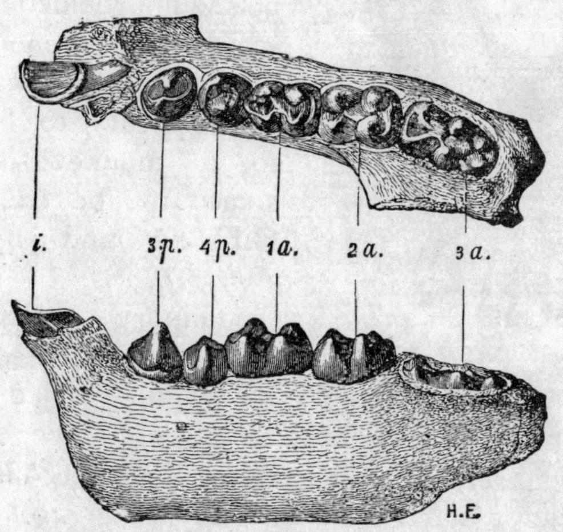

FIG. 47.—Lower Jawbone of *Oreopithecus bambolii* from the Miocene deposits of Tuscany, seen in profile and from above. Natural size.

i, incisors; 3*p*, 4*p*, premolars; 1*a*, 2*a*, 3*a*, molars. After A. Gaudry.

under the name *Dryopithecus fontani*.[2] This find caused a great sensation, for Edouard Lartet and Albert Gaudry declared that *Dryopithecus* was more closely akin to Man than any other known monkey.[3] But in 1890, the discovery of a more complete and better preserved lower jawbone (Fig. 48) enabled

[1] Gervais, P., *Zoologie et Paléontologie générales*, 2nd series, 1876, p. 9.

[2] Lartet, E., "Note sur un grand Singe fossile qui se rattache au groupe des Singes supérieurs" (*Comptes rendus de l'Académie des Sciences*, xliii., 28th July 1856).

[3] Gaudry, A., *Les enchaînements du monde animal. Mammifères tertiaires* (Paris, 1878, p. 236).

Gaudry[1] to show that this fossil monkey was in reality lower in the scale of life than the large living anthropoids. Fragments of even less significance, most often no more than isolated teeth, have been found in other European countries. *Anthropodus* from the "bonherz" in Swabia, examined by Schlosser, *Gryphopithecus* from Hungary, described by Abel, both known only from minute fragments, seem also to be connected either with *Dryopithecus* or with closely related forms.

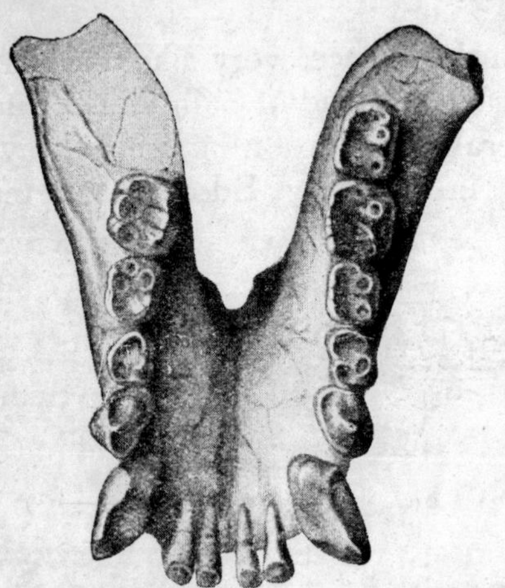

Fig. 48.—Lower Jaw of *Dryopithecus fontani*, from Saint-Gaudens. Three-quarters natural size. Palæontological Gallery of National Nat. Hist. Museum of France.

The Upper Miocene at Eppelsheim yielded a femur to which the name *Pliohylobates* (= *Paidopithex*) was given, and this appears to have belonged to a monkey considered by some to be related to the Gibbons and by others to *Dryopithecus*.

Gaudry's excavations at Pikermi, in Greece, from 1853 to 1860, yielded the remains of twenty-five individuals of *Mesopithecus pentelici*, and enabled him to reconstruct the first known skeleton of a fossil monkey (Fig. 49). The famous palæontologist was able to determine the affinities of *Mesopithecus*, and to show that it represented a form combining two modern types: for whilst its limbs were like those of a Macaque, its teeth resembled those of the Sacred Indian Ape (*Semnopithecus*).[2]

Fossil Monkeys of the Siwalik Hills.

The Siwalik Hills, chains of hills or mountains in Northern India, foothills of the Himalayas, are well known to palæontologists on account of the rich faunas of fossil vertebrates contained in their layers, the age of which varies from Mid

[1] Gaudry, A., "Le Dryopithèque" (*Mém. de la Soc. géolog. de France, Paléontologie*, No. 1, Paris, 1890).

[2] Gaudry, A., *Animaux fossiles et géologie de l'Attique* (Paris, 1862).

Miocene to Upper Pliocene. These faunas include several species of monkeys which have been studied by two English palæontologists, Lydekker and Pilgrim.[1]

Some are Dog-faced Monkeys (*Cynomorphs*), two are Sacred Indian Apes (*Semnopithecus*), two, Short-tailed Baboons, in addition to a Macaque, and a *Cercopithecus*. Others, more interesting in connection with the present discussion, belong to several genera of Anthropoid Apes (*Simiidæ*), and these all lived towards the end of Miocene times.

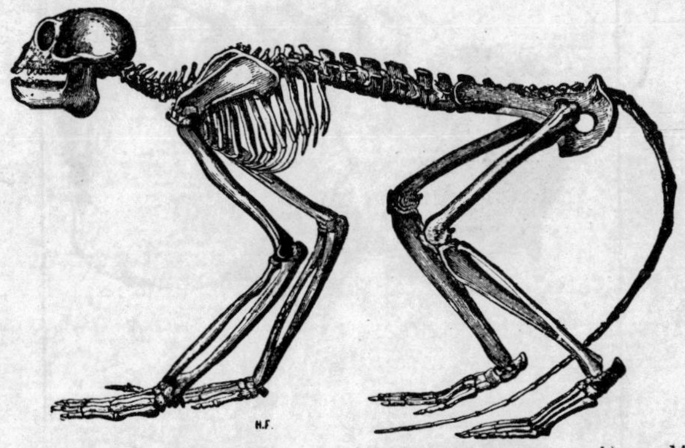

FIG. 49.—Skeleton of Fossil Monkey, *Mesopithecus pentelici*, from Upper Miocene at Pikermi in Greece. About one-sixth natural size. (After A. Gaudry).

First of all, there is the European genus *Dryopithecus*, rediscovered by Pilgrim in the Siwalik Hills, where it is represented by three species, one of them a giant of its kind (*D. giganteus*). If to these three species be added the European forms, the genus *Dryopithecus* seems to embrace a group of Anthropoid Apes having manifold variations and forming a special branch now extinct. Certain of its smaller branchlets appear to have been fairly closely related to modern Chimpanzees and Gorillas, with, however, certain rather more primitive characters. Others would seem to have been more closely related, judging at least from the characters of their dentition, to the human stock, as Lartet and Gaudry first believed. A similar opinion has been expressed during the last few years

[1] Lydekker, R., "Indian Tertiary and Post-Tertiary Vertebrata" (*Palæontologia Indica*, series x., vol. iv., 1886). Pilgrim, G. E., "New Siwalik Primates" (*Records Geological Survey of India*, xlv., 1915).

by Gregory in America and by Sera in Italy.[1] *Dryopithecus* would seem then to have been an ancestral and synthetic form.

Further there is the genus *Palæosimia*, which may be considered a direct progenitor or collateral relation of the Orangs (*Simia*); and also the genus *Palæopithecus*, first associated by Lydekker with the Chimpanzees, which seems, however, to have some affinity with *Dryopithecus*, and according to Gregory, may be closely related to the Gorillas.

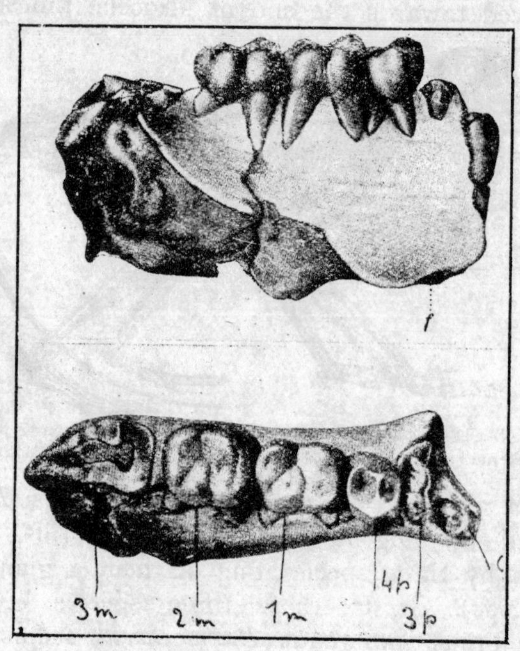

FIG. 50.—Portion of Right Lower Jaw of *Sivapithecus indicus* from the Siwalik Hills. Natural size. (After Pilgrim.)

Sivapithecus.

There is finally the curious genus *Sivapithecus*, recently discovered and described by Pilgrim, who has not hesitated to place this new fossil Primate among the Hominians. It is only known from some isolated teeth and two jaw fragments, by means of which Pilgrim has

[1] Gregory, W. K., "Studies on Evolution of the Primates" (*Bull. of the American Museum*, xxxv., 1916). Sera, G. L., "La Testimonia dei fossili Antropomorfi per la questione dell' origine dell' Uomo" (*Atti della Società italiana di Scienze naturali*, lvi., 1917).

attempted the restoration of the lower jaw (Figs. 50-52). The general form of the latter more resembles the human form than the jaw of any other anthropoid ape, living or fossil. It has, indeed, the canine of an anthropoid ape, but the true molars in their general appearance are more human in type than those of any known ape.

Pilgrim has no doubt that the characters of the lower

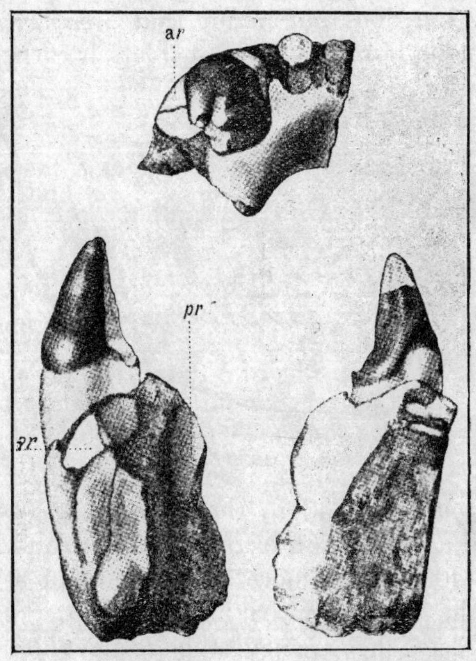

FIG. 51.—Fragment of Left Lower Jawbone of *Sivapithecus*, with the canine in place. Natural size. (After Pilgrim.)

jawbone of *Sivapithecus* justify the conclusion that this fossil belonged to the direct progenitors of the human race, a conclusion the importance of which contrasts with the slightness of the evidences on which it is based. Lydekker, who was the first to study the fossil monkeys of the Siwalik Hills, at once raised doubts as to the value of the new genus, and questioned whether it might not be the lower dentition of his *Palæopithecus*, of which only the upper dentition was known. For my part,

I have called attention to the fact that, apart from the presence of a strong canine, the characters which truly distinguish apes from Man occur in the skull and the extremities—that is to say, in parts of the skeleton which in the case of the genus *Sivapithecus* are still unknown to us, and in consequence its place in the classification cannot yet be established with any certainty.[1]

More recently, Gregory has keenly criticized the restoration made by his English colleague and has opposed his conclusions. He does not admit that *Sivapithecus* should be classified with the Humans, and considers it rather to be closely

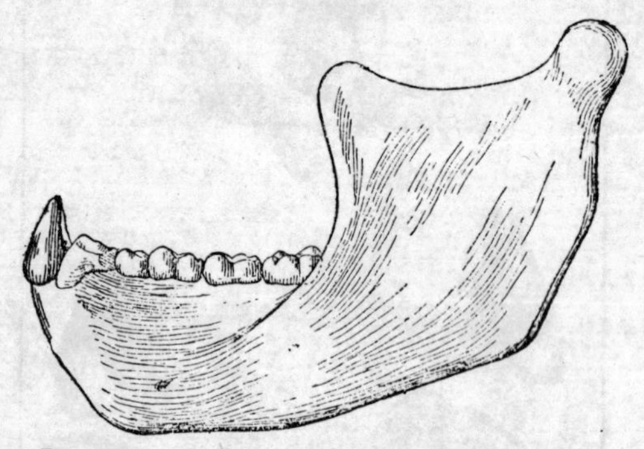

FIG. 52.—Restoration of the Lower Jawbone of *Sivapithecus indicus*. Half natural size. (After Pilgrim.)

allied to *Dryopithecus* and to the Orang. In his turn he also has attempted a restoration of the fossil jaw, and has succeeded in creating a form much more like the jaw of a female Orang than that of the most primitive human being.

Pilgrim's researches are nevertheless most interesting. They show that, during the Miocene period, Asia was inhabited by Anthropoid Apes, with characters diverging in all directions, and perhaps even in a certain degree, as in the case of *Dryopithecus* and *Sivapithecas*, towards the human type. In any case, they have made known to us a new type, exhibiting, in its lower jaw and its dentition, certain characters representing a new morphological transition between the Anthropoids and Man. These are valuable results, on which I shall have occasion elsewhere to lay stress.

[1] For more detail, see the critical analysis which I have made of Pilgrim's Memoir, in *L'Anthropologie*, vol. xxvi., 1915, p. 397.

Pliocene and Quaternary Monkeys.

Returning to Europe, we find that the Pliocene deposits contain fairly numerous remains of monkeys, but now, even more than in previous ages, we have to deal only with animals closely akin to the genera and even to the species of the present day: Macaques from Montpellier and the Val d'Arno; *Dolichopithecus* from Perpignan, akin to *Mesopithecus* from Pikermi; and *Semnopithecus* from Montpellier and Italy.

During the Lower Pleistocene period, monkeys very near to the modern Macaques or even identical with them, lived in various parts of Europe—in England, in Wurtemberg, in the Pyrenees, and in Sardinia. Here the difference is reduced to nothing more important than a difference in the geographical distribution of the species.

Until recent times, we had no knowledge of Anthropoid Apes from European Pliocene or Quaternary deposits. When we come to discuss the fossil remains found at Piltdown in England, and attributed to a man of extremely primitive type, *Eoanthropus*, we shall see that there is occasion to make an important distinction; the jaw of *Eoanthropus* is perhaps only the jaw of a Chimpanzee, for which I have suggested the name *Troglodytes dawsoni*, and which an American geologist proposes to call *Pan vetus*. It seems that a tooth found in the alluvials at Taubach, near Weimar, and attributed by Wehring to Man, is likewise the tooth of a Chimpanzee.

Conclusions.

We see that palæontological discoveries concerning monkeys are comparatively scarce. We must confess that, in contrast to the case in other orders of mammals, for example the Perissodactyles or the Proboscidians, palæontology has rendered but feeble assistance in the work of reconstructing the history of the Primates. Here zoology has supplied many more facts than palæontology, for nature at the present day shows more variations and a far greater abundance of transitional forms than in past time—at least so far as the past has been revealed to us.

Palæontology, nevertheless, does contribute a few to the number of transitional forms, and, moreover, unfolds a scheme of evolution conforming almost exactly with the zoological order.

To begin with, it shows us the first Primates in the form of small creatures, confined to a northern Americano-European continent, and of so generalized a type as to render it difficult sometimes to distinguish them from certain small contemporary mammals which must be regarded as the starting-point of other orders, such as the Insectivores. Further, it helps us to group these first types in the world, and to understand their correlative differentiation. It helps us to comprehend, for example, how certain of them were able to surmount the Lemur stage, and to give birth in South America to the Flat-nosed (Platyrrhine) Monkeys, which have since remained independent and isolated. Palæontology has discovered at Fayum the remains of several animals with synthetic characters, and so has revealed certain ancestral forms, in several cases of Long-nosed (Catarrhine) Monkeys, and in one case of the Anthropoid Apes. Thus it proves the great antiquity of the splitting off of the different ancestral branches. Much later, after a period of differentiation regarding which we have no light whatever, we see appearing in the Old World certain monkeys closely related to living monkeys, and representing transitional forms between their various types. There is the *Mesopithecus* of Pikermi, which links the Macaques to the Indian Apes and their kind (*Semnopithecus*); there is *Oreopithecus*, which seems to be connected both with the dog-faced monkeys (Cynomorphs) and the man-like monkeys (Anthropomorphs); there are the species of *Dryopithecus*, as well as *Palæopithecus*, and *Sivapithecus*, in which are found associated, characters now scattered among several species of Anthropoids.

It must be confessed, however, that palæontology has not yet revealed any indisputable transitional form, any material proof of a hereditary connection between the ape form and the human form, for we cannot attribute any conclusive value to the *Homunculus* of Ameghino, or the *Anthropodus* of Schlosser. The significance that the latter scientist attributes to *Propliopithecus* is quite hypothetical. *Sivapithecus* is still only a monkey, although it shows very interesting tendencies towards the human structure. How *Pithecanthropus* must be regarded in this respect, we shall soon see.

Palæontological evidences relating to the higher Primates are not only very rare, they are also extremely poor. Apart from *Mesopithecus* and *Dolichopithecus*, of which we have almost complete skeletons, but which are relatively of little interest from the point of view concerning us here, since their intimate relationship is with modern tailed monkeys, we possess only minute fragments, at most some incomplete jawbones. This poverty of material is not only regrettable in itself; there is a danger that it may lead to serious error. It may be said that practically every day we learn, often at the expense of our pride as palæontologists, that incomplete evidences must be interpreted with great caution; that the famous law of the correlation of characters, formulated by Cuvier, of which it has been so often said that it enabled him to reconstruct an entire fossil by means of one single bone of its skeleton, is very often at fault; that nature takes a kind of pleasure in varying combinations in the most unforeseen manner, and that she produces associations of characters expressly made to lead astray such naturalists as still retain absolute faith in the attractive Cuvierian legend.

Poverty of Palæontological Material.

In reality, in order fully to appreciate the nature of a fossil animal, and to assign to it its true place in the group to which it belongs, mere fragments of bone are in most cases insufficient. It is necessary to have skeletons, or at least complete skulls, especially when, as in the present case of very compact groups, there are numerous closely related forms which differ from each other only in fine degrees of structure, and that not always in the same parts of the skeleton. Perhaps, indeed, these shades of difference would not seem to us of so great importance were they possessed by creatures more distantly related to us than the apes.

Therefore, even if a chance discovery were to put us in possession of a fragment of a creature forming a stage in the descent of Man, it might happen that we were not in a position to recognize its true nature from the single fragment. And on the other hand, owing to the occurrence of physiological convergence, another fragment of the same kind might show characters which we should be tempted to regard as human characters, or as tending towards such, whereas examination of fragments less incomplete would save us from this error.

Such observations as these are not encouraging; they condemn palæontologists to a fate to which they have long since become resigned, but which is none the less hard.

Be this as it may, the advance of our science is dependent on chance discoveries or on the importance of systematic explorations. The latter, however, are very expensive. Public authorities, at least in France, have not yet shown as much interest in this kind of work as it deserves; they are more concerned, and increasingly so as time goes on, with sciences of a utilitarian character. Further, palæontological explorations planned with a view to discovering remains of fossil monkeys are particularly uncertain in their results. Even in beds very rich in bones of mammals, the remains of Primates are extremely rare, and this because of the more or less arboreal life which these animals lead, of the relatively small number of individuals, and of conditions which enable their skeletons more easily to escape fossilization. For three years Filhol, on behalf of the French National Museum of Natural History, carried out considerable excavations in the Miocene bone-layers at Sansan, in the hope of discovering therein skeletons or portions of skeletons of *Pliopithecus* or *Dryopithecus*. He came across not the smallest fragment. The material difficulties are thus very considerable, and the slight resources hitherto placed at the disposal of palæontology are quite insufficient to overcome them.

CHAPTER IV

PITHECANTHROPUS—THE "APE-MAN"

We have just stated that palæontology has not yet revealed the existence of a fossil ape which perceptibly lessens the gulf between the living apes and Man. In spite of certain indications deduced from the structures of the teeth, we have no justification, up to the present, for declaring that there have existed fossil apes more highly organized than the large living Anthropoids. Readers already familiar with the great questions of palæontology will be sure to object that, in so saying, I have taken no account of *Pithecanthropus*.

The discovery of this celebrated fossil was one of the most sensational made in the domain of natural science during the course of the nineteenth century. Regarded as representing the long-sought transition stage from Ape to Man—the "Ape-Man," *pithecos-anthropos*—it has given rise to many lengthy discussions. I have thought it necessary to devote a special chapter to it, not only because of its importance, but also because many naturalists separate *Pithecanthropus* from the Apes and range it among the human beings, or Hominians.

Story of the Discovery.

In Java there have long been known certain deposits, partly of volcanic origin, which contain bones of fossil mammals. In 1890, Eugène Dubois, a Dutch army doctor, was commissioned by his government to explore these layers. His work lasted for several years and was most fruitful in results; for besides enormous quantities of bones of various animals, there were found the remains of a large Primate—a skull-cap, a femur, and two teeth. In 1894, Dubois published a monograph[1] on his discovery, and its title summarizes its principal conclusion, that this newcomer into the

[1] Dubois, E., "*Pithecanthropus erectus*, eine menschenaehnliche Uebergangsform aus Java." In 4to with 2 plates. (Batavia, 1894.)

world of fossils represents the "intermediate form between the Anthropoids and Man," implied by the doctrine of evolution; it is the "precursor of Man." The following is the diagnosis which the author gives of the new creature; it is of both generic and specific significance:—

"PITHECANTHROPUS ERECTUS.—Skull much larger, absolutely and also relatively to the body mass, than in the large apes, but less bulky than in Man; cerebral capacity about two-thirds that of Man. Inclination of the nuchal plane of the occiput much greater than in the large apes. Dentition different from that of the latter, although of archaic form. Femur of human dimensions and suited for walking in the upright position."

Dubois' memoir was everywhere read, commented upon, and discussed. Palæontologists and anthropologists all the world over published their opinions or impressions, so that the list of works relating to *Pithecanthropus* soon became considerable.[1] At the International Zoological Congress held at Leyden in 1896, in spite of certain divergences of opinion, there was unanimous agreement in recognizing the great interest of the Dutch scientist's discovery, and in regretting the insufficiency of the relics brought to light. On every hand the wish was expressed that new explorations might soon be undertaken to supply additional information.

In 1906, Mme. Selenka, the widow of a German zoologist, and author of important researches on the embryology of the Anthropoid Apes, organized, at great expense, a scientific expedition to Java to prosecute new investigations. These

[1] The following are the most important of these works: Manouvrier, L., "Discussion du *Pithecanthropus erectus* comme précurseur présumé de l'Homme" (*Bull. Soc. d'Anthrop. de Paris*, 1895); "Deuxième étude sur le *Pithecanthropus*" (*Ibid.*); "Réponse aux objections contre le *Pithecanthropus*" (*Ibid.*, 1896). Virchow, R., "*Pithecanthropus erectus* Dubois" (*Zeits. f. Ethnologie*, xxvii., 1895). Dubois, E., "*Pithecanthropus erectus*, eine menschenaehnliche Uebergangsform" (*Congrès intern. de Zool.*, Leyden, 1896). Houzé, M., "Le *Pithecanthropus erectus* Dubois" (*Bull. Soc. d'Anthrop. de Bruxelles*, xiv. and xv., 1896). Marsh, O. C., "On *Pithecanthropus erectus* from the Tertiary of Java" (*American Jour. of Sci.*, i., 1896). Martin, R., Weitere Bemerkungen zur Pithecanthropus-Frage (Zurich, 1896). Schwalbe, G., "Studien über *Pithecanthropus erectus* Dubois" (*Zeits. f. Morphol. und Anthrop.*, i., 1899). Berry, R. J. A., and Robertson, A. W. D., "The place in Nature of the Tasmanian aboriginal as deduced from a study of the calvarium" (*Proc. Roy. Soc.*, Edinburgh, xxxi., 1910).

lasted eighteen months. The precise spot where the remains of *Pithecanthropus* had been exhumed was easily found, thanks to a little commemorative monument which Dubois had had erected. Around this site the soil was dug out to a depth of 12 metres, and more than 10,000 cubic metres of earth removed. The expedition was thus enabled to garner important collections of fossil animals, which now rest in the Museums of Berlin and Munich; in addition, numerous observations on the geology of Java were made; but the principal object was not attained—not the least relic of *Pithecanthropus* could be found. The scientific results of this campaign were recorded in a volume published in 1911 by the efforts of Mme. Selenka and M. Blanckenhorn.[1]

FIG. 53.—The Island of Java. The shaded square corresponds to the part shown in more detail in Fig. 54.

The most important of the results, worked out by numerous specialists, deal with the geology and palæontology of the region of Trinil. The interest of these new observations lies in the determination of the age of the deposit, a question of prime importance, which must first be considered.

Study of the Deposit.

The deposit in question occurs at Trinil, a village situated near the town of Ngawi, on the banks of the river Solo, or Bengawan, at the foot of the great volcano Lawou-Koukousan, whose terminal cone, still active, rises to a height of 3254 metres (Figs. 53-55). This volcano, like all the volcanoes of Java, rises from a substratum of Tertiary deposits of marine origin. At its base occur layers of sand, cinders, and volcanic lapilli, rearranged by fierce streams of water descending from the volcano, as well as tuffs, and clays of more or less fluviatile origin;

[1] Selenka, L., and Blanckenhorn, M., *Die Pithecanthropus-Schichten auf Java. Geologische und palæontologische Ergebnisse der Trinil-Expedition, etc.*, vol. in 4to, with plates (Leipzig, 1911).

these are spread over great expanses, sometimes attaining 350 metres in thickness, and resting unconformably on more ancient marine formations. The section illustrated here (Fig. 56), shown in somewhat diagrammatic fashion as in the Selenka publication, gives the details of the beds forming the bank

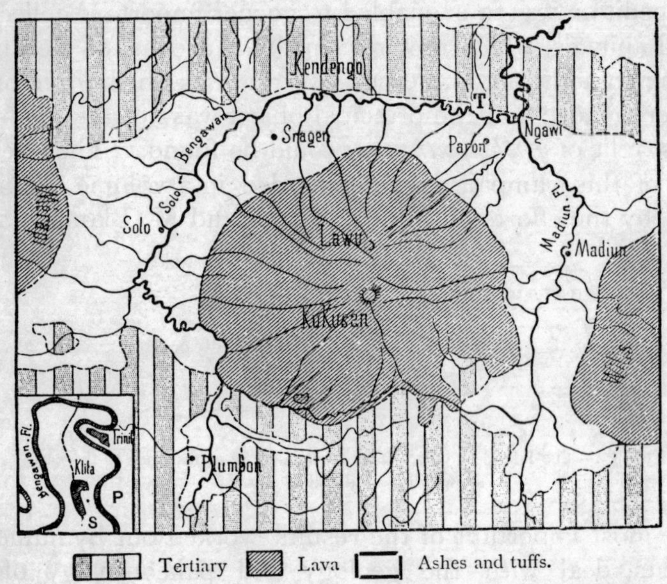

FIG. 54.—Geological Map of the volcano Lawou-Koukousan and its surroundings. Scale, 1 : 1,250,000. (After Verbeck.)

Trinil is 5 kilometres to the south of Paron, at point T. Inset on left of map, site on scale of 1 : 1,000,000. P, point where *Pithecanthropus* was found.

of the river Solo, at the very spot where the remains of *Pithecanthropus* lay.

The layers contain many fossils of all kinds: plant impressions; molluscan shells, *Unio, Corbicula,* and *Melania,* belonging to species, with one exception, still living in the country; vertebrate bones, of fishes, reptiles, and mammals. Mammals formed a very rich fauna; Elephants, including one or two species of the primitive genus *Stegodon*; Rhinoceros and Tapir; a Hippopotamus with archaic characters; ruminants, antlered and horned, among them a Pliocene genus, *Leptobos,* still showing relationship to the Antelopes; carnivores, especially of the cat tribe; a giant Pangolin, a monkey (Macaque), and, finally, *Pithecanthropus*. This fauna includes several new genera, and almost all its species are different from existing species. Its affinities with the faunas of India show that at

FIG. 55.—The *Pithecanthropus* bed on the bank of the river Solo. The spot of the discovery is indicated by a white cross. (From a photograph in Mme. Selenka's work.)

the period when the animals at Trinil lived in Java, this island was united to the Asiatic continent.

It was in the layer containing lapilli ("*bone layer*" in section,

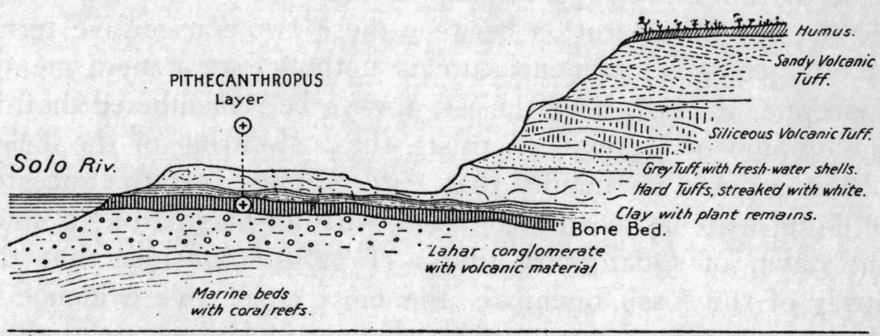

FIG. 56.—Geological section of the *Pithecanthropus* bed.

Fig. 56) that the remains of *Pithecanthropus* were found by Dubois, all at the same level: first, in September 1891, one of the teeth; a month afterwards, the skull-cap, 1 metre distant from the tooth; the following year, in August 1892, the femur, 15 metres from the skull, and a little later, a second tooth, 3 metres from the skull.

The Age of Pithecanthropus.

Dubois, founding his conclusions on the similarity between the fossil fauna of Trinil and that of the upper deposits of the Siwalik Hills in India, attributed the *Pithecanthropus* layer to the Upper Pliocene. The underlying marine conglomerate, containing only 53 per cent. of existing species of molluscs, he regarded as Miocene. These conclusions have been disputed by several members of the Selenka expedition, particularly by the geologists Volz, Elbert, and Carthaus, who, following the conchologist Martin, regard the marine deposits as Pliocene, and the formation of the river sands and of the lapilli as Quaternary. This is also the opinion of Schuster, who studied the imprints of fossil plants from the same bed. Blanckenhorn summarizing the findings of his various collaborators, is of opinion that the *Pithecanthropus* layer may be ascribed to that period of transition which up till now has been termed the *Elephas meridionalis* stage of the Upper Pliocene.[1]

The differences of opinion between Dubois and his opponents are more apparent than real. They rest chiefly on differences in the definition of the limits of the Tertiary and Quaternary. The summit of the Upper Pliocene and the base of the Lower Quaternary are obviously the same thing, for the line of separation between these two consecutive terms of our geological nomenclature is nothing but a mere mental concept. It must, nevertheless, always be remembered that, in the opinion of German scientists, the post-dating of the Trinil bed is opposed to the idea that *Pithecanthropus* is the ancestor of Man, who has been recognized, with all his attributes, since the dawn of Quaternary times. I myself consider that the study of the fossil mammals, the most conclusive evidence in the present case, argues rather in favour of Dubois' opinion.

[1] Some wished to go further. As the fossil plants seem to indicate a somewhat colder, and in particular a moister climate than the present climate of Java, they have attributed the formation of the fossil-bearing deposits to a rainy epoch, and endeavoured further to correlate this epoch with the first Glacial Period in Europe. Comparisons of this kind between deposits situated at such distances are exceedingly rash.

Let us now study the bone-remains of *Pithecanthropus*.[1] These remains are completely fossilized, as are the bones of the animals which accompany them. Their density is considerable: the femur weighs almost double the weight of the femur of a modern man of the same size.

The Skull-Cap.

The skull-cap (Fig. 57) measures 0.185 metres in length and 0.130 metres at its greatest breadth, which gives a cephalic index of 70, and places the skull in the dolichocephalic group.[2] In spite of its large dimensions, it presents at first sight a simian aspect, due especially to flattening in a vertical direction.

We may estimate the capacity of the whole skull at about 850 cubic centimetres. As, in normal Man, even amongst the primitive races, this capacity very rarely falls below 1000 cubic centimetres, and as in the largest anthropoid apes it scarcely ever exceeds 600 cubic centimetres, the volume of the skull of *Pithecanthropus* is intermediate between that of the highest Apes and of the lowest Man. The weight of the brain must have been about 750 grammes.

The bones of the cranial vault are so fused that the sutures can no longer be distinguished. The anterior supra-orbital portion of the frontal bone exhibits a continuous ridge, a kind of rim similar to that which occurs in Gibbons and Chimpanzees. Behind this rim the forehead is very narrow and receding, more receding than in the Chimpanzee. The frontal bone has a slight keel on the median line, but the skull lacks any trace of the sagittal crest possessed by the largest anthropoid apes, the Orang and the Gorilla. On the contrary the

[1] To the four relics already mentioned: skull-cap, two upper molars and a femur, there must be added a premolar found after Dubois' departure. This scientist is also said to have discovered later, a good many kilometres distant, a fragment of the chin region of a jawbone. These two latter pieces have not been described.

[2] The cephalic index is the ratio of the greatest breadth of a skull to its greatest length, multiplied by 100: $\frac{\text{breadth} \times 100}{\text{length}}$. Elongated or *dolichocephalic* skulls have the smallest indices (below 75); short or *brachycephalic* skulls have the largest indices (above 80); the medium group is called the *mesocephalic* or *mesaticephalic* group (between 75 and 80). In other words, when a skull is said to be dolichocephalic, it means that its breadth equals about 75 per cent. of its length, while the breadth of a brachycephalic skull equals at least 80 per cent. of its length, and so on.

temporal lines are not very prominent and are widely separated from one another, as in the Gibbons, Chimpanzees, and Man,

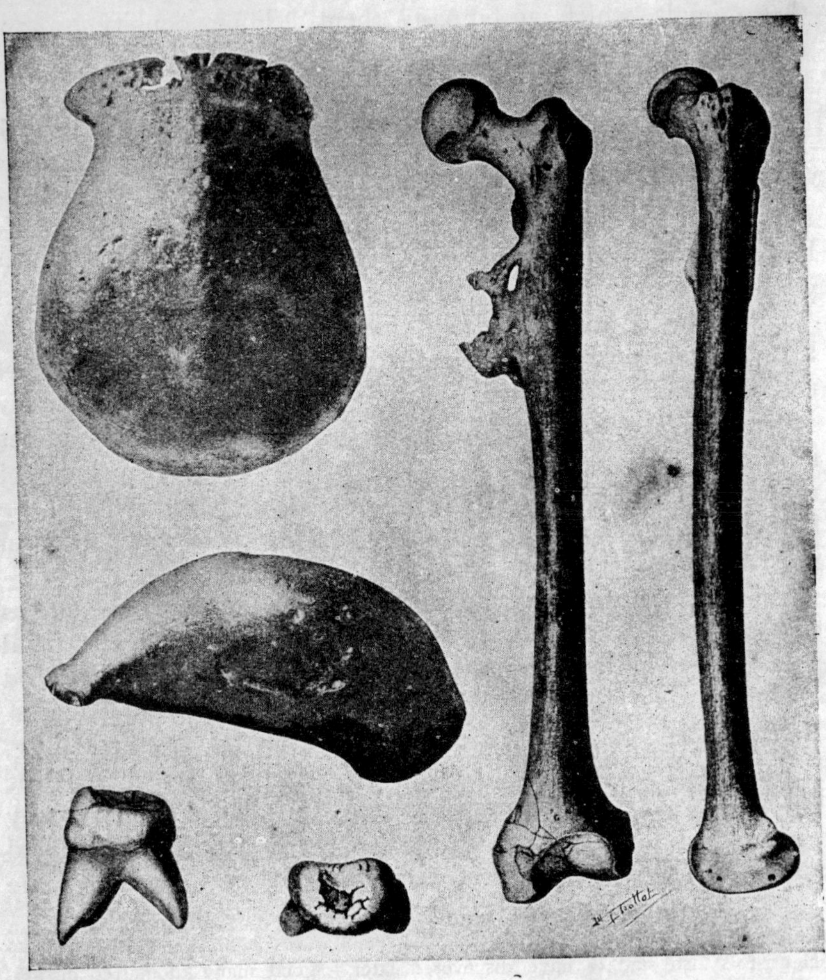

FIG. 57.—The Bone-Remains of *Pithecanthropus*.

The skull-cap, seen from above and in profile, about one-third natural size. The femur, front view and profile, about one-fourth natural size. The third upper right true molar, in profile and the crown, natural size. (After Dubois.)

and this indicates relatively slight temporal muscles and mandibular apparatus. According to Keith,[1] the fronto-malar region is that of an ape and in no respect resembles that of a Man.

[1] Keith, A., *The Antiquity of Man*, p. 264 (London, 1915).

The nuchal plane, formed by the occipital bone, is more inclined than in the anthropoid apes, and less inclined than in Man. Following Dubois, various anatomists, particularly Manouvrier, have rightly laid stress on the presence of the protuberance, here somewhat less but still continuous, which in apes unites the occipital, temporal, and supra-mastoid crests (*crête inio-mastoïdienne* of Topinard, or *crête temporo-occipitale* of Manouvrier). This protuberance is always widely interrupted in Man, even in types which appear to us most primitive—even in microcephalics (Fig. 58).

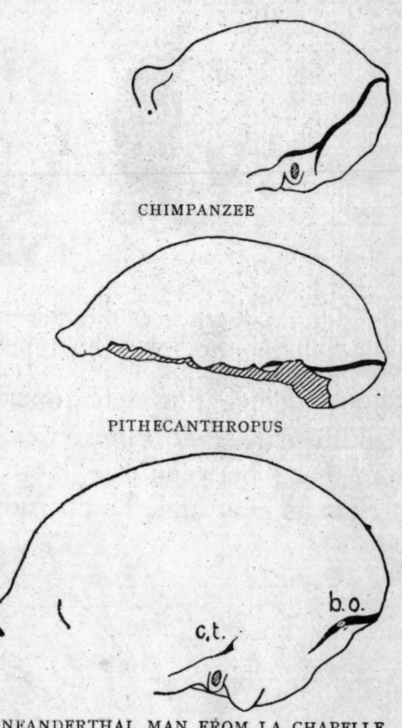

FIG. 58.—Sketch showing by a thick dark line the continuity of the occipito-temporal ridge in the skulls of a Chimpanzee, and of *Pithecanthropus*. In the skull of a man as primitive as La Chapelle-aux-Saints type, the two elements, temporal crest (c.t.), and occipital ridge (b.o.) are already discontinuous.

Taken as a whole, these structures are very similar to those of Chimpanzees and Gibbons. Dubois has said that the skull of *Pithecanthropus* might be compared to a Gibbon skull enlarged to twice its size. Figs. 59 and 60 show that, in its principal characters, the Trinil skull-cap is really intermediate between that of an ape, like the Chimpanzee, and that of a man of really low status, such as Neanderthal Man.

The Brain. A cast of the interior of the skull-cap enabled Dubois to gain an idea of the brain that had been contained therein. The convolutions, less simple than those of the Gibbons, seem already to be of the human type, an opinion shared by an eminent English specialist, Elliot Smith. But if the centres of sensation are well developed, as in apes, the association centres are much less developed than in Man. The frontal region, the site of the higher faculties, the noblest from the psychical point of view, is particularly reduced in size. The inferior frontal convolution, double that of a Chimpanzee or of an Orang, is

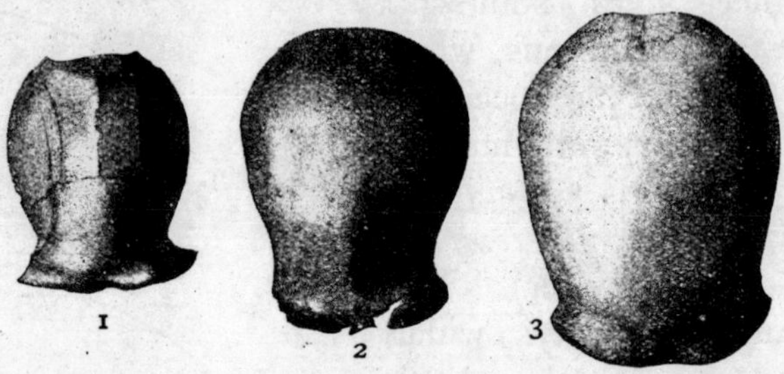

FIG. 59.—Skulls: (1) of a Chimpanzee; (2) of *Pithecanthropus*; (3) of Neanderthal Man, seen from above and on the same scale (about one-fifth natural size).

only half that of a European. In his brain, as well as in his skull, *Pithecanthropus* then may truly be considered as intermediate between the large apes and Man. Dubois declares that he may have had a rudimentary articulate language.

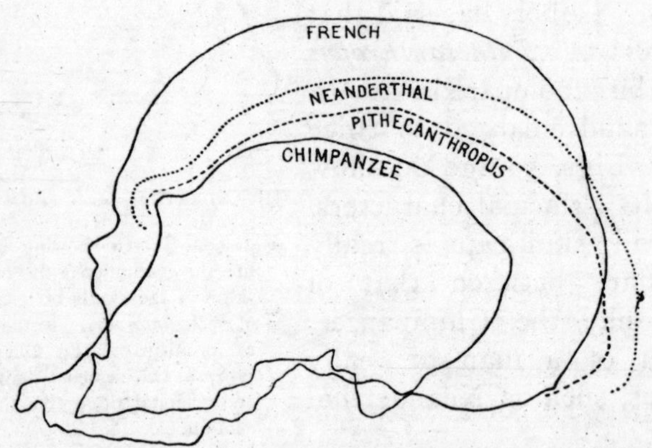

FIG. 60.—Superimposed profiles of the Skulls of a Chimpanzee, of *Pithecanthropus*, of Neanderthal Man, and of a modern Frenchman. About one-third natural size.

The Teeth.

The teeth found are three in number: a premolar which has never been figured; two last upper true molars (wisdom teeth), one of which, almost intact, is from the right side (Figs. 57 and 61), while the other, from the left side, is much worn and must have belonged to an older individual. These teeth were found separately, at various intervals of time and space. Each is of larger proportions than the corresponding human tooth, even if they be compared with the largest dentitions, such as those of the Australians. The true molars have strong, very divergent roots, a simian character; on the other hand, their crowns are relatively more developed in a transverse than in a longitudinal direction, and this may be regarded as a human character. In short, these teeth differ both from those of Man and from those of living or fossil apes (Fig. 61); but their greatest

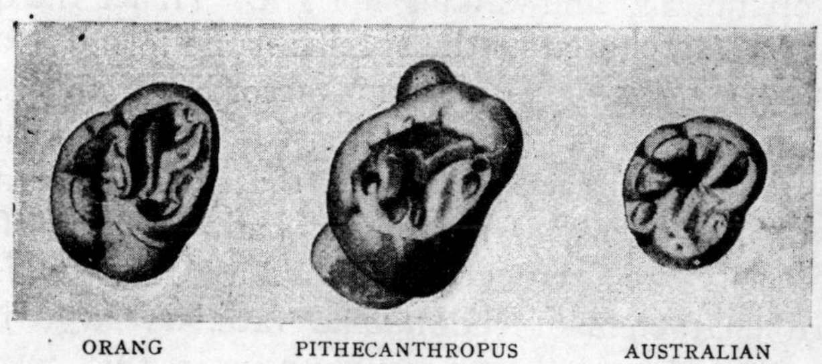

ORANG PITHECANTHROPUS AUSTRALIAN

FIG. 61.—Last true upper molars ("wisdom teeth") from the right side of an Orang, of *Pithecanthropus*, and of an Australian aboriginal. One and a half times natural size. (After Gregory.)

similarities are with those of the Orang, and, according to Gregory,[1] of *Dryopithecus*. They are thus definitely simian.

The Femur.

The left femur (Fig. 57) is complete. It measures 0.455 m. in length; if it really belonged to a Man, he must have been 1.65-1.70 metres in height. It exhibits in the upper portion, on the level of the lines of bifurcation of the femoral crest (*linea aspera*), a large irregular development of bone (exostosis). This accidental pathological structure has aroused much discussion, the net result of which is to show that it has no bearing on the zoological nature of *Pithecanthropus*.

In its whole structure, this femur is so human that, had it been found alone, there would have been no hesitation in attributing it to a Pliocene Man. It has been studied with very special care, and has been found to exhibit certain simian

[1] Gregory, W. K., *Studies on the Evolution of the Primates*, p. 320.

characters; but it is not proved that these characters may not be found in human femora. It is remarkable for its almost perfect straightness, a character which the femora of Gibbons and even of cynomorphic apes exhibit in the highest degree, but, according to Martin, it is more distorted than in the Gibbons. Dubois has called particular attention to certain differences in the manner of insertion of the muscles, especially the large adductor, and this would indicate a more marked aptitude for climbing than Man possesses.

In any case, this femur points to its possessor having the faculty of standing and walking upright. Hence the qualifying adjective *erectus*, chosen as the specific name for a creature who was *pithecanthropus*—an Ape-Man—according to his skull.

Interpretation of the Facts.

Such are the facts. If we possessed only the skull and the teeth, we should say that we were dealing with a large Ape; if we had only the femur, we should declare we were dealing with a Man. Of the two principal characters of the human race, its large brain and upright attitude, the latter would seem in this case to have been wholly acquired before the former. Duckworth has pointed out that this is not consistent with the ontogenetic development of Man.[1]

A first and important question thus arises: Did the skull-cap, the teeth, and the femur, found separately and at more or less considerable intervals of time and distance, belong to the same being? Dubois considered himself justified in asserting that they did, because no remains of large Primates have ever been found in Java, except in this spot at Trinil, and the simultaneous presence of several species appears very improbable. Further, the various bones were scattered at quite inconsiderable distances from one another.

These are certainly good arguments, but they are not conclusive. Some doubt remains, and will still remain, until new and more fortunate explorations put us in possession of less imperfect remains found in close association. The more recent discovery at Piltdown in England of the jawbone of a Chimpanzee, associated with the skull of a Man (see Chapter

[1] Duckworth, W. L. H., *Prehistoric Man*, p. 6 (Cambridge, 1912).

VI.), is not calculated to reduce the uncertainty regarding the unity of the discoveries at Trinil.

The fact having been taken for granted that the Trinil remains belonged to one and the same creature, attempts at restoration were undertaken. Dubois and Manouvrier published reconstructions of the cranium and even of the whole skull (Fig. 62). These attempts, coming from medical men, and

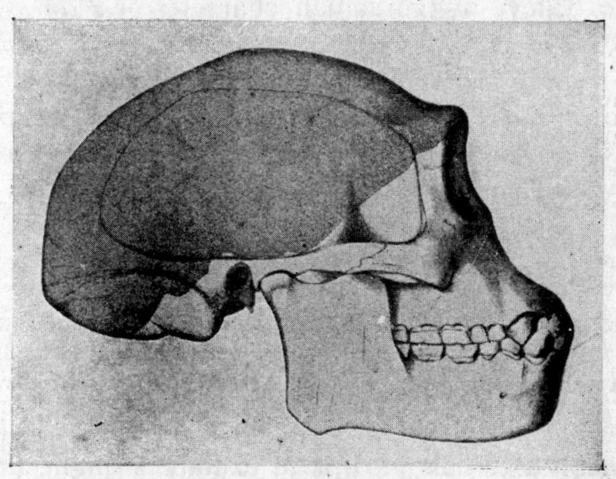

FIG. 62.—Reconstruction of the Skull of *Pithecanthropus*, after Dubois. One-third natural size. The shaded surface indicates the only portion known.

being based principally on human anatomy, are far too hypothetical, since we possess no data for the reconstruction of the base of the skull, the whole face, and all the apparatus of the lower jaw. It is astonishing to find a great palæontologist like Osborn also publishing attempts of this kind. Dubois ventured still further in the realm of imagination when he exhibited at the International Exhibition of 1900, in the Dutch Indies pavilion, a painted model of *Pithecanthropus* as he appeared in life!

Various interpretations have been given of the facts which I have just summarized as briefly but as accurately as possible. According to many scientists, Dubois' opinion, that we have here transition form between the Anthropoid Apes and Man, is justified. Of such, I may mention in France, Manouvrier and Verneau; in America, Marsh and Osborn; in England, Duckworth, Sollas, and Keith; in Australia, Berry and Robertson;

in Germany, Nehring, Schwalbe, Haeckel, and others. Other German authors, Virchow, Krause, Waldeyer and Ranke, the Italian Sergi, and the Swiss R. Martin, believe that *Pithecanthropus* was simian in nature. Topinard in France, Houzé in Belgium, Lydekker, Turner and Cunningham in England, are inclined to regard it as a Man.[1]

These differences of opinion are more apparent than real. Those who believe in the simian character of *Pithecanthropus* really look upon it as an ape superior to all living apes, and those who believe in its human character regard it as inferior to all known men, living or fossil. Wherever we place the Trinil fossil, according to its morphological characters, in the series between Ape and Man, as at: P, P′, P″,

APE......P......P′......P″......MAN,

the fact remains that in all its characters known to us, this fossil stands in an intermediate, or if terminological exactitude be preferred, an interposed, position. This is a positive fact admitted by all competent naturalists.

To consider only the most important relic, the skull-cap; unquestionably this falls into place exactly, I might almost say ideally, between that of the large apes, like the Chimpanzee, and of a man of archaic characters, such as the Neanderthal Man.

But it must be distinctly stated, and in this case repeated, that resemblance does not always imply descent. Even if, in the sum of his known characters (poor at the best), *Pithecanthropus* actually forms a structural link between the large apes and Man, it does not necessarily follow that he must be regarded as a genealogical link, and this distinction is not, as has been asserted, merely a question of words.

Genealogical Relationships of Pithecanthropus.

In order to come to a decisive conclusion regarding his true genealogical relationships, we should require to possess at least the complete skull and lower jawbone of *Pithecanthropus*; for all the reconstructions, with their more or less marked anthropomorphism, which have been advanced by different

[1] I merely mention without comment the absolutely unfounded opinion that it was an imbecile or microcephalic human being, and the other equally wild assertion that *Pithecanthropus* was the result of a cross between a Man and an Ape.

authors, will never help to solve the problem. In the present state of our knowledge, I do not think that we are yet in a position to believe that there was any direct descent between *Pithecanthropus* and Man, such as the genealogical tree prepared by Dubois would indicate (Fig. 63, I).

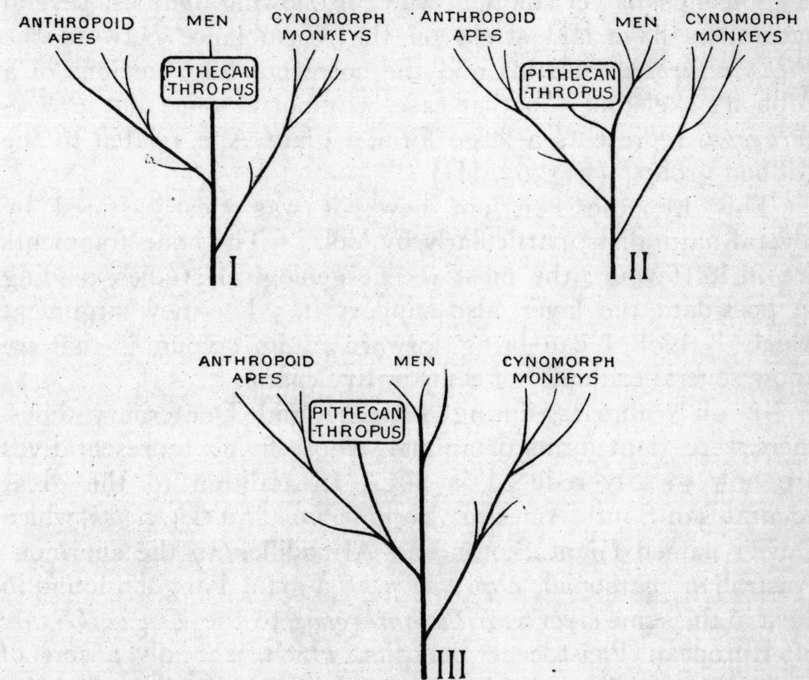

FIG. 63.—Diagrams showing the place of *Pithecanthropus* among the Primates : I, according to Dubois ; II, according to other naturalists ; III, according to Boule.

It is certainly more satisfactory to admit that the evolutionary branch to which the famous Javan fossil belongs, was different from the human branch. Naturalists have no longer any doubt that we are related to the apes ; but it is of some interest to try to define this relationship, especially when we meet with a creature apparently more akin to us than any other. Dubois rightly pointed out that if *Pithecanthropus* is, so to speak, only our granduncle instead of our grandfather, he is none the less an Ape-Man representing a stage in human descent. The majority of scientists to-day adhere to this view. They consider *Pithecanthropus* to be an extinct lateral twig of the human branch. As such he is regarded by Keith, Gregory and Osborn, whose general views are shown in the second diagram (Fig. 63, II).

It is possible, however, to interpret these genealogical relationships in yet another way. Following Dubois, several naturalists have laid stress on the resemblance between the *Pithecanthropus* remains and the corresponding portions of a Gibbon's skeleton. In that case, why not assume that *Pithecanthropus* represents a large form, a giant Ape, related to the Gibbon group? (Fig. 63, III).

This hypothesis is not new; it was clearly stated by several naturalists, particularly by Volz.[1] The bone fragments are in its favour; the most recent geological studies, tending to post-date the layer, also support it. The new argument which I think I can bring forward in its favour, is that we know several examples of comparative cases.

In all countries, during Pliocene and Quaternary times, there were giant forms of animals whose living representatives are now greatly reduced in size. In addition to the great edentates of South America, *Megatherium* and *Glyptodon*, which Cuvier named Giant Sloths and Armadillos, to the enormous Australian marsupial, *Diprotodon*, to a giant Pangolin found in Java in the same layer as *Pithecanthropus*, to the *Trogontherium* of European Pleistocene deposits, which is really a sort of giant Beaver; in addition, also, to the whole series of large running birds of Madagascar and of New Zealand recently extinct, examples among the Primates themselves are not awanting.

Pilgrim found in the Siwalik Hills the remains of a monkey which he named *Dryopithecus giganteus*. *Megaladapis* of the recent geological formations in Madagascar is none other than a giant Lemur. *Archæolemur* and *Hadropithecus*, from the same layers, are also Lemurs of larger size than the living forms; but they show morphological characters of a higher order, denoting a tendency towards the higher ape type,

[1] Volz, W., "Das geologische Alter der Pithecantropus-schichten bei Trinil" (*Neues Jahrbuch für Mineral* . . . Festband, 1907).

for the tendency towards greater perfection is not exclusively confined to the human branch.[1]

We may therefore consider that *Pithecanthropus*, discovered in the same zoological region as the modern Gibbons, may have been a large species either of the genus *Gibbon*, or rather of a closely allied genus related to the same group. This form might have been superior to its congeners, not only in size but also in other morphological characters, and particularly in cerebral capacity, a character of the first importance in which *Pithecanthropus* truly approaches the human stock.

Just as *Archæolemur* and *Hadropithecus* show the tendency of certain of the lower Primates to ascend towards the higher simian forms, in the same manner *Pithecanthropus* appears to show a tendency to ascend from one form of anthropoid ape to a higher form, analogous to and parallel with the human form. It would thus represent a branchlet, more highly specialized than the neighbouring branchlets of the most highly developed Gibbon branch, and it must soon have died out, perhaps because of this very specialization. *Pithecanthropus*, then, does not belong to the ancestral line of the genus *Homo*. The more or less "human" characters of its skull-cap, and even of its femur, can only be looked upon as characters due to convergence and not to descent.

This way of regarding *Pithecanthropus*, which I cannot too often repeat is after all only hypothetical, seems to me at least as rational as any other. It is indeed more in keeping with the general facts of palæontology, which show that phylogenetic branches become more and more thickly clustered and more and more profuse in variation, as well as more and more independent. In accord with the new light which science sheds, we see the different branches lengthening towards the base, preserving the while their distinctness, and we see also how their union with the mother branches or with the main stem is projected farther and farther away, very

[1] I must, however, observe that, according to Elliot Smith, the brain of *Archæolemur* shows certain characters of specialization in a retrograde direction. So far as I know, the brain of *Hadropithecus* has not been studied.

often quite beyond the points which our researches have reached.

Not only at its origin, but even towards the end of its evolution, the human branch would thus have as neighbours, and in a sense as rivals, other branches of the higher Primates, the offsprings of a common stem. Various ape-like forms, starting from the first anthropoid stages, would seek, under the influence of the same environment or of the same needs, to evolve towards types better adapted to new conditions, that is to say, towards types of greater perfection. Several of these forms may have been able to surmount the stage where the living Anthropoids seem to stand, and to have acquired some of the higher characters that Man possesses to-day; but the direct descendants of our primitive ancestors alone would seem to have reached the end of this race towards the goal of progress.

This interpretation of the memorable Javan discovery does not lessen its interest. I am tempted to say that, on the contrary, it increases it, since the human line, while still retaining its independence, seems thus morphologically less isolated than formerly from the neighbouring lines. It leads us to the admission that in other days there existed Anthropoids higher than living Anthropoids, but inferior to the fossil Men known to us, who themselves were inferior to living Man. The physical relationship between Ape and Man here asserts itself from a new point of view.

In any case, in whatever manner we interpret them, the facts revealed by Dubois' discovery remain of supreme interest to science.

CHAPTER V

THE PROBLEM OF TERTIARY MAN

Eoliths

As soon as the existence of Quaternary Man was definitely proved, the desire was stimulated for evidence of the existence of a Man in still earlier times—of Tertiary Man.

It must be allowed that no research could be better justified.

The Problem.

The teaching of palæontology regarding the evolution and progressive development of mammals in general, warrants us in believing and stating that Man or his immediate precursor must have lived towards the end or even from the middle of the Tertiary era.

We have seen (p. 32) that, during the Eocene and Oligocene periods, mammals still differed so greatly in their build from modern types, that palæontologists often experienced great difficulty in placing them in the orders or even in the families established by zoologists for the classification of living forms. Representatives of the order of the Primates had already been found, but these forerunners had not outgrown the evolutionary stage now represented by the lower forms of the group, the Lemurs and the tailed monkeys. From this we gather that the existence of Man, at least of Man possessing all the main characteristics of his kind,[1] would seem to have been all but *impossible* during the first half of Tertiary times.

During the Miocene period, the majority of the modern families of animals were clearly differentiated. Many of the Miocene genera are extinct, but there were, even then, certain

[1] I make this reservation because Men may have existed, potentially, in some primitive form of Primate. We have already considered the rôle attributed to *Propliopithecus* from Fayum (see p. 82).

representatives of genera which still exist. Anthropoid apes were numerous. The existence during this period of a Man, or rather of a *pre*-Man, is quite *possible*.

During the Pliocene period, almost all the modern genera were represented; many species are akin to living species, and may be considered as the ancestors of the latter. The existence of a human being, or even of a true *Homo*, is quite *probable*.[1]

From the palæontological point of view, then, no *a priori* objection can be taken: physically, Man is closely akin to the anthropoid apes which existed, well defined, from at least the Lower Miocene. Why should he not also be, like them, geologically very ancient? His late survival may be explained, according to de Quatrefages, by his intellectual superiority, which would enable him to resist the various causes which led to the destruction of other species, and to adapt himself better than these to new conditions.

Prehistoric archæology also supports this view. We now know that from the beginning of Quaternary times, Man occupied, if not the whole globe, at least a large part of the surface of the globe, practising everywhere a somewhat rude but remarkably uniform stone industry, more or less similar to the industries known as Chellean or Saint-Acheulean (Fig. 64). It would even appear, as we shall presently see, that he was already represented by several physical types. This of necessity implies a previous stage, a Tertiary stage. The origin of Man must certainly date from a past still more remote than that from which we actually know his presence.

Therefore, not only is there no *a priori* reason for denying the existence of Tertiary Man, or, to put it otherwise, of an immediate ancestor of Quaternary Man, but, on the contrary, all the evidence tends to affirm such existence. A creature,

[1] Thirty years ago, I wrote (*Revue d'Anthrop.*, 1889, p. 217): "The majority of the genera of living mammals are very ancient. From the Upper Miocene onwards we have true Cats, true Hyænas, true Deer, the true Rhinoceros; that is to say, representatives of living genera. Even certain species are closely akin to living species. Examples become more frequent during the Miocene period, and I see no reason why the genus *Homo*, with all its characteristics, should not have been contemporary with the genera of animals which I have just enumerated. I am perfectly convinced that palæontologists will some day find the bone-remains of our Tertiary ancestors."

already in possession of the chief physical and even mental attributes of the human stock, must have lived somewhere during the Pliocene period, and perhaps even during the Miocene.

But these theoretical considerations, convincing as they may sound, are not sufficient to solve the problem. The true question is this : Are we really in possession of material proofs

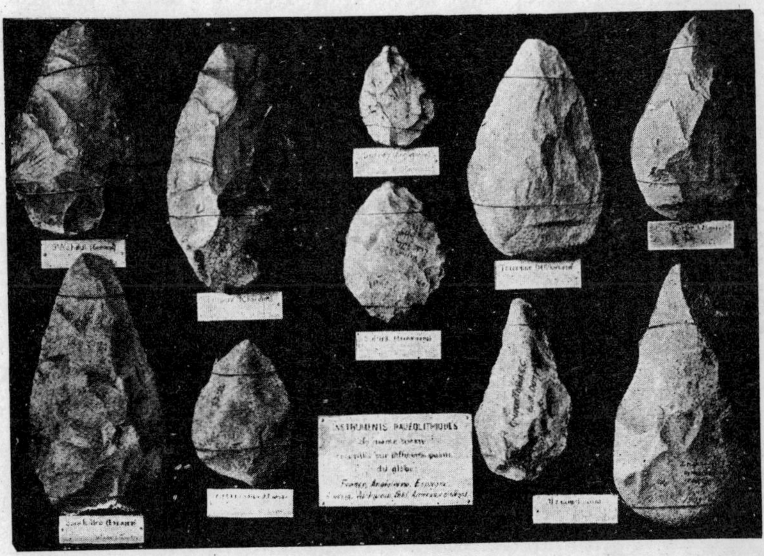

FIG. 64.—Palæolithic implements of similar forms from different parts of the globe. France (Saint-Acheul, Tilloux, Toulouse), England, Spain, Algiers, South Africa, and North America. One-third natural size. Palæontological Gallery, French National Museum of Natural History, Paris.

of the existence of a Tertiary Man, of bone-remains, products of a primitive industry, or traces of any kind whatsoever? I regret to state that not one of the discoveries hitherto brought forward to settle this question appears to me to be conclusive.

In face of such a statement, the reader will probably wonder why, in this book on fossil men, I devote a whole chapter to a being of whom we have no knowledge. I do so because, while "the question of Tertiary Man" is not solved, it none the less insistently arises, and because some of the evidences in which I myself place no faith have nevertheless their partisans. The discussion is not yet settled; and so the arguments must not be rejected in mass, but, on the contrary, must be brought forward and properly presented, in order that they may be the better submitted to serious criticism. In this way a sound foundation will be laid for future researches.

Historical Review.

The first of the discoveries referred to followed very soon after the victory of Boucher de Perthes. In 1863, Desnoyers, librarian of the French National Museum, announced that he had found in the gravels of Saint-Prest (Eure-et-Loir), regarded as Pliocene, bones of large animals bearing lines or incisions which could only be considered the work of man.

In 1867, at the International Congress of Prehistoric Archæology and Anthropology, in Paris, Abbé Bourgeois exhibited flints from the Oligocene deposits at Thenay (Loir-et-Cher). According to him, certain of these flints showed fractures, small pittings, and traces of splitting by fire, while others had been deliberately dressed (Fig. 65, 1-3). At the same time, his fellow-worker, Abbé Delaunay, showed some bones, carved in the fashion of those from Saint-Prest, which had been obtained from Miocene shell-beds of marine origin, at Pouancé (Maine-et-Loire). Soon after, all the great beds containing mammalian fossils: Oligocene in the Bourbonnais, Miocene in the Orleans district, in the Gers, and at Pikermi in Greece, yielded bones which were alleged to have been deliberately broken, scratched, or incised.

In the same year, 1867, Professor Issel, of Genoa, announced the discovery of a human skeleton in the Pliocene layers of Savona in Liguria, and the Americans sent word of the discovery of a human skull in their gold-bearing alluvials at Calaveras in California, which also were regarded as Tertiary.

In 1875, Professor Capellini, of Bologna, described from marine deposits of Pleistocene age at Monte-Aperto, in the province of Siena, the bones of a fossil whale showing incisions of a particular character, which could be attributed only to man.

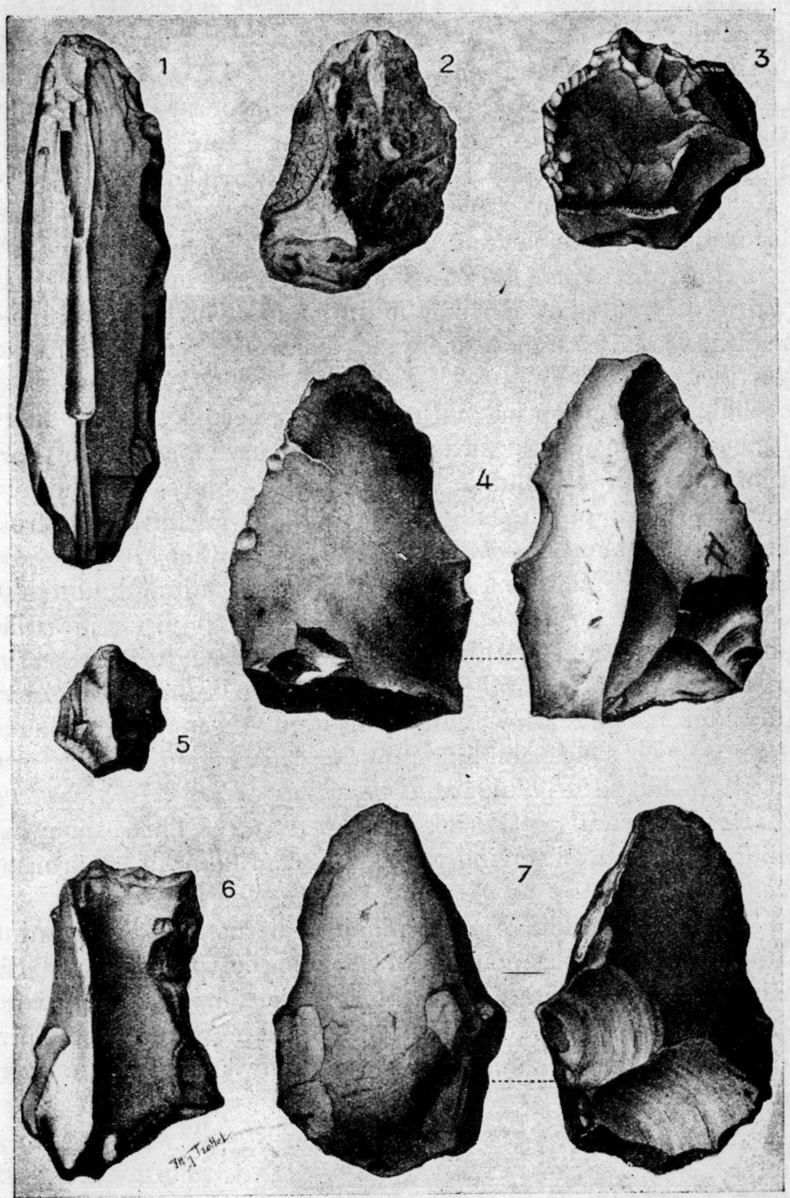

FIG. 65.—Tertiary Flints regarded as products of Primitive Human Industry. 1 and 3, split flints from Thenay; 2, burnt and fractured flints from Thenay (Loir-et-Cher); 4, split flints from Otta (Portugal); 5, 6, 7, split flints from Puy Courny, in Cantalia. Natural size.

About the year 1878, the Portuguese geologist, Ribeiro, recorded certain *dressed* flints from a Miocene deposit at Otta, in the neighbourhood of Lisbon (Fig. 65, 4); whilst the Cantalian geologist, Rames, sent to the International Exhibition a case of flints of the same kind, collected in alluvials of the Upper Miocene at Puy Courny, near Aurillac (Fig. 65, 5-7).

During a decade these discoveries gave rise to many discussions in Reviews, Societies, and Congresses devoted to prehistoric archæology.[1] The existence, based on certain "dressed stones," of Tertiary Man or of a Tertiary forerunner of Man, was ably supported by de Quatrefages and Gabriel de Mortillet.

The latter scientist, although he rejected all the other discoveries, became the ardent and zealous supporter of those from Thenay, Otta, and Puy Courny. Since, on the one hand, his ideas on the subject of evolution precluded him from dating the genus *Homo* as far back as the Miocene and Oligocene, and since, on the other hand, he had no doubt as to the deliberate carving on the flints exhibited, G. de Mortillet conceived the existence of a creature intermediate between the large apes and Man, and this he first named *Anthropopithecus*, and later *Homosimius*. He even went so far as to found three species corresponding to the three deposits; he proposed to locate at Thenay, *Homosimius bourgeoisi;* at Otta, *H. ribeiroi;* and at Aurillac, *H. ramesi.*[2] This somewhat childish method of solving a serious scientific problem aroused lively criticism.

The final result of interminable discussions was rather antagonistic to the new theory, and the partisans of the belief in the deliberate dressing of Tertiary stones became fewer and fewer. In the end, while each protagonist clung to his own opinion, discussion gradually died down and a lull succeeded.

[1] The bibliography relating to these early discoveries is so voluminous that I could not give it here. See Mortillet, *Le Préhistorique*; and particularly Reinach, S., *Antiquités nationales*, "I. Alluvions et Cavernes." Numerous references will be found on succeeding pages.

[2] Mortillet, G. de., *Le Préhistorique*, p. 105 (Paris, 1883), 3rd ed. by Gabriel and Adrien de Mortillet, p. 96 (Paris, 1900).

About 1889 a reawakening occurred. The venerable English geologist, Prestwich,[1] signalled its commencement by his publications on the supposed implements found at Ightham in England. They dealt with flints showing indistinct traces of working, and with splinters or natural fragments slightly retouched, much older than those of Saint-Acheul, and "dating from the pre-glacial period." As the flint-containing deposits of the Kentish plateau may well be regarded as Pliocene, there was once more involved the question of Tertiary Man, who, in his archæological aspect, reappeared this time on the soil of Great Britain.

The Eolith Theory.

The work of Prestwich at Ightham made a great impression in England, for he ranked as a geologist of high standing. His lead was followed; so that before long, stones fashioned after the same manner were being found almost everywhere. English enthusiasts discovered them in other parts of the British Isles, for example in the Norfolk Forest-bed, and even in the Upper Tertiary of India; while Continental workers described them from numerous layers in France, Belgium, and Germany. Among amateur collectors of dressed flints there developed a perfect frenzy for seeing in every broken pebble and jagged edge products of human industry. Thus the question of "Tertiary flints" became again the order of the day, decked in new name, to wit "the Eolith theory"; it compelled the attention of all prehistorians; in every country it had warm supporters. Presently an enthusiastic Belgian geologist, M. Rutot, put himself at the head of the movement by a perfect avalanche of publications.[2]

To the two periods of the stone industry long recognized, the Palæolithic and the Neolithic, G. de Mortillet had added a more ancient period which he called *Eolithic*, and which was meant to include "all that related to the Tertiary

[1] Prestwich, J., "On the Occurrence of Palæolithic Flint Implements in the Neighbourhood of Ightham," (*Quarterly Journal of the Geol. Soc. of London*, xlv., 1889).

[2] The majority have been reviewed in *L'Anthropologie*, viii. to xvii., *passim*. See particularly, Rutot, A., "Le Préhistorique dans l'Europe centrale. Coup d'œil sur l'état de nos connaissances relative aux industries de la pierre." (*Congrès d'archéol. et d'histoire* at Dinant, in 1903. Namur, 1904).

period."[1] According to Rutot, who adopted the name invented by G. de Mortillet, but somewhat distorted its original meaning,

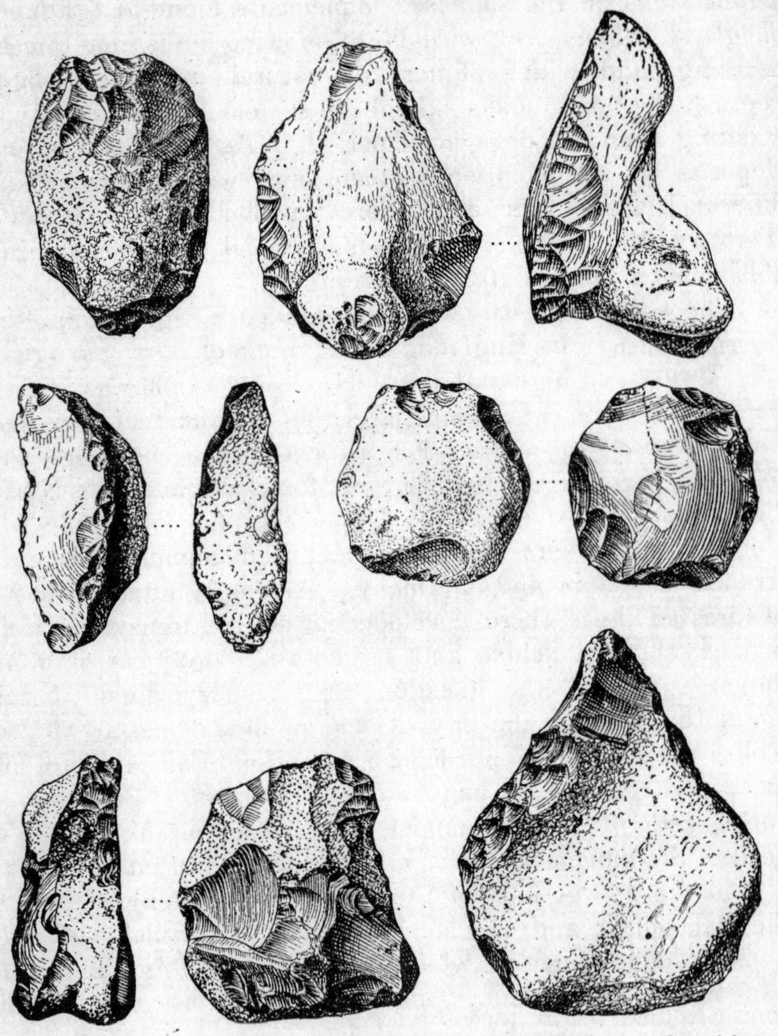

FIG. 66.—Eoliths from the "Reutelian" of Belgium. The chipped stones figured here, according to Rutot, are considered by him to be hammers, grinders, scrapers, and discs. Half natural size.

the Eolithic period furnishes no stone dressed to a *deliberat* *shape*, but only *natural forms, directly utilised.* We must giv

[1] *Le Préhistorique*, 1st ed., p. 18.

the name *eoliths* to these rude and primitive implements, these stones which were simply used. Eoliths may be recognized by signs of retouching, that is to say by slight marks of chipping, localized or disposed as if they had been knocked off deliberately and systematically, in order to adapt the piece of flint to a definite end. "The special appearance," said Rutot, "that experts rightly name 'dressed' is only to be attributed to human and essentially deliberate action, and all eoliths, whether from the Aquitanian [Oligocene] or Mosean [a name given by Rutot to a Quaternary level in Belgium] or from the intermediate stages of Miocene, Pliocene, and Lower Mosean, which exhibit retouching produced by usage, must be included among the authentic relics of primitive industries."[1]

Eoliths, thus regarded, are to be found in great quantities in Quaternary gravels, along with implements of definite and standard form. M. Rutot goes so far as to describe, from the gravels of the north of France and of Belgium, several "industries" of this kind—"Reutelian," "Mafflian," "Mesvinian," and others (Fig. 66).

But such objects are also to be met with in much more ancient deposits: chipped stones from the Oligocene at Thenay, from the Miocene at Otta and Puy Courny, flints from the Pliocene in France, England, Egypt and India, are eoliths. And here the question becomes much more serious, for the partisans of the new theory believed that a few chipped or fractured stones were enough to prove the existence, in geological times so remote, either of Man or of his immediate forerunners. After having made numerous converts in Belgium, Germany, England, and even in France, after a clamorous career which gained it notoriety everywhere, the eolith theory succumbed under the repeated blows of certain geologists and prehistorians. Accurate observations, made in every quarter by serious naturalists, showed that the play of the physical forces of nature may produce eoliths, and that it is impossible to distinguish the so-called human eoliths from eoliths wholly independent of any intelligent or conscious interference. The numbers of converts were almost as great as, although less striking than,

[1] *Bull. de la Soc. d'Anthrop. de Bruxelles*, vol. xx., 1902, p. 66.

the numbers of the original adherents. Nevertheless, in recent years, a form of the eolith theory has been resurrected in England.

FIG. 67.—Rostro-carinate ("Keeled") Flint found at the base of the Crag at Ipswich. Two-thirds natural size. (After Sir Ray Lankester.)
K. the keel. The shaded parts show the original surface of the flint nodule.

In October 1910 Mr J. Reid Moir of Ipswich announced that he had taken from the bottom layer of a Pliocene marine

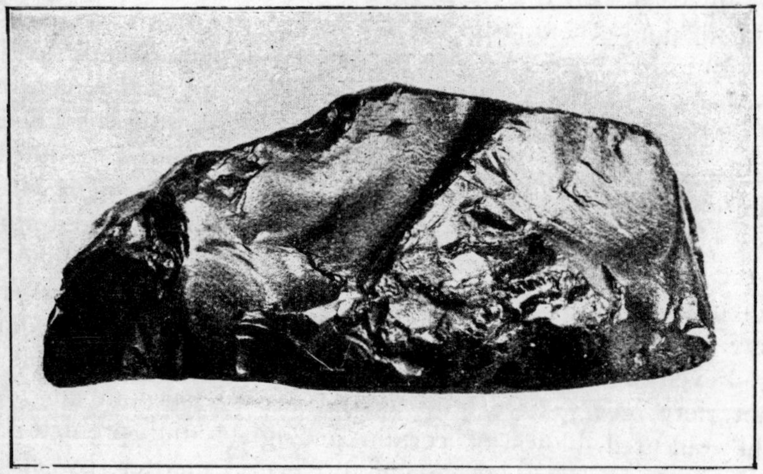

FIG. 68.—The most perfect example of Keeled Flint (*type specimen*) from the Crag at Norwich, side view. (After Sir Ray Lankester)

deposit, called the Red Crag of Suffolk, flints proving the existence of Man prior to this Pliocene deposit.[1] This discovery made a great impression on the mind of the eminent biologist

[1] Moir, J. Reid, "The Flint Implements of Sub-Crag Man," (*Prehistoric Society of East Anglia*, i., 1911).

Sir Ray Lankester, who took an active interest in the matter.[1] Moir and Ray Lankester avoid bringing eoliths forward again, that is to say, pebbles simply used and bearing only the marks of this usage. It is now a question of implements of a very definite form, of a type quite as characteristic as, say, the Saint-Acheulean types, but far older, the "rostro-carinate" or eagle-beak type. The Chellean, Acheulean, and Moustierian flints are essentially flattened, in leaf form. The rostro-carinate type of the Crag is essentially narrowed in a lateral direction (Figs. 67 and 68). Further, it is true that they are also associated with scrapers, hammers, and large picks.

The idea that these flints are artefacts has been contested in England by Sollas. I have myself discussed at length the works of Ray Lankester and Moir, giving numerous reasons which lead me to reject their interpretations.[2] I consider that the flints from the Suffolk Crag are only varieties of eoliths, and I have no hesitation in attributing them to the action of natural forces, which I shall presently discuss, rather than to the intervention of an intelligent being.

Discussion of Evidences. Human Bone-Remains.

After this long historical introduction, there must follow a discussion of the evidences to which appeal has been made. We have just seen that the testimony brought forward in favour of the existence of Tertiary Man is of two kinds: (1) human bone-remains; and (2) products or traces of deliberate work.

(1) The facts belonging to the first category, being from their very nature the most conclusive, should of themselves be sufficient to settle the question. Unfortunately no discovery of human bones, which, up to the present time, has been made in Tertiary deposits, has been able to withstand criticism. The only discoveries deserving examination here, are those made at Savona and Castenedolo in Italy, and at Calaveras in America.

[1] Lankester, Sir Ray, "On the Discovery of a Novel Type of Flint Implements . . ." (*Philosophical Transactions*, Series B., vol. 202, 1912).

[2] Boule, M., "La Paléontologie humaine en Angleterre" (*L'Anthropologie*, xxvi., 1915).

Savona.

The skeleton from Savona, in Liguria, was found in 1852, during the reconstruction of a church, in a shelly clay of marine origin, rich in fossil oysters of Pliocene age; but it was not exhibited by Issel till fifteen years after, at the Paris Congress in 1867. The remains are rather badly preserved, so much so that Issel himself described them under the vague title of "Anthropoid."[1] The distinguished professor of Genoa University was convinced of the Pliocene age of the Savona skeleton; nevertheless, he acknowledges that the circumstances of the discovery were not such as to inspire confidence. He admits that, unfortunately, no naturalist was present "to confirm, by precise and scrupulous observation, that the deposit had not been resorted, and that the bones had really been buried at the same time as the oysters."[2]

Castenedolo.

The bones from Castenedolo, near Brescia in Italy, belong to several skeletons of men, women, and children, and were found on various occasions in a shelly bed of sand and clay, of marine origin and of Pliocene age. The first discovery was made in 1860, the second in 1880; they were described by Professor Ragazzoni,[3] and immediately aroused a lively discussion. The anthropologist Sergi has been, and continues to be, their chief supporter, and he, in order to explain the presence of human skeletons in deposits of marine origin, was obliged to assume the shipwreck of a whole family.[4] In 1889, the discovery of a new human skeleton was the subject of an official report by Professor Issel, who came to the conclusion that the human bones and the Pliocene deposits were not of the same age.[5] In 1896, a pupil of Ragazzoni, Professor Cacciamali, took up the question again, and, after a laborious and thorough inquiry, declared that there must still be doubt in the matter. At Castenedolo, as in Savona, we are very probably dealing with more or less recent burials.[6]

1 Issel, A., *Ligura preistorica*, p. 140 (Genoa, 1908).

2 *Congrès internat. d'archéol. et d'anthrop*, first meeting, Paris, 1867, p. 76.

3 Ragazzoni, "La collina di Castenedolo . . ." (*Ateneo di Brescia*, 1880).

4 Sergi, G., "L'Uomo terziario in Lombardia" (*Archivio per l'Antrop. e la Etnol.*, xiv., 1884).

5 *Bulletino di paletnologia italiano*, xv., 1889, p. 89.

6 Cacciamali, G. B., "Geologia della collina di Castenedolo" (*Ateneo di Brescia*, 1896).

The Calaveras Skull.

The Calaveras skull is too famous for me to pass it over without a few words.[1] It is said to have been found in 1866, at a depth of 30 feet in the pit of a mine dug through superimposed lavas and auriferous gravels of Pliocene age, in the Bald Hill near Altaville, in Calaveras County, California. This origin is more than doubtful. All the circumstances suggest that some hoax was perpetrated—"a practical joke of the miners," as the geologist Marcou declared.

The incomplete skull (Fig. 69) now belongs to the Peabody Museum in Cambridge, U.S.A.: it is partially fossilized and covered with a thin calcareous concretion. But the various American scientists who have studied it, are unanimous in affirming its resemblance to the skull of the modern Indians of that region. Its resemblances to a skull found in a Californian cave, likewise covered with a limy coating, are such that the anthropologist Hrdlička declares we can ascribe both remains to a single race, and what is more, to the same tribe. It is very probable that they have a similar origin, and that, like the other, the Calaveras skull also came from a cave. It cannot be very ancient in the geological sense of the word.

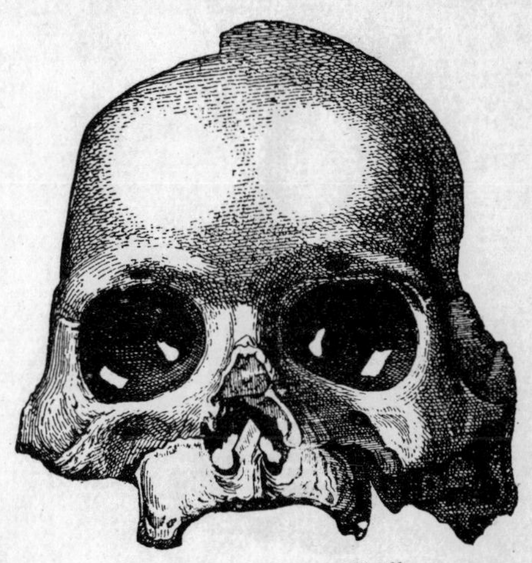

FIG. 69.—Calaveras Skull.

Ameghino also believed that he had found, in South

[1] The bibliography is large, for this discovery has been much discussed. The original report is by Whitney, "Auriferous Gravels of the Sierra Nevada," Cambridge, U.S.A., 1879. We must also note: Holmes, W. H., "Review of the evidence relating to auriferous gravels Man in California" (*Smithsonian Report*, 1899). Wilson, Th., "La haute ancienneté de l'homme dans l'Amérique du Nord" (*L'Anthropologie*, xii., 1901). Hrdlička, A., "Skeletal Remains . . . in North America" (*Smithsonian Institution*, *Bureau of Ethnology*, Bull. 33, 1907). See also an article by Nadaillac, "Le Crâne de Calaveras," in the *Revue des questions scientifiques*, 1900.

America, bone remains of Tertiary Man. He was hopelessly mistaken, as we shall see in the last chapter of this work.

There is no good reason for dwelling further on these various discoveries, since they have been almost entirely abandoned, even by anthropologists who believe they possess other material proofs of the existence of Tertiary Man.[1]

FIG. 70.—Bone of *Halitherium*, a Sirenian fossil, bearing incisions; from the Miocene shell deposits of the Maine-et-Loire. Natural size. (After Farge.)

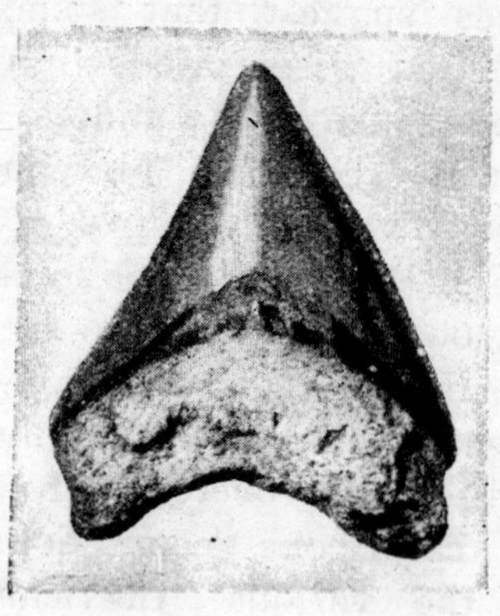

FIG. 71.—Tooth of a large Shark (*Carcharodon megalodon*) from Miocene shell deposits in Touraine, probably the cause of the incisions depicted opposite. Half natural size. Palæontological Gallery, French National Museum of Natural History.

2. Proofs other than those derived from bones themselves may be furnished by traces or products of deliberate handiwork. They may be classified in four groups: (*a*) Bones of animals bearing traces of human action; (*b*) Hunters' traps;

[1] The later supporters of these discoveries accuse their opponents of allowing themselves to be swayed by purely theoretical considerations, such as that the resemblance and even the identity of the supposed Tertiary bone-remains with those of modern Man, is contrary to the doctrine of evolution.

I cannot, for my part, accept this reproach. The argument from general considerations certainly seems to me to be not without a certain value, but I do not consider it of itself sufficient, for I am far from denying *a priori* that the modern human type may be very ancient. The essential criterion must be definite proof that the relics brought forward are really of the same age as the deposits from which they were exhumed, and I affirm that this contemporaneity has not been satisfactorily demonstrated. The whole preliminary question rests on this and on this alone.

(*c*) Burnt flints, splintered by fire; (*d*) Flints used or worked by percussion.

Animal Bones, Scratched, Incised, etc.

(*a*) The first group now forms, in Déchelette's phrase, a "kind of obituary of ephemeral hypotheses" in which no longer any one believes. The fractures, marks, scratches, incisions and perforations found on great numbers of fossil bones, of all ages and from all kinds of deposits, are due either to physical forces, such as desiccation, the accumulation of surrounding deposits, heavy pressure, or to the gnawing of carnivores. The incisions or grooves, so common on the bones of marine animals, such as *Halitherium* of French shell-deposits (Fig. 70), or *Balænotus* from the Italian Pliocene, all of which are found in deposits laid down by the sea, were made by the sharp teeth of great sharks which lived in this sea (Fig. 71). This is no longer disputed, and I can only refer those of my readers who wish fuller information on the subject to the excellent critical account by G. de Mortillet.[1]

Pseudo Pit-traps.

(*b*) In 1888, O. Fisher discovered bone-remains of *Elephas meridionalis* in a chalk-fissure at Dewlish in Dorset, in the south of England. In 1905 he endeavoured to show that this was artificial. It was said to be a true trench or trap, excavated by a human contemporary of the Southern Elephant to capture this Proboscidian, just as the natives of Equatorial Africa capture living Elephants: it was supposed to be the work of a Pliocene trapper.[2]

This hypothesis has recently been revived by its author. An inquiry conducted by experienced geologists, under the auspices of the Dorset Field Club, clearly demonstrated the natural character of the fissure, due wholly to erosive action, and the absence in this so-called trench of any trace of Man's handiwork.[3]

[1] *Le Préhistorique*, 1st ed. 1883, pp. 34 *et seq.*

[2] Fisher, O., "On the occurrence of *Elephas meridionalis* at Dewlish, Dorset" (*Quarterly Journal of the Geol. Soc.*, lxi., 1905, p. 35).

[3] *Proceedings Dorset Natural History and Antiquarian Field Club*, xxxvi., 1915, p. 209.

The Cracked Flints from Thenay.

(*c*) In the Oligocene conglomerate of Thenay, in the district of Loir-et-Cher, there are sometimes to be found, together with eoliths, flints the surface of which is covered with fine cracks, like that of old earthenware, or "decorticated"—covered with fine pitting. Similar appearances may be produced by heating flints and then cooling them rapidly; they are, then, cracked and splintered by fire. In fact, certain native races, the Mincopie of the Andaman Isles and the Australians, treat with fire any stones they wish to split, before shaping them. Abbé Bourgeois and G. de Mortillet supposed that Man or the "Anthropopithecus" of Thenay did the same.[1] There would seem to have existed, therefore, during the Oligocene, a creature sufficiently intelligent to be able to light a fire and to submit flints to its action, with a definite purpose in view.

All the evidences are against such an interpretation of facts which have long supplied material for discussion at scientific congresses. The conditions of the deposit where the flints were found do not support it. Cracked stones have been found in other and more ancient Eocene deposits, and these can be ascribed to no intelligent agency. Many observations have proved that different natural causes may produce the characters of the Thenay flints; changes of temperature of meteorological origin, especially in the conditions of a tropical climate, like that of the Oligocene; spontaneous forest fires; the action of thermal waters; or simply, as in our present climate, the somewhat prolonged action of atmospheric agents. Laboratory experiments have given similar results. It is therefore quite unnecessary to invoke human agency to explain the cracked flints of Thenay.[2] It intervenes in this case as

[1] Bourgeois, Abbé, "L'Homme tertiare, Étude sur les silex travaillés" (*Congrès intern. de Paris*, 1867, p. 67); "La question de l'Homme tertiaire" (*Revue des questions scientifiques*, 1877), etc. Reprints of Abbé Bourgeois' various publications regarding Thenay will be found in Dr F. Houssaye's *L'œuvre de l'Abbé Bourgeois*, Paris, 1904. Mortillet, G. de, *Le Préhistorique*, 1st ed.

[2] "Discussion at the Congrès de l'Association française à Blois" (*Matériaux*, 1884). Arcelin, A., "Silex tertiaires" (*Ibid.*, 1885). D'Ault-Dumesnil, "Note sur de nouvelles fouilles faites à Thenay" (*Ibid.*, 1885). Mahoudeau, P. H., and Capitan, L., "La Question de l'Homme tertiaire à Thenay" (*Revue de l'Ecole d'Anthrop.*, xi., 1901), etc.

a kind of *Deus ex machina*. No longer does any one believe in it; the evidence in the Thenay case has been finally pigeon-holed.

Flints used or worked by Percussion. Eoliths.

(*d*) The last category, by far the largest, comprises the stones shaped or simply used as found, and constitutes the eolithic industry.

Whilst G. de Mortillet considered the *bulb* or *shell of percussion*, which is formed in consequence of a blow sharply directed upon a particular point,

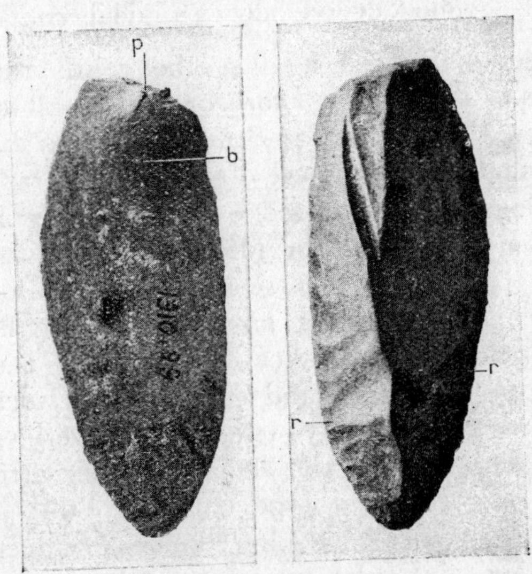

FIG. 72.—Splinter of Moustierian Flint from the Grotte du Placaid, retouched at the point. Three-quarters natural size.

On the inner surface, detached from a mass of flint: *p*, point where struck; *b*, percussion bulb. On the outer surface; *r*, retouching.

as the criterion of deliberate shaping, he attributed to retouching only secondary importance (Fig. 72). M. Rutot, as we have seen, does not attach much weight to the percussion bulb, but only to retouching with a view to use. We shall see presently that this second character is no more conclusive than the first.

Discoveries of stones, apparently shaped or simply used, have been made in deposits belonging to all geological periods, from Eocene to modern times. Enumerating from the most ancient to the most recent, the principal layers are :—

EOCENE, even Lower Eocene.—Duan (Eure-et-Loir), Clermont (Oise),[1] Escheu (Somme), etc.

OLIGOCENE, MIDDLE.—Boncelles, province of Liége, Belgium.[2]

OLIGOCENE, UPPER.—Thenay (Loir-et-Cher).[3]

MIOCENE, UPPER.—Puy Courny (Cantal),[4] Otta, in Portugal.[5]

PLIOCENE.—Kentish plateau, Crag of Suffolk, Norfolk Forest Bed (England), Burma, Thebes (Egypt), etc.

QUATERNARY.—Almost everywhere in alluvials containing flints.[6]

The objections raised against the eolith theory are of various kinds. They apply more or less to all the deposits, so that it would be tedious and unnecessary to examine them in detail. I shall describe these objections generally.

Palæontological Objections.

Palæontology is not opposed on principle to the existence of Pliocene or Miocene eoliths, for it admits, as we have seen, that in Miocene times, representatives of the genus *Homo* may have lived on the earth. But palæontology takes exception to the theory of the existence of Oligocene, and especially of Eocene eoliths.

Eoliths have almost always irregular and very varied forms, in which the supporters of the theory of their artificial origin endeavour to see the prototypes of palæolithic implements; hammers, implements for striking, knives, scrapers, piercers,

[1] Breuil, H., "Sur la presence d'éolithes à la base de l'Eocène parisien" (*L'Anthropologie*, xxi., 1910).

[2] Rutot, A., "Un grave problème" (*Mém. de la Soc. belge de Géol.*, xxi., 1907).

[3] Bibliographical references on previous page.

[4] Rames, J. B., "Géologie du Puy Courny" (*Matériaux*, 1884). Quatrefages, A. de, "Introduction à l'étude des races humaines" (1887). Boule, M., "Temps quaternaires et préhistoriques du Cantal" (*Assoc. franç. Congrès de Toulouse*, i., and *Revue de L'Anthropologie*, 1889, p. 216). "Le Cantal miocène" (*Bull. de la carte géolog.*, 1896). Verworn, M., *Die archæolithische Cultur in den Hipparionschichten von Aurillac* (1905). Mayet, L., "La Question de l'Homme tertiaire" (*L'Anthropologie*, xvii., 1906), etc.

[5] Ribeiro, C., *Descripçao de alguns silex e quartzites lascados incontrados nas camadas de terreno terciairo* (1871). "L'Homme tertiaire en Portugal" (*Congrès internat. de Lisbonne*, 1880). Cartailhac, E., *Les âges préhistoriques de l'Espagne et du Portugal* (Paris, 1886).

[6] The bibliography of the Pliocene and Quaternary deposits is too large to be given here. Consult contents of *L'Anthropologie*.

etc. (Fig. 66, p. 118). But they are always the same, whatever their antiquity. They show no gradations with the passage of time, as Rutot himself allows. From the lowest Tertiary to the Quaternary, they preserve the same rough and defaced aspect. No progress, therefore, could have been made during this immense lapse of time. Such a fact is opposed to the laws of evolution and of general palæontology.

Either the creatures who shaped these stones were Apes, as Gaudry, convinced of the superiority of *Dryopithecus* over all living apes, supposed, and their skeletons have yet to be discovered; or else we have to deal with a Man whose state of intellectual stagnation during so long a period of time is incomprehensible. On the other hand, the permanence and repetition of the forms is explained by the hypothesis of an origin due to natural causes, the effects of which are always the same.

Geological Objections.

Since my first visit to Thenay, thirty-eight years ago, I have constantly thought that eoliths have no worse opponents than the geological conditions of the layers in which they are found.

On a heap of stones of any kind, on a beach, in a ballast pit, on garden walks spread with gravel, nay, even on a heap broken by a road-mender, nothing is easier than to make a selection of specimens obviously reproducing the same shapes, and these it is quite possible to classify in groups according to their shapes, so that, arranged on a table or in a drawer, they produce an impression of order and deliberate repetition. This is apparently the chief reason of the success of collectors of eoliths in the eyes of visitors whose geological knowledge is seldom of a kind to enable them to avoid such a pitfall.

Studied in their beds, the same stones have quite a different significance. We then see that the most exacting care in selection has been necessary to obtain these deceptive series; that the stones, said to be deliberately dressed, far from indicating the occurrence, at the spot where they were found, of the site of a human settlement, of a shelter, or such like, on an ancient surface, really form part of a true geological formation, usually thrown violently in place, just like thousands of other elements in the same formation; that the specimens, chosen with a preconceived idea, do not differ in any essential character from their innumerable neighbours in the layer; that it is easy to find every possible transitional form, ranging from the rudest elements composing the bulk of the gravel, conglomerate or breccia, as the case may be, to those in which traces of fracture and splintering, a little more numerous or a little more definitely grouped, have gained for them the honour of being singled out and collected as representative specimens.

Such is the general and the main fact characteristic of the histories of eoliths of all ages and in every country; a fact which in itself ought to have sufficed to put naturalists on their guard, and to have ended all the discussions which have long occupied so many persons, the majority without weight either as geologists or as prehistorians.[1]

There are other geological arguments. Eoliths are found only in countries where flint occurs and never beyond layers naturally containing this material. This would not have been the case had these stones been weapons or implements, for such Man would not have failed to scatter almost everywhere on his travels or in his settlements. All eoliths are of flint, and all alluvial formations of cataclysmic or torrential origin, whatever their age, yield eolithic products, provided they contain flint pebbles. These pebbles constitute an integral part of true geological formations sometimes occupying thousands of square kilometres in extent, and these it would be absurd to attribute to Man.

Let us add that many deposits containing eoliths are of marine origin. As the English palæontologist, Boyd Dawkins, has wittily pointed out, it is extraordinary that fossil Man so often reveals himself to us in the form of an aquatic being, incapable of travelling over hill and dale.

Technical and Experimental Objections.

The study of the stones themselves does not support the eolith theory. We have just seen that, as regards the deposits, all stages of the stones may be observed *in situ*; from "used" or "shaped" forms, to rough pebbles of any description; it is

[1] Boule, M., "La Paléontologie humaine en Angleterre" (*L'Anthropologie*, xxvi., 1915).

therefore impossible to decide where eoliths begin or end. Further, along with eoliths of usable size, we usually find countless pebbles of similar and even identical shapes, but smaller, and the dimensions of such may imperceptibly diminish till they become sub-microscopic. The flint gravels of Parisian gardens and walks are full of them. It is impossible to maintain that tiny splinters could ever have been used, that is, that they are eoliths. Why attribute a different origin to the larger specimens?[1]

Effects of Natural Forces.

We now know that natural forces, quite independent of human will, can produce appearances identical with those presented by eoliths, such as the percussion bulb, and the "retouching" regarded as characteristic of deliberate dressing.

The waves of the sea produce such effects, as has been shown by the studies of Hardy at Dieppe, of the present author on the Norfolk Coast, of Fraas on the Rügen beach, of Sollas in the Isle of Wight, of P. Sarasin at Nice, and so on.

The same may be said of effects caused by sudden changes of temperature, observed by many travellers in the East, by Stanislas Meunier (experiments on the action of frost), by Penck (on the action of glaciers), and so on.

Of similar significance also are various manifestations of pressure, exercised through direct blows, accumulations, landslides, earth-falls, local foldings, etc. Many observers, Arcelin, Commont, and Breuil in France, Worthington Smith, Warren, and Haward in England, have called the attention of prehistorians to the effects of such varied happenings, the close relationship of which to certain eolith-containing deposits they have observed.[2]

They have shown that pressure, or compressions due to the weight of the deposits, produce, even in the depths of geological

[1] Sarasin, P., "Einige Bemerkungen zur Eolithologie" (*Jahresbericht der Geogr. Ethnogr. Gesellsch. in Zürich*, 1909).

[2] See particularly: Warren, S. H., "On the Origin of Eolithic Flints by Natural Causes" (*Jour. Anthrop. Inst.*, xxxv., 1905). "A natural eolith factory beneath the Thanet Sand" (*Quart. Journ. Geol. Soc.*, lxxxvi., 1921). Breuil, H., "Sur la présence d'éolithes à la base de l'Eocène parisien" (*L'Anthropologie*, xxi., 1910). Haward, F. N., "The Chipping of Flints by Natural Agencies" (*Proc. Prehistoric Soc. of East Anglia*, i., 1912). The origin of the "Rostro-carinate Implements" (*Ibid.*, iii., 1918-1919).

layers, results similar to those due to blows—sometimes percussion bulbs, and sometimes the chipping off of more or less numerous small flakes or retouches, which are often distributed in so regular a manner as to simulate the results of deliberate action (Fig. 73). Blocks of flint have frequently been found broken into several pieces, which while bearing all the characters of eoliths, remain *in situ*, each piece close beside

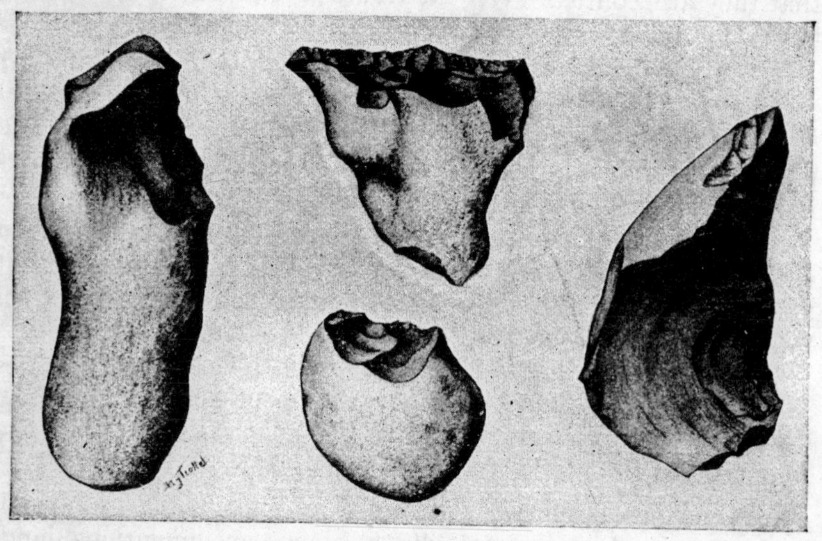

FIG. 73.—Natural Eoliths from the base of the Eocene deposits at Clermont (Oise). Two-thirds natural size. (After H. Breuil.)

the other or even still in actual contact (Fig. 74). Breuil's observations on the Lower Eocene deposits in the neighbourhood of Clermont (Oise) are of particular interest in this respect. In a large number of cases, flint nodules are found fissured, split and dispersed in small pieces, and these might readily be transformed into eoliths by subsequent re-sorting, in the course of which they might undergo shocks due to the dynamic forces of streams or of the waves of the sea.

A Modern Eolith Factory.

With regard to the play of natural forces, certain experimental facts have been quite conclusive. There is, near Mantes, a factory which produces cement by mixing chalk with plastic clay. These substances are mingled with water in round tanks, the water being kept in constant movement by means of a horizontal wheel, placed above the surface of the liquid, on the spokes of which are hung harrows of cast-iron which plunge into the water and chalky mass to a depth of 0.20 m. from the bottom. The circumference of the wheel rotates at a speed of about 4 metres in a second, the speed of the Rhone in flood. The speed of the liquid is, of course, less because of the friction. The water moves in an eddy, which carries along not only the particles of

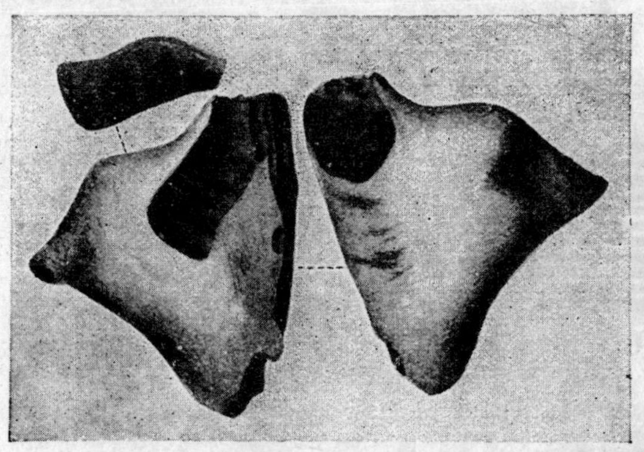

FIG. 74.—Flint split in three fragments which remained in contact with each other. Eocene eolith beds of Clermont (Oise). Two-thirds natural size. (After H. Breuil.)

chalk or clay, but also a certain number of flint nodules which, overlooked by the quarrymen, have been deposited along with the chalk in the tanks (Figs. 75-77).

During the course of twenty-nine hours these flint nodules undergo thousands of reciprocal shocks, and their clinking, plainly audible, reminds one of the sound of pebbles rolled by a turbulent stream. When the machinery stops the flints remain at the bottom of the tank, covered by a chalky deposit. They are taken out, washed, and put in heaps, in order to be used in the making of concrete.

Now these pebbles, which have been submitted in the tanks to the mechanical action of an artificial whirlpool comparable to the dynamic action of a natural torrent, present all the characters of ancient river gravels. The majority become round pebbles identical with those in ballast-pits, but, as in all ancient flint alluvials, be they Oligocene, Miocene, Pliocene

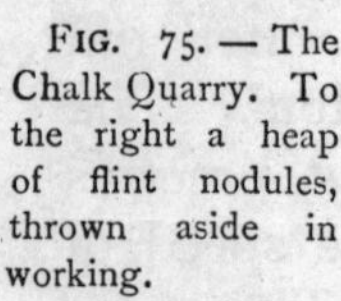

FIG. 75. — The Chalk Quarry. To the right a heap of flint nodules, thrown aside in working.

FIG. 76. — The Diluting Tanks, where the flints come into violent contact with each other.

FIG. 77.—Heaps of Eoliths taken from the tanks.

FIGS. 75-77 —Works at Guerville near Mantes (Seine-et-Oise), a factory turning out both cement and eoliths. (Photographs by E. Cartailhac.)

or Pleistocene, many specimens show "retouching." Moreover, the splintering is not always distributed all over these pebbles as if by chance, but there may be localized flaking on

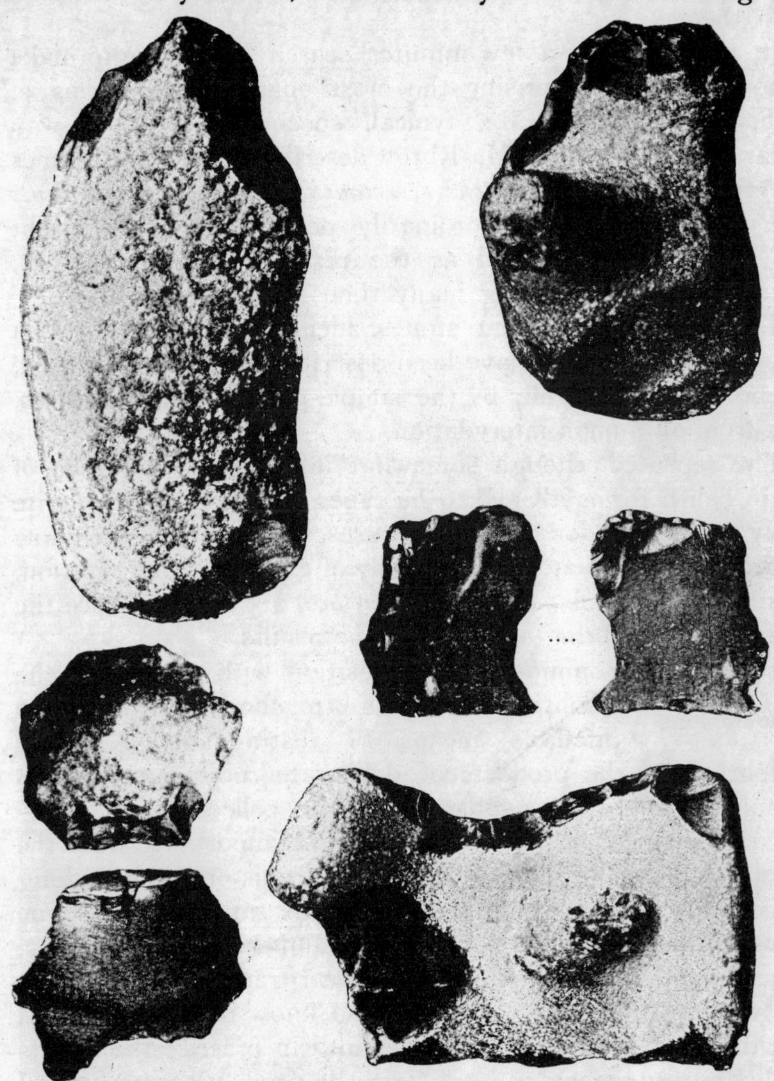

FIG. 78.—Eoliths from the cement works at Mantes.

Specimens similar to those which Rutot describes under the name of hammers, scrapers, scratching knives, retouchers and notched flints. Below to the left-hand corner, flint with plane of fracture and percussion bulb. Natural size.

only one surface of the flint, grouped, as if of deliberate purpose, at definite points, that is to say actual "retouching" in the sense given to this word by the supporters of the eolith theory.

In the course of a few minutes' search I was able to make a fine collection comprising the most characteristic forms of eoliths, forms said to be typical, specimens in every way similar to those which M. Rutot describes under the names of *strikers*, *scrapers*, *scratchers*, *retouchers*, and *notched flints* (Fig. 78). Certain extraordinarily perfect specimens might well have been described as the results of finished work, of "methodical retouching many times repeated."[1]

These facts prove that stones, identical with those from ancient alluvials which have been described as worked or used, may be shaped naturally by the simple play of physical forces, apart from all human intervention.

The repeated, though somewhat infrequent occurrence of certain eolith forms alleged to be types of implements, is quite simply explained; for the same causes, acting under the same conditions and on materials originally of the same configuration, as untouched nodules of flints often are, ought to produce the same effects and bring about the same results.

Conclusions.

This, in my opinion, is the position with regard to the "eolith problem." In short, we have no infallible means of distinguishing natural accidents from the products of deliberate rudimentary workmanship. Where the pebbles have been collected in Quaternary deposits, this distinction is of slight importance from the philosophic point of view. All expert prehistorians have long known that there are to be found in the same layers as fine palæolithic implements, much greater numbers of simpler flints, mere splinters, bearing or seeming to bear traces of working, or, shall we say, of usage. But they also know the difficulty and often the impossibility of distinguishing in practice the results of deliberate rudimentary work from the productions of natural agencies.

The question is much more serious when we have to deal with pebbles from Tertiary deposits, in which there has never

[1] Boule, M., "L'Origine des Éolithes" (*L'Anthropologie*, xvi., 1905). Obermaier, H., "Zur Eolithenfrage" (*Archiv für Anthropologie*, 1905).

been discovered the least trace of fossil Man. We must then be doubly careful, and make absolutely certain, in the first place, that stones having all the characters of eoliths could not have been fashioned by natural forces. Now, all the facts I have just stated show that eoliths collected in Tertiary deposits do not provide indisputable evidence of the existence of Man at the periods when these deposits were formed.[1]

This existence, it cannot be too often repeated, is possible in Europe, it is probable in some other undetermined spots of the globe; but we have not yet, scientifically speaking, the right to assert it; up to the present we possess no material proof. The opposite view, in the present state of our knowledge, can only be a matter of opinion. "I am very proud of the antiquity of my family," said Sir John Evans, long ago, "but I want some other proof of it than a percussion bulb."

Since the first edition of this work appeared, Mr Reid Moir has continued his researches in the neighbourhood of Ipswich. He has found, in the Red Crag at Foxhall, flints which really seem to present the characteristics of deliberate shaping, and his views in this respect are shared by various prehistorians on the continent as well as in England. The author of this work is, however, not yet convinced, and continues to reserve judgment. (For further details see Appendix, page 469.)

[1] Supporters of the eolith theory of the existence of Tertiary Man have sometimes violated their modesty in comparing themselves with Boucher de Perthes. But surely the position is entirely different. Boucher de Perthes had to struggle especially against preconceived ideas, against a deliberate determination to deny the antiquity of Man. To-day no competent naturalist doubts the existence of Tertiary Man; we no longer dispute anything but the value of the material data put forward in support of his existence.

CHAPTER VI

THE MEN OF THE CHELLEAN OR LOWER PLEISTOCENE AGE

THE chief geological, palæontological, and archæological characters of the oldest Pleistocene formations, corresponding very closely to the *Chellean* period of prehistorians, have already been discussed (see p. 50). We know that these characters indicate a warm climate, rich vegetation, and a fine mammalian fauna—in fact a most favourable environment for human occupation. And, indeed, the presence of Man is disclosed by an industry, admittedly very primitive, which has abundantly scattered its stone products, some of them most carefully wrought, in various lands almost all the world over (see p. 112). It is not implied, however, as we shall see later, that these universally distributed implements are everywhere of the same age.

The Industry of the Lower Pleistocene.

Influenced by G. de Mortillet, prehistorians have long believed in the extreme simplicity and uniformity of Chellean implements. "The industry of this period," it has been said, "is of the simplest, most rudimentary character ; it comprised a single stone implement, a flint or fragment of other rock, dressed on both surfaces by hammering, and rendered more or less amygdaloid by means of blows generally of some force. This implement varies much in form, size, finish, and in material ; but, once known, it is always easily recognized."[1]

The variations are indeed very considerable, as the accompanying drawings (Fig. 79) show, and as is still better shown by the specimens in the d'Acy collection in the Museum at Saint-Germain. Further, we now know that along with the amygdaloid flints, which G. de Mortillet has named

[1] Mortillet, G. de, *Le Préhistorique*, 1st ed., p. 133.

"hand-hammers" (*coups de poing*) to indicate the chief use he attributes to them, and which, in his opinion, were implements of general utility, the industry of very ancient Palæolithic Man comprises other forms of implements, differently worked, with retouching more localized on one surface only. Such forms occur more or less commonly in all layers, especially in those which represent ancient human settlements, rather than in the alluvial deposits of more or less turbulent streams, where only the most bulky objects have been able to withstand the effects of reshuffling. Many have been figured by d'Acy and Commont. Now such specimens exhibit the closest resemblance to types considered by archæologists as characteristic of the succeeding Moustierian period (Fig. 79, Nos. 5, 6, 7). Thus, Moustierian forms, which in France are specially characteristic of the period of the Mammoth fauna, have sometimes been found, as at Grimaldi, along with the warm fauna of the Lower Pleistocene, that is to say of the Chellean period of archæologists.[1]

In France, archæologists divide the Lower Pleistocene into two periods. The oldest, the Chellean, is said to be characterized by thick and rough amygdaloid stones shaped by the chipping off of very large flakes, and having in consequence cutting edges forming a zig-zag (Fig. 79, No. 1). The following period, called the *Acheulean*, marks the transition from the Chellean to the Moustierian; the amygdaloids are less thick, occasionally flattened (the *limandes* of the Somme quarries), more regular and symmetrical in form, and

[1] Boule, M., *Les Grottes de Grimaldi*, *Géologie et Paléontologie*, i. (1906). This fact has greatly surprised prehistorians who believe in the chronological stability and infallibility of the various types of palæolithic implements. Reflection, however, shows that there is here nothing extraordinary. The human industry named *Moustierian* is one of the most primitive imaginable; it is certainly simpler than that of the fine amygdaloids from Chelles or Saint-Acheul, since it only comprises flakes retouched at the edges. It is therefore not astonishing that we should find it from the commencement of the Palæolithic period; and, moreover, it is extremely probable that there may have been, as early as the Lower Pleistocene, various archæological facies, distributed according to the regions inhabited by Men who themselves may have differed from each other. The idea of the homogeneity of the human populations of the Chellean period, based on the wide distribution of stones wrought in an almond or amygdaloid shape, is a common theory with nothing to recommend it but its simplicity.

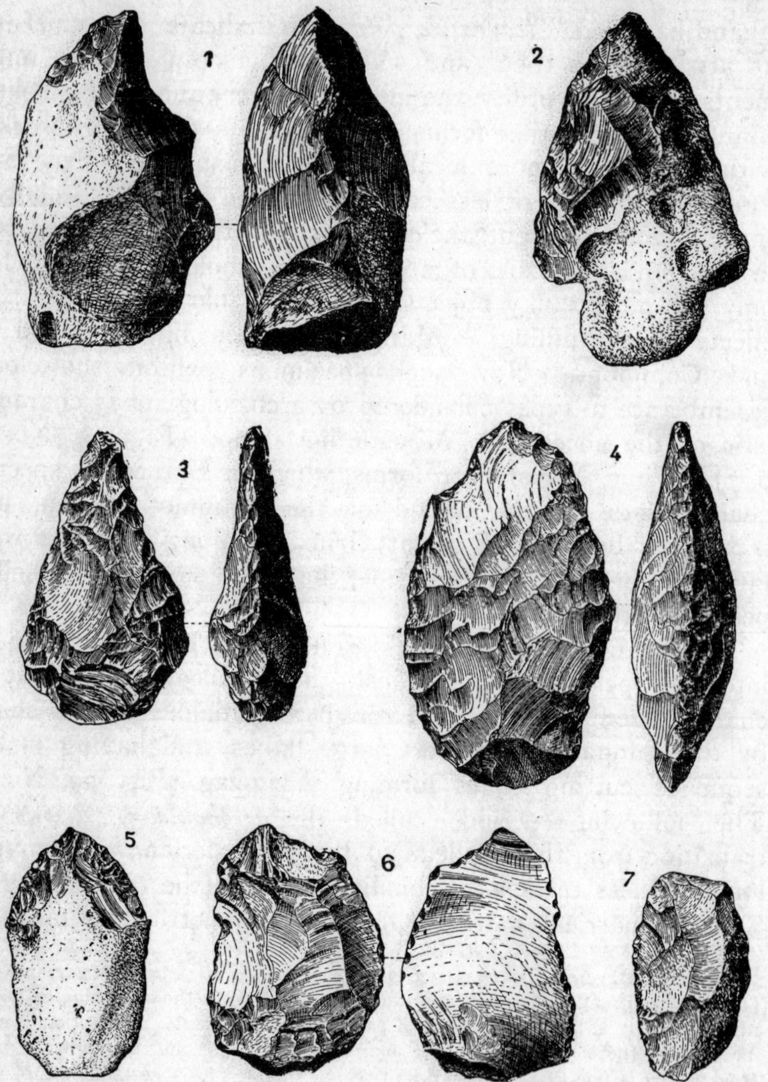

FIG. 79.—The Stone Industry of the Lower Pleistocene (Chellean and Acheulean from Saint-Acheul (Somme)). One-third natural size. (After Commont.)

1. Rude implement, with large flakes, of Chellean make; seen full face and in profile. 2. Primitive hand-hammer (*coup de poing*), the base or heel of which, intended for grasping, is formed by part of the original flint nodule. 3. Amygdaloid of Chellean type, seen full face and profile; the cutting edges still present a zig-zag line. 4. Flatter amygdaloid, Acheulean type, seen full face and in profile; the cutting edges are less sinuous. 5. Flint dressed on one side, and at the upper extremity retouched into a point. 6. Another flint, dressed on one side only. 7. Flint dressed in form of a scraper.

showing finer and more careful work, which gives the cutting edge a straighter outline (Fig. 79, Nos. 3 and 4). Some authors even distinguish a *pre-Chellean* period, preceding the Chellean and characterized by an extremely rude stone industry without true amygdaloids.

It is extremely probable that the men of Chelles and Saint-Acheul utilized and worked wood, but they do not seem to have made use of bone.

However that may be, the great quantity of the stone products of these early ages, *which must have extended over a great period of time*, would lead us to expect many discoveries of contemporary human bones.

In reality, such fossils remained almost unknown till 1908.

Remarks on Human Bone Discoveries.

Certain discoveries in previous years have sometimes been brought forward as belonging to the Chellean period, but my first duty must be to show that they are either of uncertain age or belong to a more recent period.

The oldest skull-fragment, from Cannstadt, near Stuttgart in Wurtemberg, dates back to 1700. The source of this specimen is quite obscure, its antiquity is more than doubtful.[1] It owes its reputation largely to de Quatrefages and Hamy, who selected it as a prototype of their "early fossil race."[2]

The second discovery was made, in 1844, amongst the flood-sorted volcanic cinders from the volcano at Denise, near Le Puy, in the district of Haute-Loire. It comprised some portions of skulls and other human bones. The antiquity of these remains has been disputed. In the course of my geological studies in Velay I devoted much care to the question of their age, or rather of the age of the deposit in which they were found. Having examined the layer several times, I am of opinion that this deposit dates back to a very remote period of Quaternary times.[3] But perhaps it would be

[1] See Obermaier, H., "Les restes humains quaternaires dans l'Europe centrale" (*L'Anthropologie*, xvii., 1906, p. 63).

[2] Quatrefages, A. de, and Hamy, E. T., *Crania ethnica*. The history and bibliography of this and the following discoveries will be found in this work.

[3] Boule, M., *Description géologique du Velay*, Paris, 1892, p. 219; *L'âge des derniers volcans de la France*, Paris, 1906, p. 31.

well not to discard the hypothesis of an artificial burial. The study of the bones, outlined by Sauvage, has still to be completed from the morphological point of view, as well as from the physical and chemical points of view.

The year 1863 was marked by the discovery of the famous jaw from Moulin-Quignon, the status of which I have already defined in the chapter dealing with the historical aspect (p. 22).

The skull from Olmo, discovered near Arezzo in Tuscany, in 1863, was not recorded by Cocchi till 1867. But so much uncertainty surrounds its geological age that it is impossible to place much weight upon it in such a work as this.[1]

With regard to the skeletons from the alluvial soils of Clichy (1868) and of Grenelle (1870), I merely mention the rashness of the anthropologists who have recently attempted to prove their geological antiquity. Such an attempt, made forty years after the discoveries by individuals who had never seen the deposits, cannot be of much value. The most elementary notions of caution demand of human palæontology that it should base its speculations only on evidences of irreproachable origin. This has not been the case with regard to the skeletons from the Parisian alluvials.[2]

In 1888, portions of the skull and several other parts of a human skeleton were found in the Pleistocene gravels of Galley Hill, near Northfleet in Kent. The gravels are situated about 30 metres above the level of the Thames, and contain many Palæolithic flints of different shapes, especially of lanceolate form. The bones are said to have been discovered at a depth of eight feet from the surface.

Seven years later, in 1895, they were described by E. T. Newton, the palæontologist.[3] Keith[4] has since asserted that the Galley Hill skeleton does not differ in any important character from that of a modern Englishman; but this does not deter him from attributing to it an age of 200,000 years.

1 The distinguished Italian anthropologist, Sergi, still considers it contemporary with *Elephas antiquus*: "Sul l'Uomo fossile dell'Olmo" (*Rivista di Antropologia*, xxi., 1916-17).

2 Mortillet, G. de, *Le Préhistorique*, 1st ed., pp. 346 *et seq.*

3 *Quarterly Journal of the Geological Society*, 1895.

4 Keith, A., *Ancient Types of Man*, London, 1911, p. 32.

More recently, several other anthropologists, in different countries, have magnified the importance of the Galley Hill discovery; it is said to prove that *Homo sapiens* there dates back to a very remote period. That is quite possible, but the proof is insufficient. The fact is that no geologist was present at the discovery of the skeleton, and none examined the layer at the time. When the bones were presented to the Geological Society of London, two highly competent scientists, Sir John Evans and Professor Boyd Dawkins, raised doubts as to their high antiquity and brought forward arguments pointing strongly to artificial burial. These doubts are but increased by the appearance of the skull, which is now exhibited in the British Museum. It is quite a distorted relic, with none of the physical characters of fossil bone remains. On the contrary, Duckworth[1] has shown that it is exactly like many comparable specimens from recent burials.

I believe that the Galley Hill discovery is of exactly the same nature as those which have been made on several occasions in the Pleistocene alluvials of the Seine, at Grenelle and Clichy, and to which we in France now attach no importance, determined as we are to consign to oblivion all osteological evidence the high antiquity of which is not absolutely assured.

Finally, some years ago, in 1912, English newspapers contained many references to a similar find brought to notice by Reid Moir, the discoverer of the rostro-carinate flints. It was said that a human skeleton had been found at a depth of 4½ feet in a layer of sand at Ipswich in Suffolk, immediately under glacial clay containing erratics. This skeleton, carefully described by Sir A. Keith, who considered it to be of quite a modern type, was said to belong to a period preceding the great glaciation of the region itself, older than the Pleistocene alluvials.[2] After having seen it at the College of Surgeons in London, and after having visited the layer at Ipswich, I was

[1] Duckworth, W. L. H., *The Problem of the Galley Hill Skeleton* (Cambridge, 1913).

[2] Moir, J. Reid, "The Occurrence of a Human Skeleton in a Glacial Deposit at Ipswich" (*Proceedings of the Prehist. Soc. of East Anglia*), i., 1912, p. 194. Keith, A., "Description of the Ipswich Skeleton" (*Ibid.*, p. 202). *The Antiquity of Man* (London, 1915).

obliged to express my opinion as follows: "The Ipswich skeleton must, as a measure of scientific caution, be entirely dismissed from the series of authentic evidences, such as have an established place and serve as a basis for speculation in the science of human palæontology."[1]

Some months later, Reid Moir loyally acknowledged that he had been mistaken. "I wish to take this opportunity," he said, "to state that those who opposed my contention as to the great age of these remains were in the right, whilst the views held by me regarding them have been shown to be erroneous."[2]

For a long time, French prehistorians, led by G. de Mortillet, considered, because of the dearth of osteological evidences from Lower Pleistocene deposits, that the Neanderthal human type was characteristic of the Chellean period. For long I opposed this assertion, stating that Neanderthal Man was essentially Moustierian, and that we were unacquainted with any bone-remains of Chellean Man. To-day the latter statement no longer holds true. Two fine and quite authentic discoveries have been made in the course of the last few years, in deposits belonging to the Lower Pleistocene, the first at Mauer near Heidelberg, and the second at Piltdown in England.

The Mauer jaw, the remains of a skull and lower jaw from Piltdown, and, perhaps, two teeth and a fragment of jaw from calcareous tufa near Weimar, are the only evidences which date or can possibly date back either to the Chellean or to the Acheulean period. I shall describe them in order, beginning with the least important, those from Weimar.

The Weimar Fossils.

The Teeth from Taubach.

The village of Taubach, in the neighbourhood of Weimar, is situated on a terrace of the Ilm, the river gravels of which are surmounted by calcareous tufa, sandy or of a close nature. For many years there had

[1] Boule, M., "La paléontologie humaine en Angleterre" (*L'Anthropologie*, xxvi., 1915, p. 38).

[2] *Nature*, 12th October 1916 [vol. xcviii. p. 109]. See *L'Anthropologie*, xxiii., 1917, p. 188.

been collected in these tufas, along with bones of large animals belonging to the warm fauna of ancient Pleistocene times, dressed flints, and bones, broken, burnt, and possibly utilized as tools. The deposit, with its traces of hearths, seemed to correspond to a station and place of encampment of a tribe contemporary with *Elephas antiquus*.

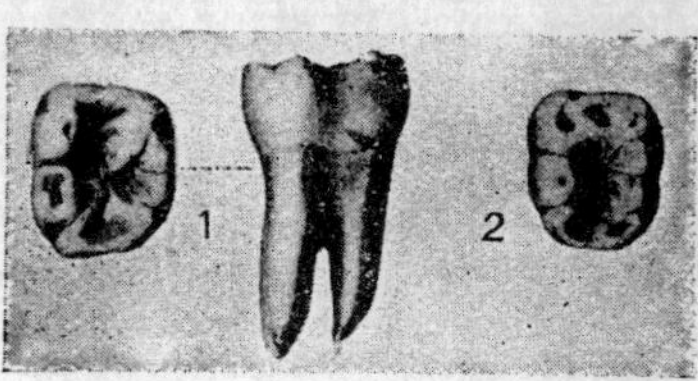

FIG. 80.—1. Tooth from Taubach (first lower left true molar), seen from the crown and in profile (inner side). 2. First lower left true molar of a Chimpanzee. Natural size. (After Nehring.)

As early as 1871, the proprietor of the deposit presented anthropologists with a human skull which he said he had found in the fossil bed, but which was identified by Virchow as Neolithic. In 1892, a naturalist, Dr Weiss, himself extracted from the deposit containing remains of the warm fauna, a tooth which was sent to Nehring of Berlin, along with another tooth, whose source, although less definite, nevertheless appeared to be the same. The first is a lower milk molar; the second a permanent lower molar, the first on the left side.

As a result of comparative study, Nehring[1] considered them to be human teeth, at the same time pointing out that they differed from the latter in their greater size, and in certain other simian characters. The true molar in particular is remarkable, on account of its length, the arrangement of the cusps of the crown, and its much corrugated enamel, all characters which give it a resemblance to the tooth of a Chimpanzee (Fig. 80).

Speaking of this tooth, Duckworth[2] has since said: "It is difficult to decide if it is a human tooth, or the tooth of a pithecoid precursor of Man." The American zoologists, Miller and Gregory,[3] do not hesitate to ascribe it to a species of fossil Chimpanzee, to which they likewise attribute the Piltdown jaw, of which we shall presently speak.

[1] Nehring, A., "Ueber fossile Menschenzähne aus dem Diluvium von Taubach" (*Naturwissenschaftliche Wochenschrift*, 4th August 1895).

[2] *Prehistoric Man*, p. 23.

[3] *Studies on the Evolution of the Primates, loc. cit.*, p. 313.

But it is possible that it may have belonged to a man related to the Neanderthal type.

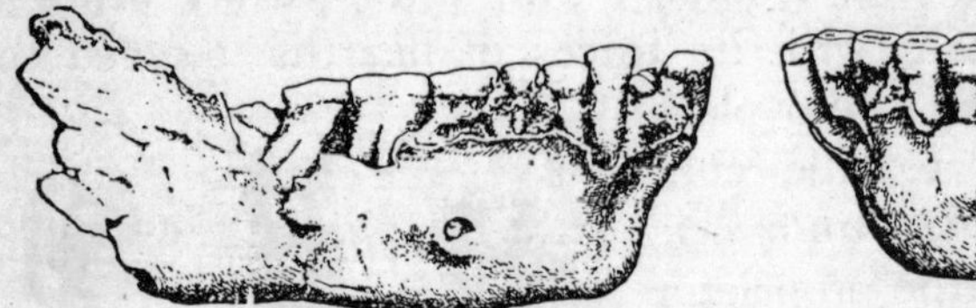

FIG. 81.—Right and left profiles of adult Weimar Jaw. Half natural size. (After Schwalbe.)

The Ehringsdorf Jaw.

Ehringsdorf is another locality in the neighbourhood of Weimar, where Quaternary formations, identical with those at Taubach, are surmounted by a deposit of loess and by a tufa or overlying travertine.

A human lower jaw was found on 8th May 1914, in the lower tufa, at 11.90 metres below the soil, and Schwalbe, the German anatomist who described it, has called it "the Weimar jaw."[1]

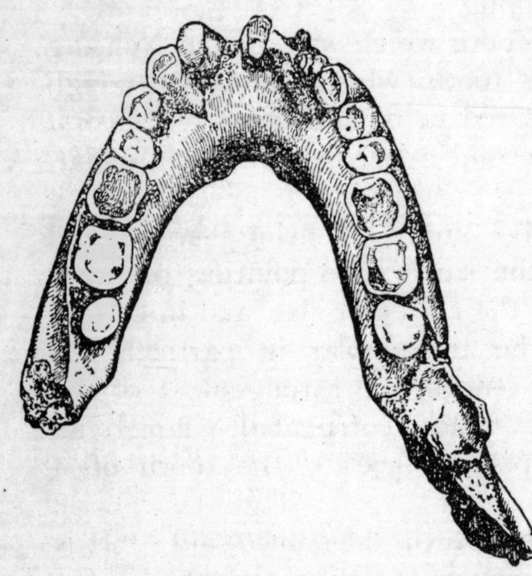

FIG. 82.—The adult Weimar jaw, seen from above. Two-thirds natural size. (After Schwalbe.)

Much mutilated and lacking the ascending ramus on the left side, it retains only a small part of the right ramus. Almost all the teeth are in position (Figs. 81 and 82).

The absence of chin is striking, the more so since it is accompanied by a very marked alveolar prognathism. It follows that the inner region of the symphysis slopes strongly from the front backwards, and so is very noticeable when the jaw, lying in a horizontal plane, is observed from above. The

[1] "Ueber einen bei Ehringsdorf in der Nähe von Weimar gefundenen Unterkiefer des *Homo primigenius*" (*Anatomischer Anzeiger*, xlvii., 1914).

chin region appears to be deprived of tubercles for the genio-glossal and genio-hyoid muscles, nor is there any mylo-hyoid ridge on the inner surface of the horizontal branches. The foramina on the chin are long and situated directly below the first true molars (almost as in *Homo neanderthalensis*). Schwalbe lays stress on the narrowness of the mandibular arch, but he supposes that the specimen may have undergone some post-mortem deformation.

The teeth are much worn; and as the premolars are less so than the canines, we must suppose that the crowns of the latter extended beyond the level of the crowns of the premolars. There is no interval (diastema) between the canine and the first premolar. A remarkable and rather unexpected fact is the diminutiveness of the last true molar; which seems to prove that the tendency of this tooth to disappear is much older than has been supposed.

Without any hesitation, Schwalbe associates the Weimar specimen with the Neanderthal type, which we shall study presently. This jaw is really human, but of a low type. It differs considerably from the Mauer specimen; there is every reason to believe that it is more recent, and that it dates from the end of the Acheulean period, while the Mauer fossil must date from the beginning of the Chellean.

In 1916 several additional bones were found belonging to the skeleton of a child about ten years of age, notably a lower jaw which is of the same type as the adult jaw referred to above. These remains were described in 1920 by Virchow, and are referred to and illustrated in the Appendix (page 470).

The Mauer Jaw.

(*Homo heidelbergensis.*)

At the end of 1908, Otto Schoetensack, of Heidelberg University, published a memoir entitled, "The lower jaw of *Homo heidelbergensis*, taken from the sands at Mauer, near Heidelberg."[1]

[1] Schoetensack, O., *Der Unterkiefer des Homo heidelbergensis aus den Sanden von Mauer bei Heidelberg* (Leipzig, 1908).

The Mauer Jaw Site.

The village of Mauer is situated 10 kilometres south-east of Heidelberg on the Elzenz, a tributary of the Neckar (Fig. 83). The bed-rock of this region is of Triassic formation, overlaid in places by a covering of Pleistocene deposits, and particularly of river sands representing the ancient bed of the Neckar. On 21st October 1907, in one of the pits worked in these sands, the "Sandgrube Grafenrain," a lower jaw was found 24 metres below the surface of the soil.

Fig. 83.—Geographical position of Mauer, near Heidelberg.

The accompanying photograph and geological section of the face of the sand-pit were made by Schoetensack (Fig. 84). The lower part of this section is formed by the "Mauer sands" the base of which contains gravels with beds of rounded pebbles, and the upper portion of which, separated from the lower by a bed of clay, consists of fine sand. The upper part of the section is formed by silt or loess, subdivided into a lower or ancient level, of a brown and sandy character, and an upper and more recent level containing small calcareous concretions.

The Mauer sands, very rich in fossils, contain shells of both land and fresh-water molluscs, which, taken as a whole, indicate a climate more continental than at present. Of thirty-five species, eight have emigrated to the East.

The mammalian fauna comprises the Ancient Elephant (*Elephas antiquus*), the Etruscan Rhinoceros, a horse which may be a transition form between the Pliocene Horse (*Equus stenonis*) and the present-day *Equus caballus*, a Wild Boar, Red Deer, Roe Deer, an Elk, a Bison, two Bears resembling the Pliocene *Ursus etruscus*, a Dog, two felines—a Lion and Wild Cat—and the Beaver. This fauna is most closely related to that of the lowest Pleistocene of France, such as we find in the oldest gravels of the valleys of the Somme (Abbeville) and of the Seine (Chelles), in the deepest deposits and settlements of the Grotto du Prince at Grimaldi, and in a few archaic formations in the caves of the Pyrenees. It even exhibits the same zoological gradations as these—a horse akin to *Equus stenonis*, a small bear nearly related to *Ursus etruscus*, and such like.

It has been suggested that the Mauer sands date from the Pliocene, but this is straining the facts of the case. The fauna of these sands hardly differs at all from that of various

FIG. 84.—Photograph and Geological Section of the Grafenrain Sand-pit at Mauer. + indicates the spot where the human jaw was found, 24 metres below the surface of the soil. (After Schoetensack.)

other European deposits, such as those at Süssenborn and Mosbach in Germany, in the Norfolk Forest-bed in England, at Solilhac and Saint-Prest in France, all of which lie on the borderland between Pliocene and Pleistocene. Yet the closest resemblances are with our "Chellean" fauna, a possibility the more to be kept in view inasmuch as it has not been disproved that the relics of the most ancient members of the Mauer sands may have been derived from deposits of greater age.

Age of the Jaw. As the topographical and stratigraphical conditions of the Mauer river formation confirm this interpretation, there is every reason to believe that the ancient owner of the human jaw extracted from the gravels was a representative of the oldest workers of Palæolithic stones. There can be no doubt that the jaw was found in the place ascribed to it, and that it is of the same age as the gravels, an

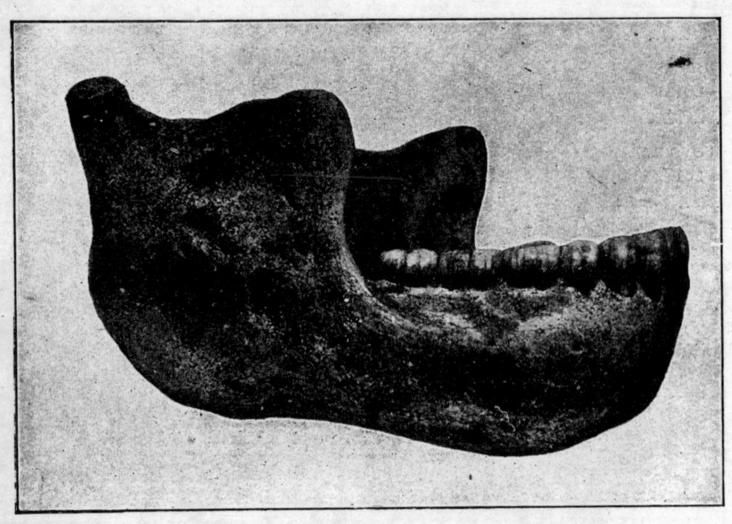

FIG. 85.—Heidelberg (Mauer) Jaw, seen in profile. Three-fourths natural size. (From a photograph by Schoetensack.)

age indicated by the animal bone remains amongst which it was lying and with which it agrees in its state of fossilization.

The Mauer discovery is of the utmost importance, since it brings us into touch with one of the rare relics of the most ancient human race known to us. Accordingly, it deserves to be studied in some detail.

Its Characters. The Mauer jaw is almost complete and is well preserved. At the very first glance it impresses one by its great size, its massive and extraordinarily powerful appearance, by the great breadth of its ascending rami, and by the complete absence of chin (Figs. 85 and 86). These are simian features, or, shall we say, stamps of the beast.

On the other hand, the dentition is altogether human ; the canines are small, and the molars possess dimensions and characters frequently found in modern Man. Therefore Schoetensack placed his fossil in the genus *Homo*, naming it *Homo heidelbergensis*, whilst Bonarelli[1] invented a new generic name for it, *Palæoanthropus*.

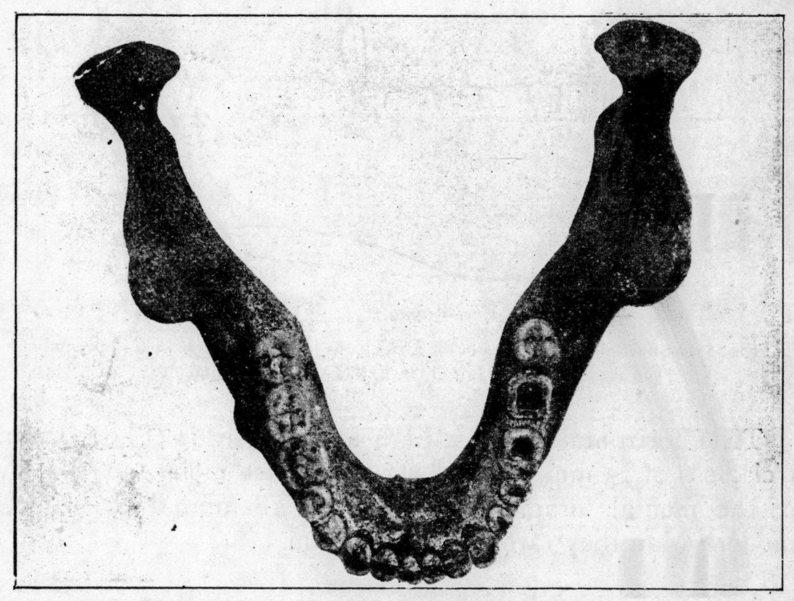

FIG. 86.—Heidelberg (Mauer) Jaw, seen from above. Two-thirds natural size. (After Schoetensack.)

Let us examine its morphology more closely.

The ascending rami are extraordinarily broad and low. They measure 60 mm. in breadth (in place of an average of 37 mm. in modern Man), and their height is 66 mm. ; so that, seen in profile, they are almost square in form. The mandibular notch (on the upper surface of the ramus) is not deep ; the coronoid process is blunt, rounded, and at a lower level than the condyle ; the latter possesses a large articular surface. Among anthropoids, the Gibbon most closely resembles our fossil in these three characters.

[1] *Rivista italiana di Palæontologia*, 1909.

The angle of the jaw has the truncated appearance noticeable in anthropoids, particularly in the Orang, as well as in *Homo neanderthalensis*, whose remains we shall examine later.

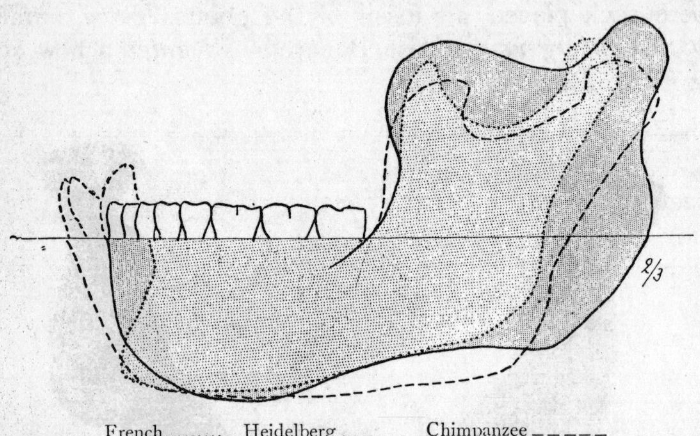

FIG. 87.—Superimposed Profiles of the Heidelberg Jaw, of a Chimpanzee's Jaw, and of a modern Human Jaw (French). Two-thirds natural size.

The horizontal rami are high and massive. They attain a thickness of 23 mm. at the level of the last molar; at the level of the mental foramen they are still 18 mm. thick, while in modern Man they rarely exceed 14 mm.

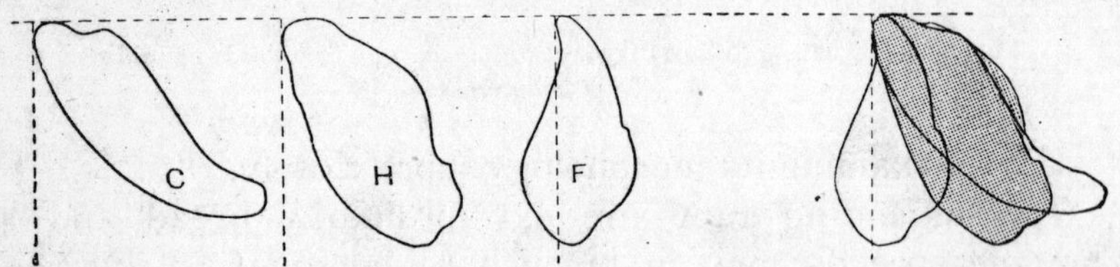

FIG. 88.—Vertical sections of various Lower Jaws at the symphysis. Two-thirds natural size. C. Chimpanzee; H. Heidelberg Man; F. Frenchman. To the right, the three profiles superimposed, that of Heidelberg being shaded.

The lower edge is concave, as in the case of many cynomorphic apes, and contrary to the case in anthropomorphic apes and Man. On the interior surface, the oblique ridge is faint, indicating weak mylo-hyoid muscles.

In Fig. 87 I have shown the Heidelberg jaw superimposed

on the jaw of a Chimpanzee and on the jaw of a modern Man (French), and this shows and summarizes clearly the differences exhibited by these three specimens as seen in profile.

The symphysis is very thick (17 mm.), its external surface has a convex and receding curve, resembling that of the apes and quite different from that of modern Man (Fig. 88). There is really no trace here of a chin. Now the chin is essentially a characteristic of the human race.

Seen full face, the region of the symphysis has, at its lower edge, a strongly marked chin pit (mental fossa); this structure occurs also in the Gibbons, and, to a less degree, in certain human jaws, either fossil or recent, as in the case of the Australians.

In modern Man the lower part of the front of the jaw-bone terminates in a definite edge, and the imprints of the insertions of the digastric muscles are apparent on the other side of this edge, that is, on its inner surface. But in the Mauer jaw the edge is replaced by a flattened area, a facette bearing the two digastric muscle impressions, which are here shorter than in the Chimpanzee, although longer than in modern Man. In this respect Neanderthal Man stands intermediate between Heidelberg Man and modern Man (Fig. 89).

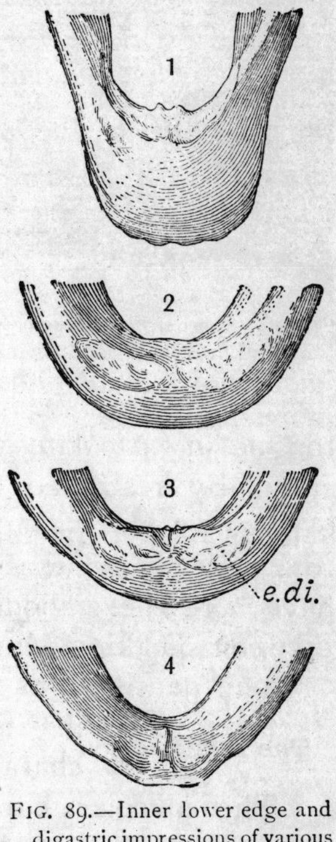

FIG. 89.—Inner lower edge and digastric impressions of various jaws. About half natural size.

1. Chimpanzee; 2. Mauer; 3. La Chapelle; 4. French. *e.di.* Impressions of digastric muscles.

The region of the symphysis further shows certain peculiarities on its inner surface. It slopes strongly from the front backwards, as does the outer or chin surface. Moreover, it is not uniformly developed: the part near the lower third, the genial region, instead of projecting as in Man, contains, as in the anthropoid apes, a hollow which takes the place of the upper genial tubercles for the insertion of the muscles of the tongue (Fig. 90).

In the characters of its symphysis, therefore, as Fig. 88 shows, the Mauer jaw is more akin to that of the large apes than to that of modern Man. We shall see later that in this respect the jaw of Neanderthal Man (*Homo neanderthalensis*) takes a place exactly between that of Heidelberg Man (*Homo heidelbergensis*) and of recent Man (*Homo sapiens*).

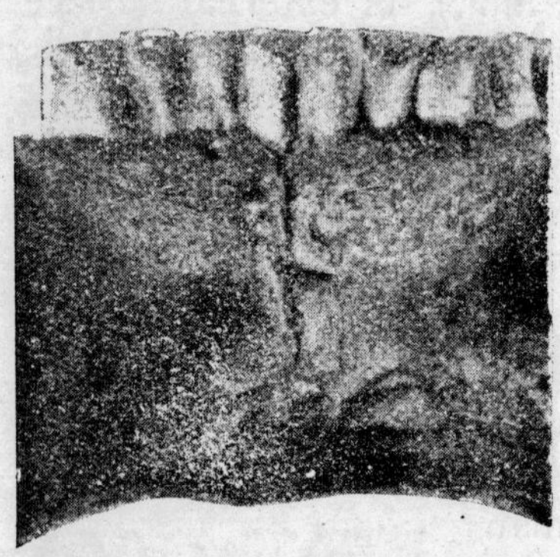

FIG. 90.—Inner surface of the body of the Mauer Lower Jaw, genial region. Photographed from a cast. Natural size.

The structure of the anterior part of the Mauer jaw shows that the space left for the tongue was greatly constricted, although less so than in anthropoid apes, and more so than in modern Man and even than in other fossil men known to us. The play of the tongue in articulate language must in consequence have been singularly restricted. It would even seem, to borrow an expression which Gaudry thought might be justifiably applied to *Dryopithecus*, that, from an anatomical point of view, we have here the embodiment of a kind of "intermediate stage between speaking Man and the howling beasts."[1]

The Dentition.

The dentition, as we have said, is definitely human, both in its entirety and in the details of its characters.

The alveolar borders, defining the dental curve, have a parabolic instead of the U-form of the great apes.

The teeth are well preserved, slightly worn on the left side; the molars on the right side have been partially broken in disengaging the relic from its matrix. The dental series is regular, continuous, without any interval (diastema) between the teeth.

The incisors are normal: their roots are slightly arched, following the curve of the region of the symphysis. The canine is not more developed than in modern Man, even of the civilized races; its crown does not perceptibly rise above the level of the neighbouring crowns. This fact is very important,

[1] "D'intermédiaire entre l'Homme qui parle et les bêtes qui crient."

for it bears on a human adaptation which is thus shown to date from a very remote period. Even at the primitive stage represented by this simian jaw, the canines were no longer used as defensive weapons.

The premolars also are quite normal. The true molars are almost as broad as long, whilst in apes they are generally longer than broad. Compared with those of civilized Man, their dimensions are considerable, and may be compared with those of many modern savages, such as the Australians (Fig. 91). On the other hand, they appear weak in comparison with the strength of the jaw which bears them.

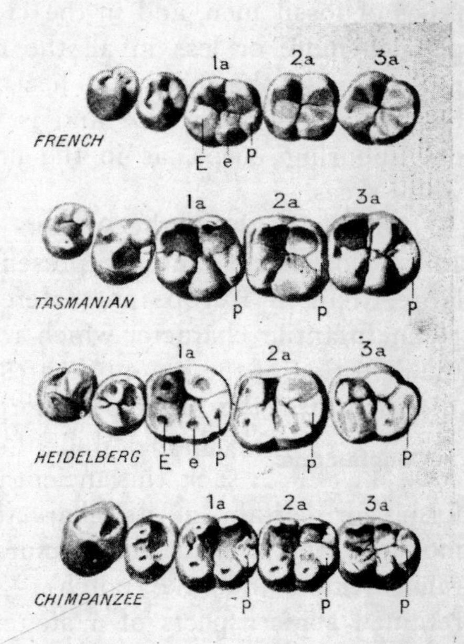

FIG. 91.—Comparative morphology of the Lower Molars of a Chimpanzee, of Heidelberg Man, of a Tasmanian and of a modern Frenchman. Natural size.

1a, *2a*, *3a*—first, second, and third true molars; *E*, *e*, external cusps; *p*, posterior cusps, highly developed in the true molars of the Chimpanzee and Heidelberg Man, less developed in the Tasmanian, and only present in the first molar of the Frenchman.

The second true molar is the largest, as is the rule in Man, but is likewise the case in apes. The last molar, or wisdom tooth, had not been cramped in its development, for the alveolar border is prolonged backward sufficiently far to accommodate a supplementary molar.

Minute study of the crowns of the true molars proves of interest, and the smallest details are important enough to be taken into consideration, for while Man and the large apes have molars which are much alike they differ from those of other monkeys.[1]

[1] With reference to this subject, see the illuminating notes of A. Gaudry: "Sur la similitude des dents de l'Homme et de quelques animaux" (*L'Anthropologie*, xii., 1901). "Contribution à l'Histoire des Hommes fossiles" (*Ibid.*, xiv., 1903).

In the anthropoid apes (Fig. 91), the crown of the large lower molars is always formed of five well-developed cusps or tubercles. In civilized Man, the fifth cusp (Fig. 91, *p*) is reduced or even completely lost, save in the case of the first true molar, which is the largest and where it is always present. In certain savage races, such as the Australians, in several types of fossil men, and in the Heidelberg Man, this fifth cusp persists more or less in all the true molars; but instead of projecting in the form of a heel, as in the apes, it has already become reduced in size and is wedged in between the two neighbouring cusps, as in the first true molar of men of the white races.

All the teeth in the Mauer jaw have large pulp cavities. In the lower races at the present day these cavities are also larger than in Europeans. Here is illustrated the persistence of an infantile character which represents a primitive stage in the dentition of the large primates.

Conclusions.

Such are the principal morphological characters of the valuable Mauer fossil. According to Schoetensack, this ancient jaw represents a very primitive form, for certain of its characters are found in the lower monkeys and even in the Lemurs (great breadth of ascending rami, weak mandibular notch). Thus it would seem to have retained some aspects of a stage through which the common ancestors of Anthropoid Apes and Man must have passed. However that may be, it is certain that, in this very remarkable anatomical specimen, in this ancient relic of one of our oldest ancestors, there is a skilfully proportioned mixture of human and simian characters.

Duckworth considers that the jaw discovered by Schoetensack may have belonged to *Pithecanthropus*—an unnecessary hypothesis. We may go so far as to say, however, that the lower jaw of the Mauer fossil, like the brain-box of the Javan fossil, actually represents an almost ideal intermediate stage between the Apes and Man.

Others have wished to go further, and to deduce from this valuable but too isolated relic more than it can signify. Reconstructions of the skull, and even the portrait of *Homo heidelbergensis* have been published. Such attempts may serve as pleasant pastimes for men of science; but they should not be allowed to pass beyond the study walls.

On the other hand comparisons between the Mauer fossil and similar fragments of other human fossils are justifiable. While reserving this examination for the following chapter, I may say, here and now, that between the lower jaw of *Homo heidelbergensis* and that of *Homo neanderthalensis*, his successor in Western Europe, there are certain similarities favouring the hypothesis of a fairly close relationship.

The Piltdown Man.

(*Eoanthropus dawsoni.*)

Historical Account.

On 18th December 1912, Mr Charles Dawson, the geologist, and Dr Smith Woodward, the eminent palæontologist of the British Museum, exhibited before the Geological Society of London human bones obtained from a very ancient Quaternary deposit. English newspapers of the day following brought this sensational news to the knowledge of the public at large, and from that time the "Piltdown discovery" has been expounded, criticized, and discussed almost everywhere. The bibliography relating to it consists, at the moment of writing, of more than one hundred articles, notes, or memoirs.

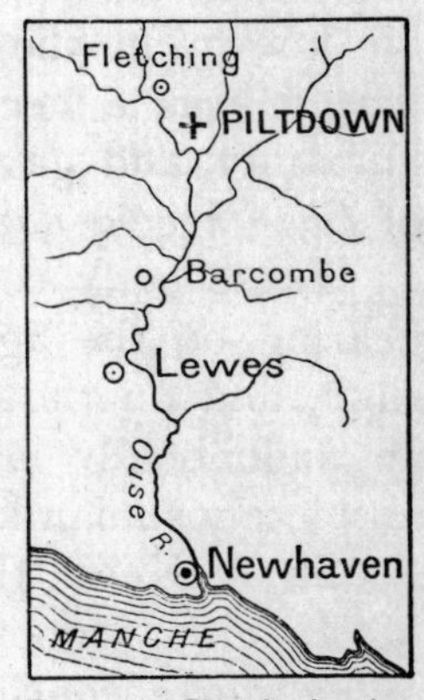

FIG. 92.—Sketch showing Geographical Position of Piltdown.

In a field near Piltdown (Fig. 92), in the parish of Fletching in Sussex, there is a small quarry where the countrymen of the district obtain stones for repairing the roads. One day one of the workmen forwarded to Mr Dawson a fragment of a human parietal bone, very ferruginous in appearance, as were the stones of the quarry. Some years later, in the autumn of 1911, Mr Dawson collected among the rubble a fragment larger than the first, belonging to the frontal region. This find was remitted to Dr Smith Woodward, who immediately recognized its great interest; and accordingly, by common consent, further explorations were carried out in the deposit during the summer of 1912.

It would seem that a human skull, entire or almost entire, had been broken by the workmen, and the fragments scattered. Some of these it was possible to recover from the rubble, while the right half of a lower jaw was obtained by Mr Dawson *in situ* in the gravel, at the very spot where the skull lay, and where, a yard away but at the same level, Dr Smith Woodward found a fragment of the occipital bone. Beside the human remains there were collected fossil animal remains and some dressed flints.

Dawson has described the geological bearings of the bed, and Smith Woodward has identified the animal bones and reconstructed the human fossil.[1] As the reconstructed skull, according to a first measurement, seemed to have a capacity of only 1070 cubic centimetres, as the fragment of jaw had a very simian appearance, and as, according to Elliot Smith, the intracranial cast revealed the most simian brain so far known in the human family, Smith Woodward believed that it was a very primitive form representing the dawn of humanity, and gave to the remains the very expressive name of *Eoanthropus dawsoni*.

These early communications were followed by discussions bearing on the age of the bed, on the reconstruction of the skull, and on the question of attributing to the same being an undoubtedly human brain-box and at the same time an absolutely simian lower jaw.

Sir A. Keith, the able anatomist of the College of Surgeons of London, opposed from the first certain of Dr Smith Woodward's conclusions.[2] According to him, the restoration of the skull was at fault: its capacity was really 1500 c.c.

[1] Dawson, C., and Woodward, A. S., "On the Discovery of a Palæolithic Skull and Mandible . . . at . . . Piltdown . . . with an Appendix by Elliot Smith" (*Quart. Journ. Geol. Soc. London*, lxix., 1913). Supplementary Note . . . (*Ibid.*, 1914). Woodward, A. S., *A Guide to Fossil Remains of Man in the British Museum*, London, 1905 [3rd. ed. 1922].

[2] Keith, A., "Ape Man or Modern Man?" (*Illustrated London News*, 16th and 23rd August 1913), etc.

instead of 1070 c.c.; so that we have here a being quite comparable to a man of our own day, to a "London citizen."

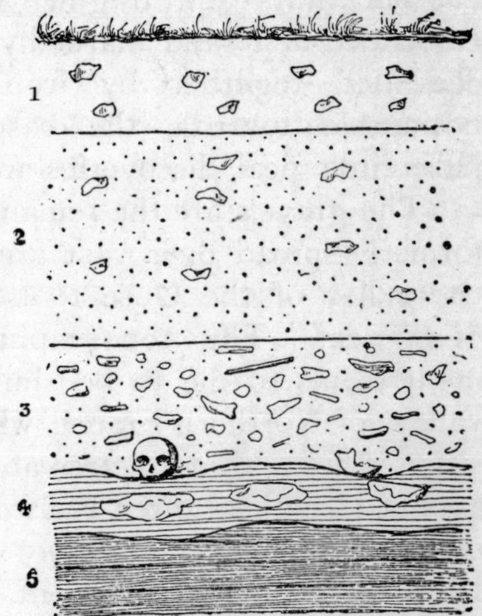

Fig. 93.—Section of the Piltdown Bed.

1. Superficial soil, 0.30 metre. 2. Pale yellow clayey sand, containing, in its flood-sorted condition, certain constituents from the underlying bed; a palæolithic implement was taken from the middle of this layer, the depth of which is about 0.75 metre. 3. Dark brown ferruginous gravel, with subangular flints and pieces of tabular iron ore. Bed of *Eoanthropus* and water-worn Pliocene fossils. Eoliths and a dressed flint. The surface of the bottom layer shows depressions. Depth about 0.45 metre. 4. Bed of sand and clay, a sort of mud formed at the expense of the underlying beds and containing large blocks of flint. 5. Strata of the Tunbridge Wells Sands (Wealden) *in situ*. (After Dawson.)

Professor Elliot Smith defended Smith Woodward's restoration.

In August 1913, a French palæontologist, a student in my laboratory, Père Teilhard de Chardin, exploring the bed in Dawson's company, found a canine tooth which was attributed to the jaw, and which reinforced the simian characters of the latter.[1]

In 1915 other human fragments were brought to light; they belonged to one or two new individuals.[2] The controversy continued and still continues, as we shall presently see.

After this short historical preface, let us examine the facts, considering first those which bear on the geological age of *Eoanthropus*, and thereafter those which relate to the fossil itself.

Geological Age of the Layer.

The geological age must be determined by stratigraphy, palæontology, and archæology.

The Piltdown gravels cover a plateau, 30-40 metres high, about 25 metres above the river Ouse; and at Piltdown, their depth varies from 0.30 metre to 1.50 metres. They rest on the Wealden sandstone at the

[1] Woodward, A. S., "Note on the Piltdown Man" (*Geological Magazine*, October 1913).

[2] Woodward, A. S., "Fourth Note on the Piltdown Gravel, with evidence of a Second Skull of *Eoanthropus dawsoni*" (*Quart. Journ. Geol. Soc., London*, lxxiii., 1917).

base of the Cretaceous formation. The gravels are composed of Weald pebbles, dark brown and ferruginous, mixed with angular flints and fragments of quartzite, all of which are embedded in a sand markedly stratified in places and strongly cemented together by iron oxides, conditions particularly noticeable towards the base of the formation, where the palæontological discoveries were made (Fig. 93).

The gravels are the remains of an alluvial formation which formerly spread over vast areas, before the last sculpturing of the valley of the Ouse took place, excavating it to a depth of 80 feet. This topographical and stratigraphical arrangement seems to me to be similar to that which occurs in the valleys of Northern France, where the middle terraces, generally situated from 25-30 metres above the river channel, have been ascribed to the Lower Pleistocene, characterized by the fauna containing *Elephas antiquus*. This is also the opinion of the English geologist, Clement Reid, who has compared the Piltdown gravels to the ancient Palæolithic gravels of the Thames valley, both having been formed subsequent to the great Glacial Period of the British Isles.

Palæontology expresses no very definite opinion on the subject. The mammal remains collected in the gravels have been attributed to two faunas of different ages. On the one hand, fragments of the teeth of a Mastodon, of *Stegodon*, and of a Rhinoceros, can only be of Pliocene age ; but as they are much worn by transporting agencies, we may suppose that they originally came from a more ancient deposit. The teeth of a Hippopotamus may have belonged equally well to Quaternary or to Pliocene times. The other remains, fragments of the antlers of Red Deer, Beavers' teeth, and the molar of a horse, would appear to be less impregnated with mineral matter and less water-worn. They may be regarded as Pleistocene and really contemporary with the gravel.

There remains the archæological aspect. The flaked flints found in the Piltdown gravel are of two kinds. A large number of very water-worn and weathered fragments may be looked upon as eoliths, and, like the water-worn teeth of the Proboscidians, as originating from an older geological formation.

Some other fragments are distinctly worked, and their sharp edges are almost as well preserved as when they were made. They have been dressed or flaked on one surface only, and their general form is in no way remarkable. We know nowadays that although flints worked on one side only were formerly regarded as characteristic of the Moustierian civilization, they are very often, sometimes to the exclusion of other forms, found also in the Chellean. Nothing could be more natural, as we have already pointed out; for implements fashioned by simple flaking are the most elementary, the most

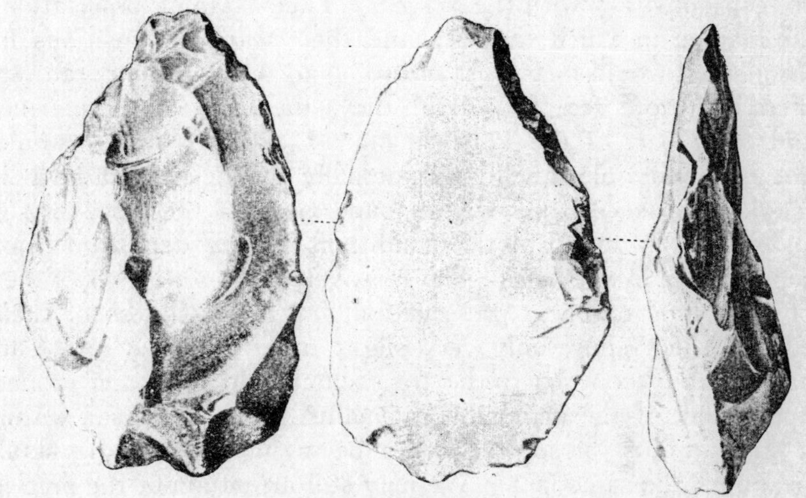

FIG. 94.—Dressed Flint from the Piltdown Gravel. Three-fourths natural size. (After Dawson.)

primitive that can be imagined; their manufacture certainly preceded that of implements dressed on both faces, and in particular that of the fine lanceolate forms of Chellean times (Fig. 94). So that in spite of the absence of amygdaloid types, it is possible, as Dawson would have us believe, that the Piltdown dressed flints are of Chellean age.

More recently there has been found in the same gravels a detached splinter of a bone of a Proboscidian, which measures no less than 0.40 metre in length and 0.10 metre in breadth. Of the artificial character of this object, which is rounded at one end, and pointed at the other, there seems to be no doubt in the minds of Dawson and Smith Woodward, who regard it as "an implement";[1] although M. Breuil looks upon it as a bone gnawed by a large Beaver or *Trogontherium*, Beaver remains having been actually found in the bed.[2]

To sum up, stratigraphy justifies us in dating the Piltdown gravels from a very remote period of the Pleistocene; the palæontological data may be interpreted in the same sense; and the archæological evidences are in no wise contradictory to this view.

Now, can we be certain that the human remains themselves are contemporary with the gravel? Their state of fossilization, the manner in which they lay, and their scattered positions in the alluvial would forbid us, according to some, from regarding them as more recent. Could they, then, be more ancient, and form part of the Pliocene fauna? *A priori*, that would not be impossible, and it was actually the belief first held in England, whence came the announcement of the discovery of a Tertiary Man. But the conditions of the deposit do not support this hypothesis. The relatively large size of the pieces of skull, their excellent preservation, and the freshness of their edges, which show only very slight traces of wear and tear, are in strong contrast to the fragmentary, decayed, and eroded appearance of the mammalian remains. The dispersal, within a small radius, of several pieces belonging to a single skull, bears out the same fact. We may still argue about the precise age of the Piltdown gravels; but there does not seem to be any doubt that the human bones were contemporary with the deposition of these gravels.

All these deductions are valid only if we take for granted the hypothesis that the superficial Piltdown formation is homogeneous as regards age and composition. Now it is possible, it even seems apparent, that this is not the case. The primary origin of the deposits may be very ancient and date from Pliocene times, as certain of the most worn fossils appear to indicate. But the first alluvial formations may have

[1] Dawson, C., and Woodward, A. S., "On an Implement from Piltdown" (*Quart. Journ. Geol. Soc., London*, lxxi., 1915).

[2] *Revue anthropologique*, July-August 1922, p. 229.

subsequently undergone successive alterations which may have mixed the more recent with the more ancient fossils. One thing is certain: that the human bones with edges still sharp cannot have undergone prolonged transportation. Moreover, the character of ferruginous mineralization loses much of its importance here, if we take into account the rapidity with which water charged with iron can change the physical aspect of objects it impregnates. Now all this region is made up of Wealden deposits formed of very ferruginous constituents.

In short, it is very difficult to gauge precisely the geological age of the human bones. With my English colleagues, I consider them as dating from the Lower Pleistocene, but I must add that this does not seem to me to have been absolutely proved.

Description of the Bones. The Skull.

The bone-remains, human or supposedly human, comprise, first, a large portion of brain-box and half a lower jaw with the first and second true molars in position. To these remains were added, later on, some nasal bones, a tooth and some fragments of another skull.

There are four pieces of the first skull, composed of nine fragments, by means of which it has been possible to reconstruct the brain-box. The largest of these pieces consists of the frontal and parietal portions of the left side. The second is a left temporal bone, almost complete and well-preserved (Fig. 95). Unfortunately these two pieces have no point of contact with the two others, which belong to the right half of the skull, and consist of the larger part of the parietal and the median region of the occipital. This blank explains the divergences which have arisen in the course of the reconstruction of the skull, a matter which will always remain somewhat problematical.

The bones are normal, without any trace of disease; they have been in no wise deformed by their mineralization.

The thickness of the bones is remarkable, from 10 mm. to 12 mm. in the parietal and frontal, and so also is the depth of the impressions of the meningeal vessels on the inside of the skull. Apart from this peculiarity, detailed examination of each of the bones hardly reveals any characters which are not perfectly human, and even more akin to those of modern *Homo sapiens* than to those of other fossil men such as *Homo neanderthalensis*.

As there is preserved in the left frontal region a portion

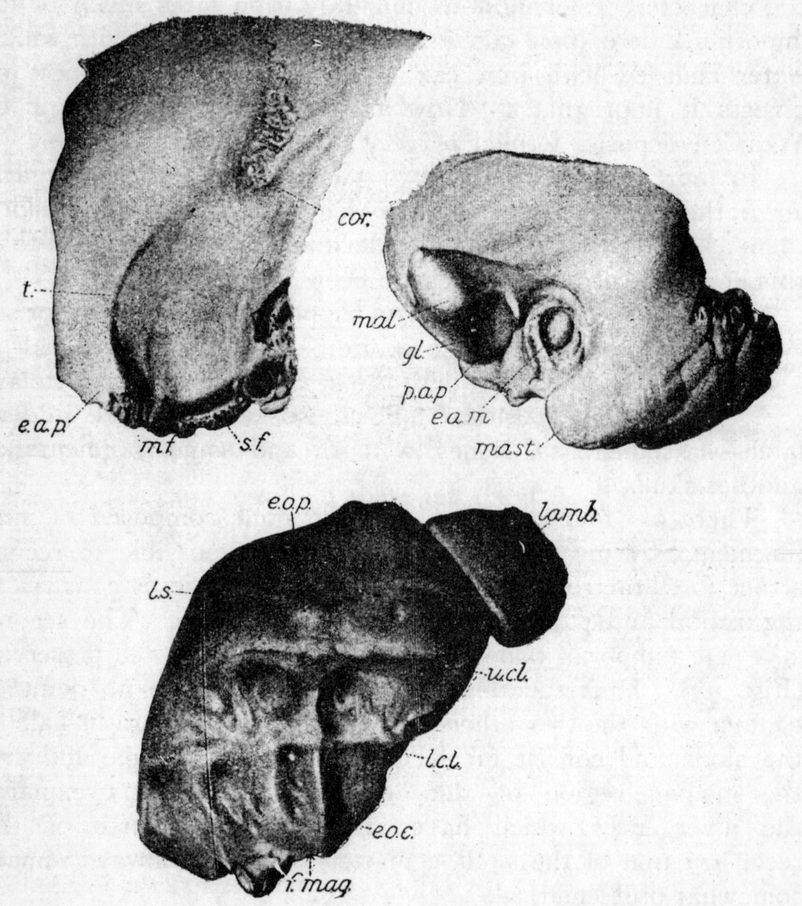

FIG. 95.—Various Fragments of the Skull of *Eoanthropus*.

Above, to the left; left frontal bone, seen from the outer surface; *cor.*, coronal suture; *e.a.p.* external angular process; *m.f.* facette for malar; *s.f.* facette for the sphenoid; *t.* temporal ridge.

Above, to the right; left temporal bone, external surface; *e.a.m.* external auditory meatus [blocked by a pebble]; *gl.* glenoid fossa; *p.a.p.* post-articular apophysis; *mal.* process for malar; *mast.* mastoid process.

Below; occipital bone, external surface; *e.o.c.* external occipital crest; *e.o.p.* external occipital protuberance; *f. mag.* occipital foramen (*foramen magnum*); *lamb.* portion of lamboid suture; *l.s.* linea suprema; *u.c.l.*, upper curved line; *l.c.l.* lower curved line.

All the figures are two-thirds natural size (after Smith Woodward).

of the external or malar tuberosity (Fig. 95), it can be definitely ascertained that the supra-orbital ridges were not more developed than in modern Man. This conclusion was confirmed in 1916 by the discovery of a second piece of frontal bone belonging to a second individual. The coronal suture is highly complex.

On the parietal bones, the curved temporal lines are very high; and the squamous suture is as arched as in a modern skull. The strong projection of the temporal ridge and the strength of the zygomatic apophysis indicate a considerable development of the masticatory apparatus. We must also note the position, almost in the middle of the parietal, of the very pronounced parietal boss, and the presence of a large flat area behind this boss.

The left temporal bone, which is well preserved (Fig. 95), is similar in all its details to that of modern races: the glenoid cavity is as deep; the tympanic bone is reduced; the mastoid process is large. Yet this bone seems to me to exhibit certain somewhat special characters—a very considerable projection of the temporal crest, and especially of the sub-mastoid tubercle, a large backward extension of the petrous region, the strength of the zygomatic apophysis, and the fairly considerable development of the post-glenoid tubercle.

The occipital bone is remarkable for its development in a transverse direction. The external occipital protuberance is placed, as is the case in modern Man, below the plane separating the cerebral hemispheres from the cerebellum, instead of above it as in Moustierian Man. It is not thus in the portion of the occipital found in 1916; here the external occipital protuberance lies above the lateral sinuses, and the neck muscles extend higher, as in Neanderthal Man. Nevertheless, the noticeable projection of the upper arches contrasts with the very slight development of the external occipital protuberance. These upper arches are continued for some distance in the direction of the temporal. The lower curved lines seem to me to be relatively nearer the upper curved lines than in modern types, conforming in this respect to the arrangement seen in *Homo neanderthalensis* and to a greater degree in Chimpanzees.

Nasal bones, found in an excellent state of preservation, are very human in character; relatively small and broad, they are rather of a Melanesian or African than of an Eurasian type.

The skull, as finally reconstructed by Dr Smith Woodward, would be mesaticephalic, almost brachycephalic, having a cephalic index of 78 and a slightly flattened vault. Its capacity, according to the latest evaluation by the same author, should no longer be 1070 but 1300 cubic centimetres, comparable, therefore, to the average capacity of many present-day savage peoples, such as the Australians, Bushmen, or Andaman islanders.

As a whole, and in spite of certain peculiarities of a primitive type and of the extraordinary thickness of bone, the skull possesses in high degree the structure of a human skull. The individual to whom it belonged, far from representing a different genus, can at the most be regarded only as the representative of a primitive race of *Homo sapiens*. Professor Ramstrom[1] of Upsala has recently drawn attention to the close resemblance of the Piltdown skull to the Aurignacian skull from Combe-Capelle, of which we shall speak later.

The Jaw.

But we have still to consider the jaw; and here the story of the Piltdown discovery becomes truly extraordinary.

The lower jaw, contrary to all the indications of the cranium, is very simian! It shows the same degree of mineralization as the skull, with which it appears to correspond fairly well in point of size. The condyle is missing, as well as the upper portion of the front half of the horizontal ramus, but in the remaining portion the two anterior true molars are in place (Fig. 96).

The ascending ramus is broad; the mandibular notch is not very deep, and the condyle must have been short. The mark of attachment of the temporal muscle is large. The mylo-hyoid groove is situated below the dental canal, instead of originating from it, as in Man. There is no mylo-hyoid ridge (internal oblique line).

[1] Ramstrom, M., "Der Piltdown-Fund" (*Bull. Geol. Instit. Upsala*, xvi., 1919).

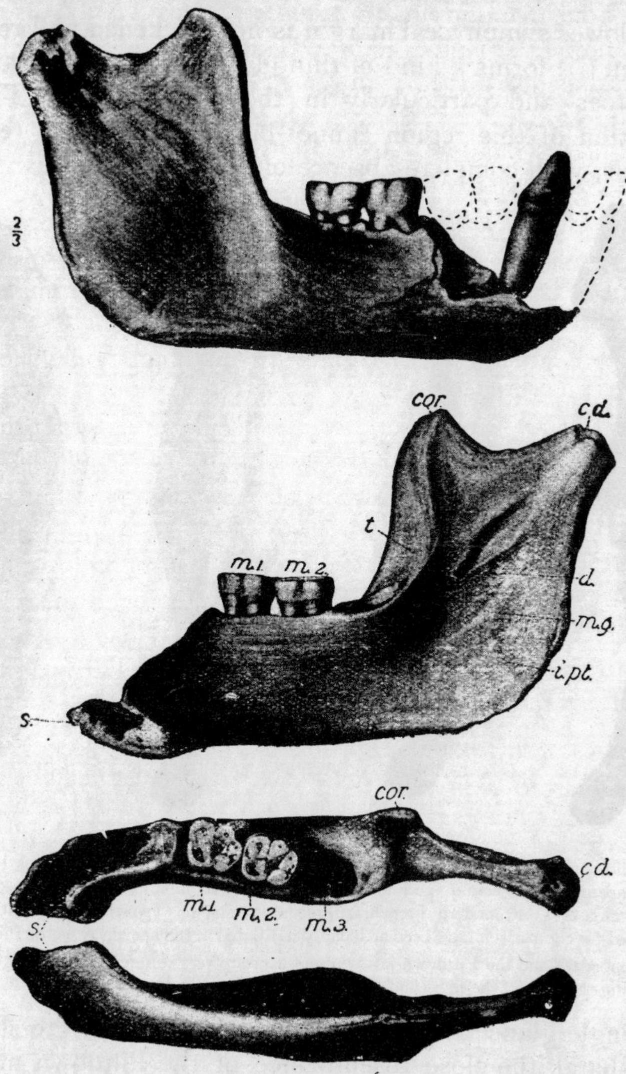

FIG. 96.—Lower Jaw from Piltdown. Two-thirds natural size. (After Smith Woodward.)

Top Fig.—Jaw seen from outer surface. The canine tooth has been drawn here as belonging to this jaw. It is now considered to be an upper canine.

Centre Fig.—Jaw seen from inner surface ; *cd.* neck of condyle ; *cor.* coronoid process ; *t.* area of insertion of temporal muscle ; *d.* inferior dental foramen ; *m.g.* mylo-hyoid groove ; *i.pt.* area of insertion of internal pterygoid muscle ; *s.* incurved boney flange of symphysis ; *m.*1, *m.*2, first and second true molars.

Bottom Figs.—Jaw seen from above and below ; *m.*3, socket of third true molar ; other symbols as above.

The lower symphyseal margin is not thickened and rounded as in Man; it forms a kind of thin plate, projecting inwards, as in the apes and particularly in the Chimpanzee. The reconstruction of this region cannot but suggest a very receding symphysis, with complete absence of chin.

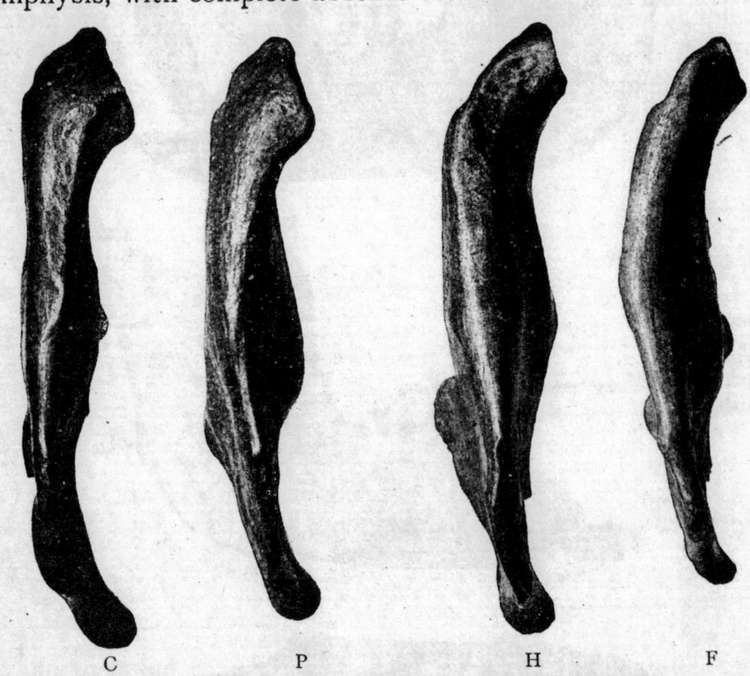

FIG. 97.—Photographs showing, in comparative view, the under margin of various lower jaws: C, Chimpanzee, from actual specimen; P, Piltdown mandible, from a cast; H, Heidelberg Man, from a cast; F, modern Frenchman, actual specimen. Two-thirds natural size. The jaws used for comparison have been artificially mutilated in the same way as the Piltdown jaw.

The Chimpanzee and the Piltdown jaws present a group very different at first sight even from that containing the jaws of the most ancient known Man and a modern European.

A single glance at Figs. 97 and 98 will suffice to show, on the one hand, the close resemblance of the Piltdown mandible to that of a Chimpanzee, and, on the other hand, the great difference which separates these two from a second small series comprising the Mauer jaw and that of modern Man.

The Teeth.

The canine tooth found by P. Teilhard (Fig. 99) differs from a human canine in its much larger dimensions, in the more raised, more conical, and more compressed form of its crown, and in the way in which it had worn, for this implies the existence, in the opposing jaw, of a similar tooth and possibly of a corresponding space or diastema.

At first Smith Woodward regarded this canine as belonging to the mandible, and described its resemblance to the canine teeth of apes, and to the lower milk canine of a human being. Gregory and Millar think that the tooth was in reality an upper left canine, very similar to that of a female Chimpanzee.

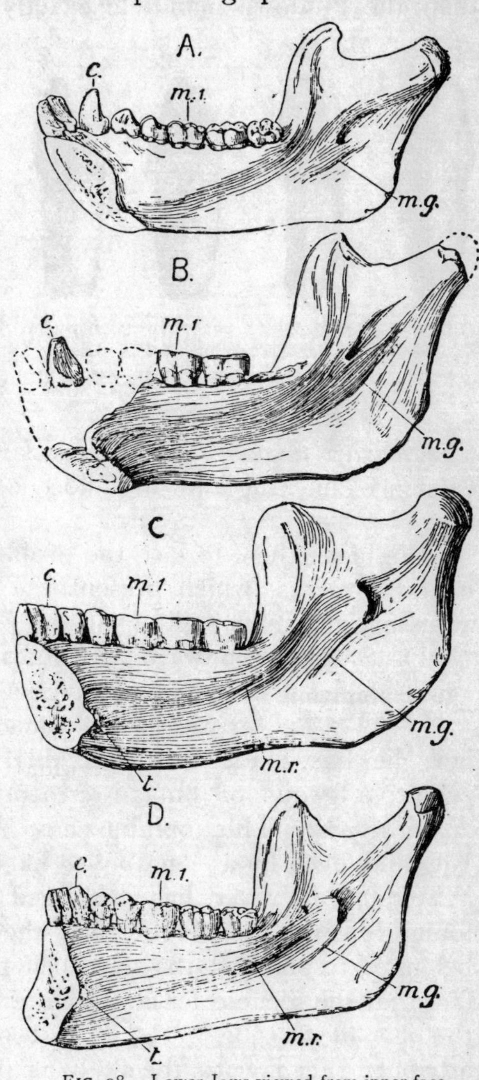

FIG. 98.—Lower Jaws viewed from inner face (half natural size).

A, Chimpanzee; B, Eoanthropus; C, Mauer (Heidelberg) Man; D, Modern Man. *c.*, canine tooth; *m.*1, first true molar; *t.*, genial tubercle; *m.g.*, mylo-hyoid groove; *m.r.*, mylo-hyoid ridge (internal oblique line). (After Smith Woodward.)

The first and second true molars, preserved in position in the jaw, are relatively long and narrow. Their neck is well marked, the roots are not attached. The crowns have five well-developed cusps, arranged as in certain anthropoids, such as the Chimpanzee, where the fifth cusp is more important and projects more markedly at the back edge of the tooth than in Man. Whatever may have been said of them, it is certain that these molars are much more simian than human, and that they wholly resemble those of the Chimpanzee.

In its general build as well as in the details of its structure, then, the Piltdown mandible exactly reproduces an ape's jaw, or, to be precise, the jaw of a Chimpanzee. I did not hesitate to say so, after having had an opportunity of studying a cast of the specimen: "Had this jawbone been found alone in the Piltdown gravels, along with the remains of Pliocene mammals, one would not have hesitated to call it *Troglodytes dawsoni*, and to declare that it testified to the existence in England in Pliocene times of an Anthropoid Ape."[1]

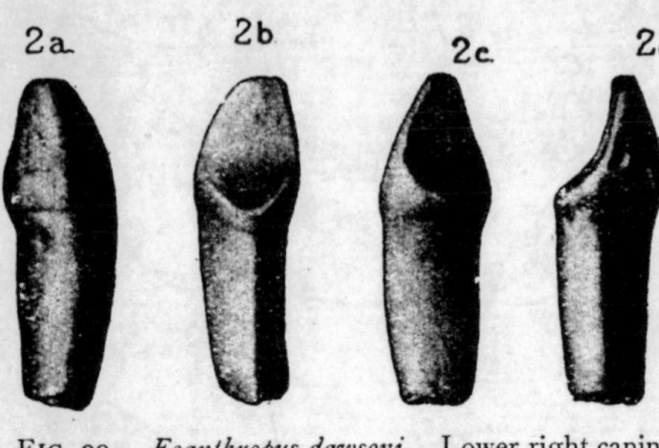

FIG. 99.—*Eoanthropus dawsoni*. Lower right canine, seen from the outer (2*a*), inner (2*b*), anterior (2*c*), and posterior (2*d*) surfaces. Natural size. (After Smith Woodward.)

We have, then, to face the problem of a small collection of bone-fragments, which present the paradoxical association of an essentially human skull with an essentially simian jaw.

Interpretation of the Facts.

Dr Smith Woodward attributes the skull and the jaw to the same individual, probably a female; and, struck by the simian characters of the lower jaw, he has made this individual the type specimen of a primitive genus of human creatures, which he has named *Eoanthropus*. This opinion has been generally accepted in England, even by an anatomist like Sir A. Keith.[2] Professor Waterston, however, has expressed doubt on this point. He found difficulty in believing that the skull and jaw could have belonged to one and the same being, and drew attention to the fact that the glenoid fossa of the temporal is not formed to fit the jaw of an ape. "It is just as possible," he said, "to attribute this jaw to the skull, as it would be to articulate a Chimpanzee's foot with the bones of an essentially human thigh and leg."[3] This scepticism gained the adherence of

[1] Boule, M., "La paléontologie humaine en Angleterre" (*L'Anthropologie*, xxvi., 1915, p. 60).

[2] *Loc. cit.*, and also *The Antiquity of Man*, London, 1915.

[3] *Nature*, 13th November 1913 [vol. xcii., p. 319].

other naturalists. The American mammalogist, G. Miller,[1] endeavoured to show, with a wealth of evidence, that the jaw was that of a Pleistocene species of Chimpanzee, and named it *Pan vetus* (*Pan* being the generic name used in America for the Chimpanzee). Another New York scientist, Gregory, whose knowledge of mammalian anatomy is comprehensive, also adopted this view.[2]

Is Eoanthropus an Artificial and Composite Creature?

We seem, then, to be faced with a double discovery: that of a Man and that of an Ape. *Eoanthropus*, therefore, would describe an artificial creature, composed of two natural creatures: *Homo dawsoni* and *Troglodytes dawsoni*.

Till now I have had a certain amount of doubt in the matter. Apart from the anatomical facts, the conditions of deposition made me hesitate. It is really very difficult to imagine the presence, at the same spot in the depth of an ancient alluvial formation, of remains belonging to two species of large Primates, and to explain, by pure chance, that these remains have the same physical characters, belong to creatures of the same stature, and are complementary pieces of the same part of the skeleton. On the other hand, the presence of an anthropoid ape in Western Europe in the Pliocene period or at the commencement of the Quaternary would not be in any way extraordinary, the less so as the Taubach teeth likewise appear to be those of a Chimpanzee.

Since the appearance of the first edition of this work certain new facts brought out by Elliot Smith and Hunter are perhaps of such a nature as to incline the balance in favour of Smith Woodward's hypothesis. But the new arguments do not appear to me to be conclusive; it is still permissible to suppose that the skull and the lower jaw from Piltdown may have belonged to two different beings (for fuller discussion, see Appendix, p. 471).

And so examination of Dr Smith Woodward's reconstruction of the skull is bereft of its interest, for it was only in

[1] Miller, G. S., "The Jaw of Piltdown Man" (*Smithsonian Miscellaneous Collections*, lxv., No. 12, 1915); "The Piltdown Jaw" (*American Journal of Physical Anthropology*, i., 1918. These two memoirs contain a lengthy critical bibliography.

[2] *Studies on the Evolution of the Primates*, p. 316.

the firm belief that the jaw and the skull-bones belonged to the same individual, that he was able to undertake this reconstruction at all (Figs. 100-102).

The shape of the brain-box, obtained by associating the fragments in their respective positions, cannot be far from the truth. The reconstruction of the face, on the contrary, is markedly hypothetical, since it could scarcely have been made without the assistance of nasal bones and a jaw which are foreign to it. It is not surprising, therefore, that it errs from lack of harmony, that it does not ring true. But that it is easy to criticize is chiefly due to the fact that here we have an example of that imprudent rashness which desires at whatever cost to glean from an unsatisfactory palæontological relic more than it could possibly yield.

General Conclusions.

The Piltdown relics are, unfortunately, incomplete. The interpretation of them, an extremely difficult task, is still doubtful regarding certain essential points. But in spite of all the uncertainties, they form a most important and one of the most instructive discoveries. Even although it be admitted that the skull and jaw are perfectly independent, it is at the same time none the less true that the fragments of the skull tell of the existence, at a period probably very early in the Quaternary era, of a man with an essentially human brain-box, and that this Man is more closely related to the ascending line of Modern Man than to that of Neanderthal Man. The beginning of our direct ancestry must thus date from a very remote past. Up to the present time, this hypothesis had been supported only by a certain number of discoveries which possessed no geological sanction, and, in consequence, were of no value as proofs. Now we have brought before us a new, carefully observed fact, of which the significance seems clear and definite, always allowing that the Piltdown skull is really as ancient as is supposed.

However it may be with regard to the question of age, which is very difficult to elucidate, we may ask if Dr Smith Woodward was justified in creating a new genus for the Piltdown fossil, for its fundamental characters are essentially human, particularly its cerebral capacity, which is as high as that of

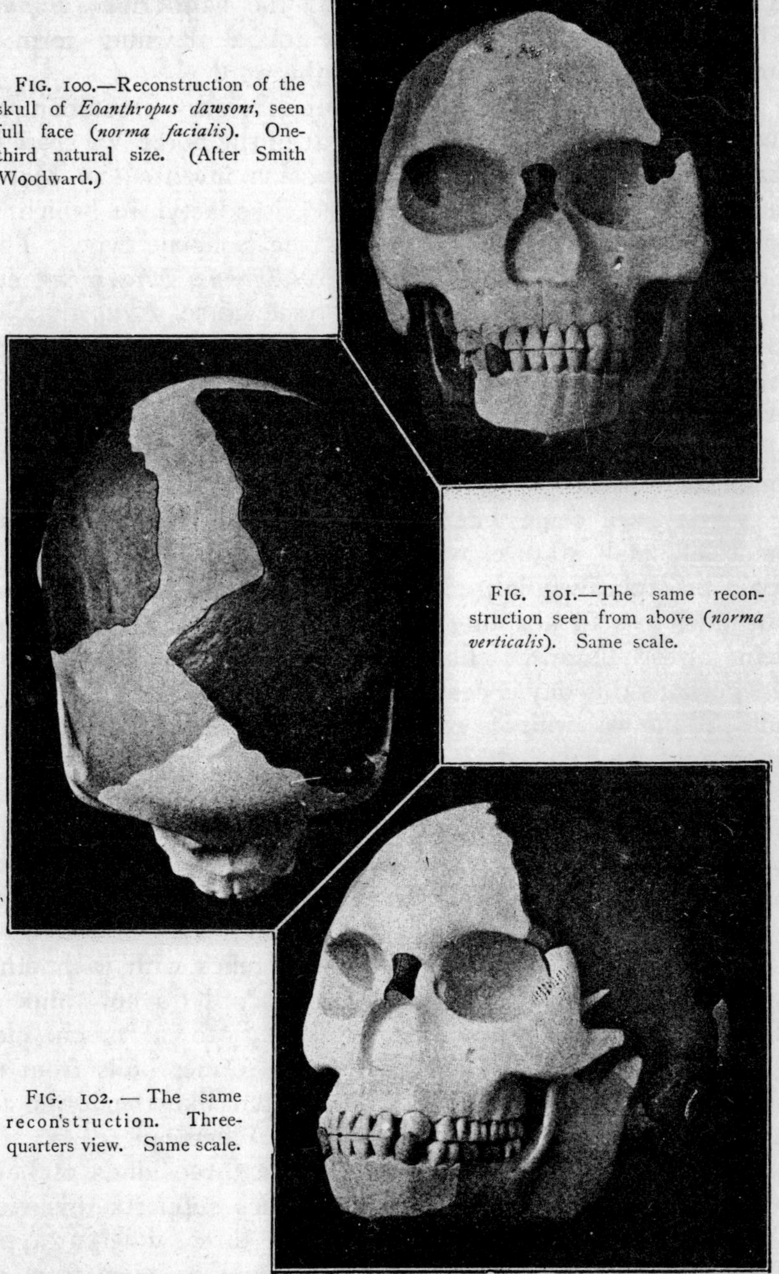

FIG. 100.—Reconstruction of the skull of *Eoanthropus dawsoni*, seen full face (*norma facialis*). One-third natural size. (After Smith Woodward.)

FIG. 101.—The same reconstruction seen from above (*norma verticalis*). Same scale.

FIG. 102. — The same reconstruction. Three-quarters view. Same scale.

many modern men. Above all, why the word *Eoanthropus?* There can be no question here of a dawning form of humanity in the zoological sense of the word.

If, from the palæontological point of view, we compare the evolution of the human race with that of the horse, we shall see that *Eohippus*, the analogous expression invented by Marsh, really applies to a very ancient Perissodactyl in which we begin to notice a tendency towards the Solipede type. Then follow other forms: *Mesohippus*, *Protohippus*, *Pliohippus*, etc., linking up this *Eohippus* with the true horse, *Equus*. Now it is quite certain that *Eoanthropus* does not bear to *Homo* the relationship which *Eohippus* bears to *Equus*. Even on Smith Woodward's hypothesis it would barely represent in the scale of Mankind, anything like the stage which *Protohippus* occupies amongst horses.

A day will come when there will be discovered a human being of small stature, with almost erect carriage, having a brain-box relatively large in comparison to the total body-bulk, but much less in absolute size than that of any other human being already known. This will be the true *Eoanthropus*.

Perhaps this day is near at hand, but it may be still very far off. While we await its coming, we must be satisfied with the knowledge we have so slowly and so painfully acquired. Our knowledge has indeed made some progress during the past dozen years, for previously we possessed no authentic remains of Chellean Man. To-day we have at least a lower jaw, the Mauer jaw, and in all probability some portions of a skull, the Piltdown skull.

May we, then, associate these two relics with each other, and attribute them to the same species? I do not think so. The Mauer discovery seems to belong to a much older geological age than the Piltdown: the former hails from the most ancient Chellean period; the latter, like the lower jaw from Ehringsdorf, cannot be older than Acheulean.

The comparative morphology of the three kinds of bone-relics yielded by these various discoveries supports this view. We are dealing with two, or probably three, different types. Thus we reach the important conclusion that, even from the commencement or from the first phase of Quaternary times, the human beings (Hominians) in our country were already differentiated. It is a fresh proof that the origins of Man are lost in a past that recedes further and further into the distance. Assuredly the Pliocene and Miocene deposits have still in store for us some strange and exciting discoveries.

CHAPTER VII

NEANDERTHAL MAN

(*Homo neanderthalensis.*)

IF the poverty of human fossil remains in the Lower Pleistocene or *Chellean* period is marked, there is a striking contrast in the wealth of human fossils in the Mid Pleistocene, which almost corresponds to the *Moustierian* period of archæologists.

Characters of the Moustierian.

We already know that this period differs greatly from the preceding one in its geological, palæontological, and archæological characters. It corresponds to the last glacial invasion, to a period of great floods and deposit of alluvium, of the formation of the greatest accumulation of superficial muds and natural cave deposits.

The Moustierian flora and fauna of Central and Southern Europe differ both from the Chellean flora and fauna, and from the flora and fauna of the present day, and indicate a much moister and colder climate. The large extinct species of mammals, the Mammoth, the Woolly (Tichorine) Rhinoceros, etc., were covered with a thick coat of hair; the majority of the species still living at the present day, such as the Reindeer, Musk-Ox, Glutton, Ibex, Chamois and Marmot now only inhabit northern countries or the highest mountains.

From the archæological point of view, there are also changes. The amygdaloid flints are smaller, less thick and finer. The predominating types are points and scrapers dressed and retouched on one face only. For the first time, we find traces of the use of bone (Figs. 103 and 113).

Here we have to deal with conditions of environment totally different from, and much more severe than those of the Chellean period. Man, obliged to protect himself against the rigors of the climate, had to modify his habitat: he sought refuge in caves, and there he lived, died, and left the bones which we exhume to-day with such excited curiosity.

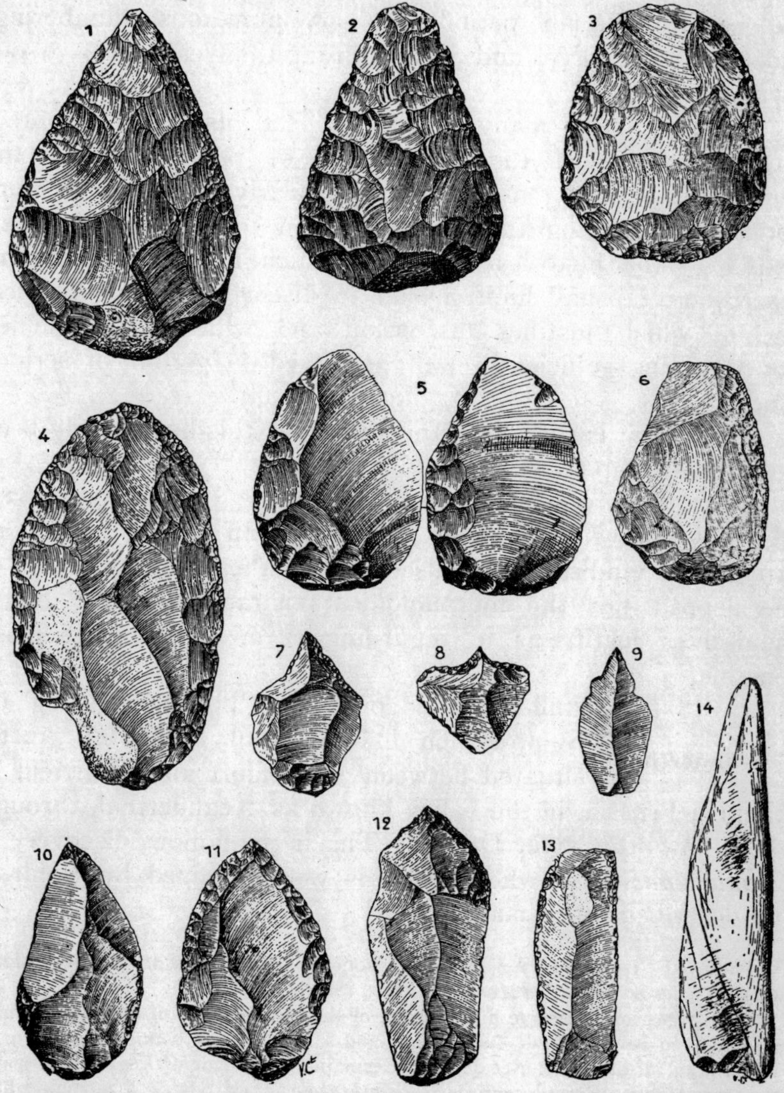

FIG. 103.—Moustierian Industry from the north of France. One-third natural size. (After Commont.)

1, 2, 3, Amygdaloid flints carefully dressed on both faces; 4, 5, 6, scrapers; 7, 8, 9, piercers; 10, 11, 12, points; 13, flake, retouched as a knife; 14, fragment of bone with striations, probably used as a compressor.

Historical Summary.

Moreover, the discoveries of human fossils attributed to the Moustierian period are very numerous, numbering, indeed, about forty, and dating from the year 1700 to our own day.

Unfortunately many must be set aside or ignored, either because of the poor state of preservation of the remains obtained, or more particularly because of uncertainty regarding the age and even the identity of their beds. In the historical summary which I think it useful to insert here, I shall limit myself to discoveries the geological age of which justifies discussion and which have yielded bone-remains sufficiently well preserved to permit of serious study.

The oldest is that at Cannstadt, and this has already been described (see p. 141).

Next in chronological order comes the Lahr find, of 1823, which was given to Cuvier (see p. 6). In spite of Hamy's[1] attempt to vindicate this discovery, neither the conditions of the deposit nor the morphological characters of the bones themselves justify us in regarding them as authentic fossil remains.

Neanderthal.

In 1856, a skull-cap and some long bones were dug up by workmen in the small Feldhofer grotto, situated between Düsseldorf and Elberfeld in Rhenish Prussia, in the valley known as Neanderthal, through which flows the River Düssel. This is the famous discovery of *Neanderthal Man*, whose remains were obtained by Fuhlrott and described by Schaaffhausen.[2]

[1] Hamy, E. T., "Nouveaux matériaux pour servir à l'étude de la paléontologie humaine" (*Congrès intern. d'Anthrop.*, Paris, 1889, p. 423).

[2] The very comprehensive bibliography of this discovery is of historical interest only. It is to be found in: Quatrefages and Hamy, *Crania ethnica:* Reinach, S., "Description raisonée du musée de Saint-Germain": Obermaier, "Les restes humains quaternaires dans l'Europe centrale" (*L'Anthropologie*, xvii., 1906), a memoir which gives numerous bibliographical references to all the deposits containing human fossils in Central Europe. Among later works published on the morphology of the skull may be singled out for mention: Schwalbe, G., "Der Neanderthalschädel" (*Bonner Jahrbücher*, Heft 106, 1901).

The greatly depressed skull-cap with its large superciliary arches (Fig. 104) keenly interested the foremost naturalists of last century. Some, along with Schaaffhausen and Huxley, saw in it the representative of a primitive human race, preserving simian characteristics; others, including the German Virchow, leaned to the view that it was rather a pathological specimen or the skull of an imbecile. For long its high antiquity was suspected, since there was no irrefutable argument in its favour. The extraordinary morphology of the skull-cap could not be successfully appealed to, for this was an isolated case. But we shall see how later discoveries have destroyed this isolation, and have justified us in affirming that the clay whence the Neanderthal bones were taken, indeed dates, like the majority of analogous formations, from the Mid Pleistocene.

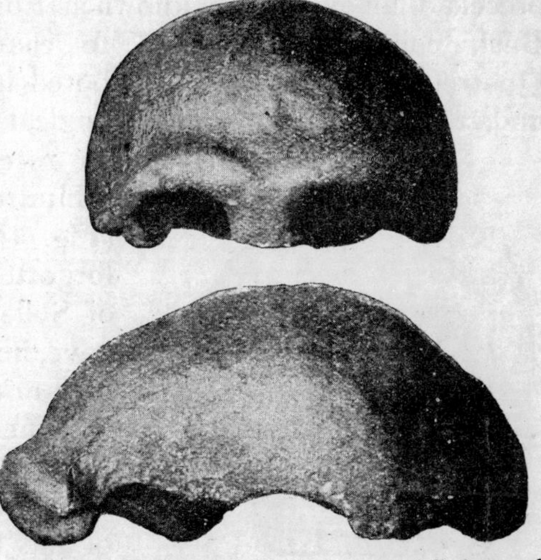

FIG. 104.—Neanderthal Skull cap, seen full face and in profile. (One-third natural size. Photograph from a cast.)

In 1859 there followed the discovery, in the Grotte des Fées, near Arcy-sur-Cure in Yonne, of a fragment of a human lower jaw, accurately dated by the bone remains of extinct animals collected at the same time.[1] This much mutilated specimen appeared remarkable for the smallness of the chin (Fig. 105).

FIG. 105.—Fragment of a Jaw from Arcy-sur-Cure. Three-fourths natural size. (After de Quatrefages and Hamy.)

[1] Vibraye, De (*Bull. Soc. géolog. de France*, 2nd series, xvii., 1860, p. 462).

Gibraltar.

In 1864, at a meeting of the British Association, an English geologist, Busk, exhibited a human skull obtained in the year 1848 from the bone-breccia of an excavation known as Forbes's Quarry at Gibraltar. Busk compared this skull to that from Neanderthal; de Quatrefages and Hamy attributed it to their *Cannstadt race*; and, thereafter, this palæontological specimen, remarkable for its receding forehead, its great orbital arches and enormous face (Fig. 106), seemed to have been forgotten. The recent researches of Sollas, Sera, Keith and others, have, however, once more brought it to notice.[1]

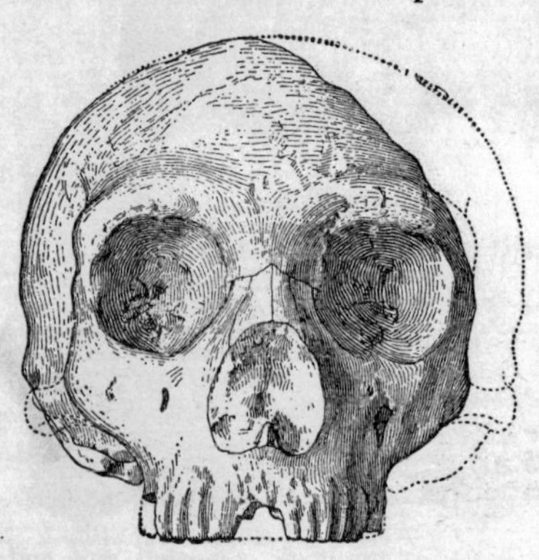

FIG. 106.—Gibraltar Skull, seen full face. Diagraph drawing from a cast. One-third natural size.

The high antiquity of the Gibraltar skull cannot be doubted. It is contemporary with the fauna of the breccia and deep deposits of the caves described by Busk,[2] and this represents a southern facies of the Mid Pleistocene fauna. The human skull and the animal bones are in the same state of fossilization; and, furthermore, we now know that Southern Spain contains a typical Moustierian industry.

La Naulette.

The year 1866 was marked by the discovery of the lower jaw from La Naulette, almost as famous as the Neanderthal skull-cap itself (Fig. 107). Found by a distinguished Belgian geologist, Dupont, in an undisturbed layer in the Trou de la Naulette, near Dinant, and accurately dated by animal remains of Mid Pleistocene age which accompanied it,[3] its morphological peculiarities, its

[1] Sollas, W. J., "On the Cranial and Facial Characters of the Neanderthal Race" (*Philosophical Transactions*, B., cxcix., 1907). Sera, G.-L., "Di alcuni caratteri . . . nel cranio di Gibraltar" (*Soc. romana di Antrop.*, xv., 1909). "Nuove osservazione." (*Arch. per l'Antrop.*, xxxix, 1909). Keith, Sir A., *Ancient Types of Man*, London, 1911, p. 121.

[2] Busk, G., "On the Ancient or Quaternary Fauna of Gibraltar" (*Trans. Zool. Soc., London*, x., 1879).

[3] Dupont, E., "Etude sur les fouilles scientifiques . . . dans les cavernes de la Lesse (*Bull. de l'Acad. roy. de Belgique*, xxii., 1866).

strength, its absence of chin, and its large molar alveoli made a great impression upon anatomists.[1]

To Hamy belongs the credit of foreseeing, if not of proving, that, on the one hand, skulls with receding foreheads and heavy superciliary arches, and, on the other hand, strong jaws without chin, known to be of this period, must have belonged to the same type, that is to say, to the same race. Dr Topinard[2] has since taken up again, in detail, the study of this jaw from La Naulette.

Some doubt overhangs the geological age of the skull-cap obtained in 1872 from the ancient alluvials at Brüx in Bohemia. Scientists are agreed in regarding this specimen as Neanderthal in type, but its poor state of preservation has rendered it of little use as evidence.

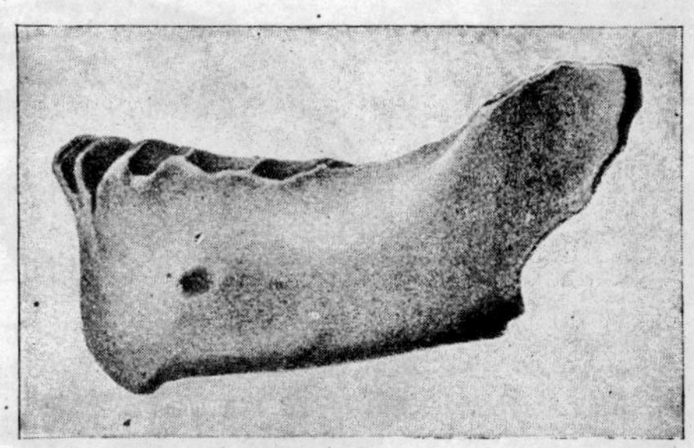

FIG. 107.—Lower Jaw from La Naulette. Three-fourths natural size. (After Quatrefages and Hamy.)

The discoveries made, in 1881 in the Schipka cave in Moravia, in 1883 in the Marcilly clay (Eure), and in 1884 in a clay pocket in the chalk at Bury St Edmunds in Suffolk, consist of mere fragments of jaws or of skulls in a state of preservation barely sufficient to warrant their classification as of Neanderthal type.

It is equally impossible to date the skull obtained towards the close of 1883 from the lœss at Podbaba, near Prague.

The Spy Discovery.

The year 1886 is remarkable for the very important discovery made by Marcel de Puydt and Max Lohest, in the Spy cave in the province of Namur in Belgium. Here the requirements of scientific precision are all fully met. The stratigraphy of the layer was definitely ascertained by a geologist; the fauna accompanying the human

[1] Pruner-Bey, "Sur la mâchoire humaine de la Naulette" (*Bull. de la Soc. d'Anthrop. de Paris*, 1866). Hamy, E. T., *Précis de Paléontologie humaine*, p. 232.

[2] Topinard, D., "Les caractères simiens de la mâchoire de la Naulette" (*Revue d'Anthrop.*, 1886, p. 385).

remains was that of Mid Pleistocene times—the Mammoth (*Elephas primigenius*), the Woolly Rhinoceros (*R. tichorhinus*), etc.; and the dressed flints were Moustierian in form.

The human bones are numerous and well preserved: two brain-boxes, some portions of the face, two lower jaws, and a large number of long bones, more or less perfect, comprise the remains of two skeletons. These valuable relics have been studied by Fraipont and Lohest, whose fine monograph[1] has added much to our knowledge of the human type known as Neanderthal. Till

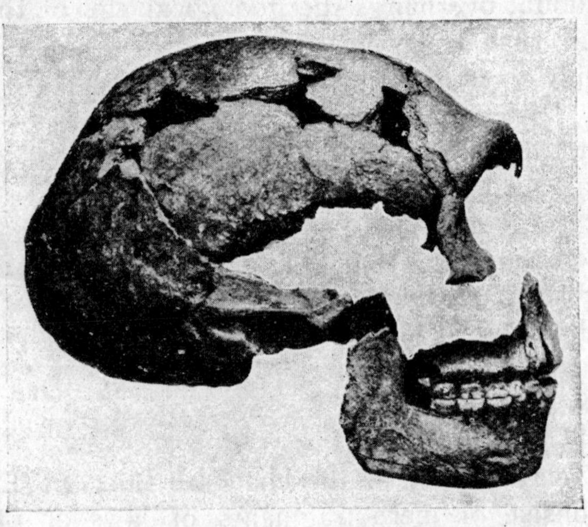

FIG. 108.—Spy Skull (No. 1) seen in profile. About one-third natural size. (After Fraipont.)

recent years the Spy skeletons were justly regarded as the most complete and the most important of ancient palæolithic remains.

In 1889, Hamy[2] examined certain human remains discovered by Piette in the Gourdan Cave, some of these being Magdalenian and others Moustierian. The latter comprised a fragment of lower jaw and a portion of face (upper maxillary and malar). Hamy, with his usual perspicacity, associated them with corresponding parts of the Gibraltar and Spy skulls.

[1] Fraipont, J., and Lohest, M., "Recherches ethnographiques sur des ossements humains découverts dans les dépôts d'une grotte quaternaire à Spy (*Arch. de Biologie*, vii., 1886. Gand, 1887).

[2] *Congrès intern. d'Anthrop.*, Paris, 1889, p. 413.

In the same year, H. Filhol described a lower jaw taken from the Malarnaud cave, near Montseron in Ariège. It was embedded in a mud containing bones of the Cave Bear and Mammoth, that is to say, in an environment and at a level corresponding to those of the La Naulette jaw, the anatomical characters of which are repeated in the Malarnaud specimen[1] (Fig. 109).

A portion of skull, obtained in 1892 in a brick-field at Bréchamps (Eure-et-Loir), has been considered by Manouvrier to belong to the Neanderthal and Spy type, the principal characteristics of which it exhibits, though in much less marked degree. To my mind it resembles rather the modern skulls known as "neanderthaloid." In any case, the conditions of deposit were not subjected to any examination, and the skull was not seen in place at the level occupied by the Moustierian flints of the brick-fields.

FIG. 109.—Lower Jaw from Malarnaud in Ariège, seen in profile. Three-fourths natural size. (After Filhol.)

A human jaw, found in 1895 at Isturitz (Basses Pyrénées), in a neighbourhood rich in bone-remains of the Cave Bear and Rhinoceros, should, according to M. Breuil, be compared to that from Malarnaud.[2]

About the same time, in the caves of Estelas in Ariège, at Aubert in Ariège, and at Sallèles-Cabardès in Aude, there were found, by F. Regnault and L. Roule, two lower jaws of children and a frontal bone with well-developed orbital arches. These may be readily assigned to Mid Pleistocene times.

[1] Filhol, H., "Note sur une mâchoire humaine trouvée dans la caverne de Malarnaud . . ." (*Bull. de la Soc. philomathique de Paris*, 1889). Boule, M., "La caverne de Malarnaud" (*Ibid.*).

[2] Breuil, H., "Les plus anciennes races humaines connues" (*Revue des sciences philosophiques et théologiques*, 1909).

Krapina.

The discovery at Krapina in Croatia is of greater import. In 1899, a professor in the University of Agram, M. Gorjanovic-Kramberger, described a Palæolithic shelter where, in an undisturbed Pleistocene layer (Fig. 110), he collected portions of from ten to twelve skulls, fourteen pieces of lower jaws, one hundred and forty-four isolated teeth, as well as numerous fragments of vertebræ, ribs, and long bones. Many of these bones, found in hearths, are calcined; indeed, the general state of preservation of all the remains leaves much to be desired. They have formed the subject of a long and painstaking monograph.[1]

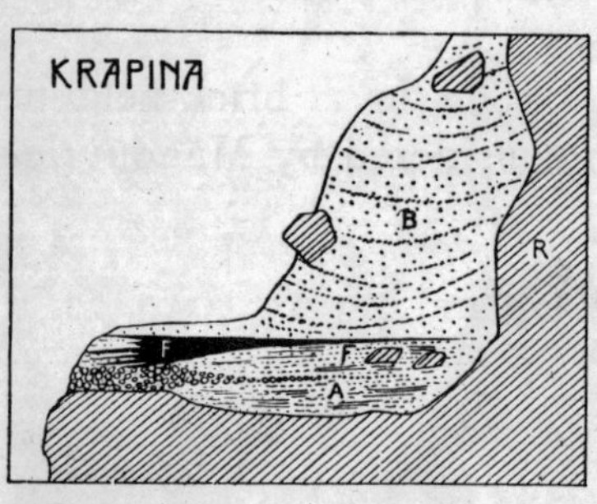

Fig. 110.—Section of the Krapina Bed. (After Gorjanovic-Kramberger.)

R, Rock, forming shelter; A, alluvial deposit of the Krapina, the bed being now 25 metres lower; F, hearths, bed where human remains were mainly found; B, products of disintegration of the Miocene sandstone forming the shelter, towards the topmost layers of which are still to be found bones of the Cave Bear.

The true age of the layer has often been misinterpreted. An attempt has been made to regard it as very ancient, on the strength of the presence of *Rhinoceros mercki* in the fauna the remains of which accompany the human bones, and this in spite of the fact that, in respect of its other members, this fauna contains nothing to distinguish it from the typical Mid Pleistocene fauna. Now, in France, *Rhinoceros mercki* persisted much longer than the Ancient Elephant (*Elephas antiquus*) and the Hippopotamus, with which it is associated in the fauna of the Lower Pleistocene. None of the really characteristic members of this fauna are to be met with at Krapina. From the palæontological point of view, therefore, this deposit cannot date further back than Mid Pleistocene times. And on this point the archæological evidence agrees with the palæontological evidence: the dressed stones belong to the recognized Moustierian types.

[1] Gorjanovic-Kramberger, K., *Der diluviale Mensch von Krapina in Kroatien*, Wiesbaden, 1906.

One of the principal results of the excavations carried out, from 1895 to 1902, at the instance of Albert I., Prince of Monaco, in the Grimaldi caves near Mentone, was the discovery of several skeletons of fossil men, accurately dated by stratigraphy, palæontology, and archæology.[1] Several of these skeletons, like those formerly found by Rivière and Abbo, do not date beyond the Upper Pleistocene period and present all the characters of the Cro-Magnon race of *Homo sapiens*. We need not dwell on these at this point. But in June 1901, the Canon of Villeneuve, director of the Prince's excavations, dug up from an older layer in the Grotte des Enfants, at a depth of 8.50 metres, two other skeletons with very different morphological characters. My learned colleague, Professor Verneau, who made a masterly study of them, established them as the type of a new fossil race remarkable for its negroid characters, *the Grimaldi race*. Examination of the bed and study of the animal bones collected immediately above and below the human skeletons, have led me to affirm that the latter date from Mid Pleistocene times, and are clearly of the same age as the Spy skeletons. But to this point I shall return in the following chapter.

Grimaldi.

In 1906, Rzehak described a fragment of jaw found in a layer "containing glacial fauna," in the grotto of Ochos in Moravia. This fragment is too incomplete to enable us to be certain of its neanderthaloid character, but it seems to have possessed a chin.

It will be sufficient simply to mention the discovery by M. Favraud of three fragments of human jaws in a Moustierian environment at Petit-Puy-Moyen in Charente. The jaws are strong, and have a receding chin and large teeth.

We come now to a series of recent discoveries of such interest that they must be described at somewhat greater length.

[1] Boule, M., Cartailhac, E., Verneau, R., and Villeneuve, L. de, *Les grottes de Grimaldi*, Monaco, 1906-1919. See lengthy summaries of this work in *L'Anthropologie*, 1906.

La Chapelle-aux-Saints.

The first is that of the skeleton from La Chapelle-aux-Saints.[1] Near the village of that name, in the Corrèze district, there is a small cave (Fig. 111), the excavation of which was undertaken by three priests,

FIG. 111.—The hill, showing entrance to the cave of La Chapelle-aux-Saints. The entrance to the cave is almost in the centre of the picture.

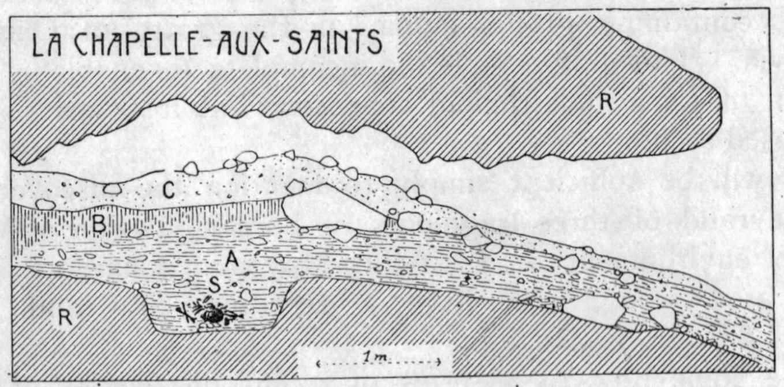

FIG. 112.—Longitudinal section of the Cave of La Chapelle-aux-Saints. (After A. and J. Bouyssonie and Bardon.)

R, rock in which the cave is hollowed out; A, archæological layer; B, clay; C, shifting sandy clay; S, human skeleton.

[1] Boule, M., "L'Homme fossile de La Chapelle-aux-Saints" (*Comptes rendus de l'Acad. des Sciences*, 14th December 1908). Bouyssonie, A. and J., and Bardon, L., "Découverte d'un squelette humain moustiérien" (*Ibid*, 21st December 1908).

MM. the Abbés A. Bouyssonie, J. Bouyssonie, and Bardon, already well known for their researches in prehistoric archæology. On 3rd August 1908, they discovered in it a quantity

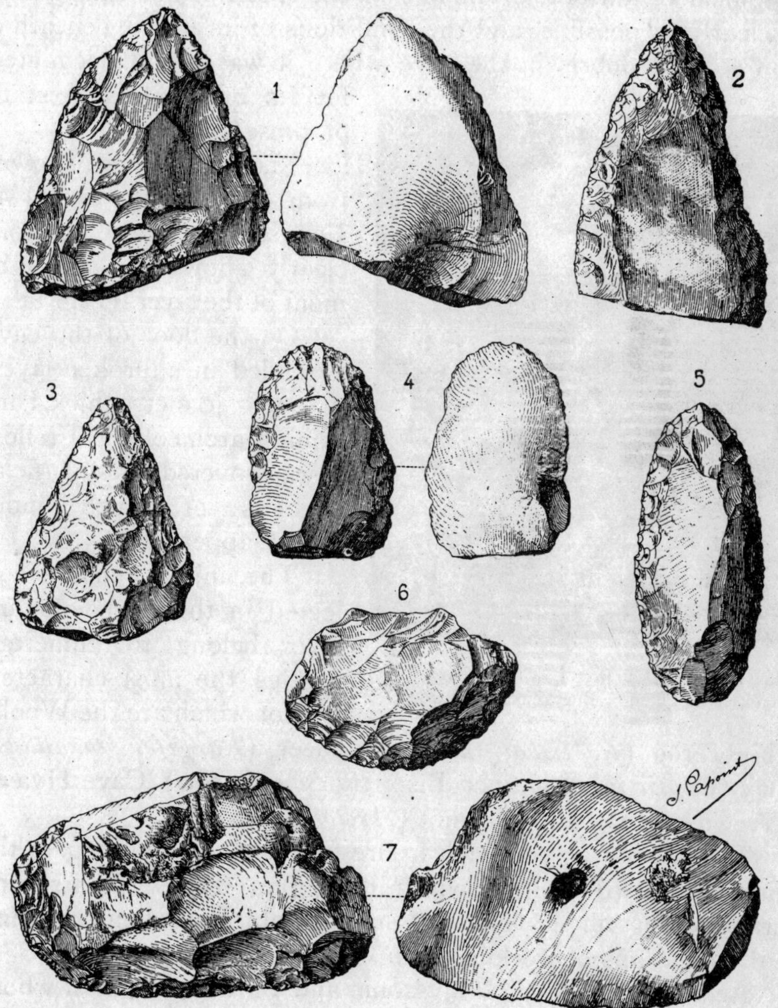

FIG. 113.—Some Dressed Flints from the Moustierian layer of the Cave at La Chapelle-aux-Saints. Three-fourths natural size.

of human bones, and these they sent to me at the Paris Museum, and, with the utmost generosity, were good enough to hand over to me.

This fortunate discovery provided the most complete and best preserved Moustierian human fossil known up to that time. It was found in unexceptionable topographical and stratigraphical conditions. The age of the skeleton was established as clearly as possible, and the conditions of its burial, beneath a thin coat of débris in the cave where it was found, accounted for its quite unusual state of preservation.

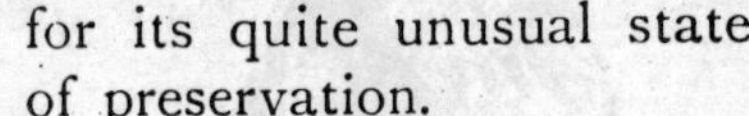

Fig. 114.—The Skull from La Chapelle-aux-Saints in its bed. Photo by Bouyssonie and Bardon.

The section here shown, from a drawing by MM. Bouyssonie and Bardon, clearly shows the arrangement of the layers (Fig. 112).

On the floor of the cave extended an unbroken layer, 0.30 to 0.40 metre thick, containing archæological relics; it was covered by superficial deposits of much more recent appearance.

The animal remains collected in this archæological layer belong to numerous species, the most characteristic of which are the Woolly Rhinoceros (*R. tichorhinus*), Reindeer (*Rangifer tarandus*), Ibex (*Capra ibex*), extinct Bison (*Bison priscus*), Cave Hyæna (*Hyæna spelæa*), and Marmot (*Arctomys marmotta*).

The layer was very rich in dressed flints, comprising in the main the two well-known Moustierian types, points and scrapers (Fig. 113). Not the smallest object of bone showing real indications of workmanship was observed.

According to MM. Bouyssonie and Bardon, the man whose skeleton they found had been deliberately buried. He lay at the bottom of a ditch hollowed out of the marly soil of the grotto, at a depth of about 0.30 metre.[1]

[1] Bouyssonie, A. and J., and Bardon, L., "La station moustiérienne de la 'Bouffia' Bonneval à La Chapelle-aux-Saints" (*L'Anthropologie*, xxiv., 1913).

The parts of the skeleton which I have been able to reassemble are the skull (cranium and lower jaw), twenty-one vertebræ and fragments of vertebræ, about twenty ribs or portions of ribs, a collar-bone (clavicle), two almost complete upper arm bones (humeri), two incomplete radii, two cubital bones, several bones of the hand, two fragments of haunch bones (ilia), two incomplete femora, two knee-caps, portions of two tibiæ, an astragalus, a calcaneum or heel-bone, five right metatarsals, two fragments of left metatarsals, and one phalanx.

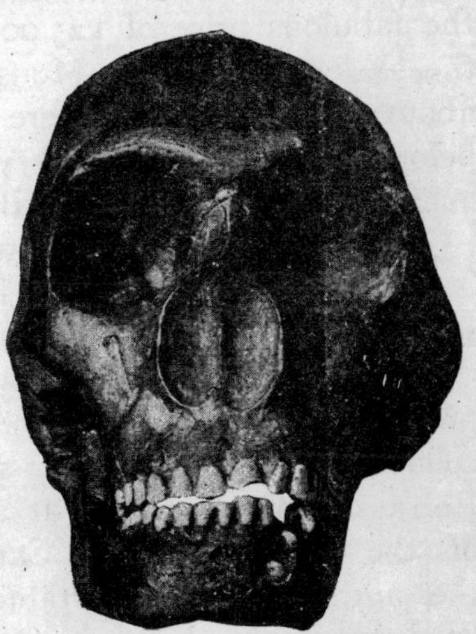

FIG. 115.—Skull from Le Moustier, seen full face, as reconstructed by the German anatomist, Klaatsch. Photographed from a cast. One-third natural size.

Le Moustier.

In January 1909, a dealer in antiquities, of Swiss nationality, who had only too long exploited, for German profit, the deposits in the Dordogne district, that is to say, the most ancient and the most valuable archives in France, revealed the circumstances under which he had discovered and exhumed a human skeleton at Le Moustier.[1]

The exhumation took place on the 10th August 1908, in the presence of a tribunal of scientists from beyond the Rhine—Klaatsch, H. Virchow, von den Steinen, Hahne, Wüst, and others (and, of course, in the absence of any French scientist). Even so the scientific value of this relic is markedly diminished by the poverty of significant stratigraphical or palæontological data, and especially by the deplorable manner in which it was extricated and restored. The reconstruction of the skull by Klaatsch, a professor of anatomy, is a positive caricature (Fig. 115).

[1] Hauser, O., "Découverte d'un squelette du type de Néanderthal . . ." (*L'Homme préhistorique*, January 1909), followed by Klaatsch, H., "Preuves que l'*Homo mousteriensis Hauseri* appartient au type de Néanderthal."

A second reconstruction, in which several of Klaatsch's distinguished colleagues were called upon to assist, has at least the merit of being more faithful. The monetary value of the skeleton from Le Moustier was, on the other hand, considered beyond compare by the "Museum für Völkerkunde" in Berlin, which paid Hauser, the dealer, the fabulous price of 125,000 francs! According to Klaatsch, the skeleton from Le Moustier is that of a young man of about sixteen years. There seems to be no doubt that it may belong to the Neanderthal type.

La Ferrassie.

At La Ferrassie, near Bugue in Dorgogne, there is a rock shelter containing many superimposed settlements, rich in Palæolithic dressed relics, which were explored for ten years by MM. Capitan and Peyrony. On the 17th September 1909, Peyrony saw some human bones projecting from the soil. At his invitation and that of his collaborator M. Capitan, several persons repaired to La Ferrassie to be present at and collaborate in the extrication of the skeleton—MM. Cartailhac, Breuil, Bouyssonie, and the author. We ascertained: (1) that the stratigraphical level is clearly the same as that at La Chapelle-aux-Saints, at the base of a layer containing Moustierian relics, and resting, according to MM. Capitan and Peyrony, on an Acheulean layer; (2) that the skeleton belonged to an individual of the Neanderthal type; (3) that the bones of this fossil Man had retained their anatomical relationships and that they lay in the very midst of undisturbed layers, without any observable trace of burial.

In 1910, M. Peyrony obtained from the same layer a second skeleton, which lay not far from the first, and indicated a slighter individual of smaller stature, very probably a woman. Finally, in 1912, he collected some portions of the skeletons of three children.[1] It would appear that there are present here the bone-remains of a whole family, killed by accident, perhaps buried beneath a land-slip. This series, which is of outstanding

[1] Capitan, Dr, and Peyrony, "Deux squelettes humains au milieu de foyers de l'époque moustérienne" (*Revue de l'Ecole d'Anthrop.*, December 1909); Station préhistorique de la Ferrassie (*Revue anthropologique*, January 1912).

value, was generously gifted to the French National Museum of Natural History by MM. Capitan and Peyrony. The clearing of the skeletons, their preparation, and the reconstruction of a skull (Fig. 116), almost as complete as that from La Chapelle-aux-Saints, were carried out in my laboratory. A detailed description of them was on the point of publication when the war broke out. Reference will frequently be made to them in the course of this chapter.

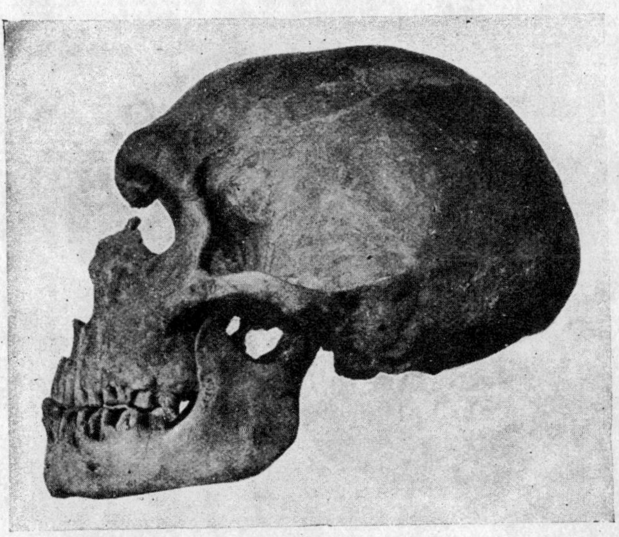

FIG. 116.—Skull of Male Skeleton from La Ferrassie, seen in profile. One-third natural size Palæontological Gallery, French National Museum of Natural History.

MM. Capitan and Peyrony likewise sent me the broken bones of a child's skull, collected by them in the Moustierian bed of the little cave at Pech de l'Azé, near Sarlat in the Dordogne.

La Quina.

It was also a long-continued exploration which led to the fortunate discovery by Dr Henri Martin in 1911, at La Quina in Charente. After several years of slow and most careful excavation, he found a human skeleton in an environment which was clearly Moustierian. The well-preserved portions of the head (Fig. 117) possess the characters of the corresponding portions of the Neanderthal skull-cap and the lower jaw from La Chapelle-aux-Saints.[1]

Sincc this period Dr Martin has discovered in the same bed a fine skull of a child aged about 8 years, and a large number of fragments belonging to 18 other individuals.[2] (See Appendix p. 472.)

In 1911, likewise, in a cave in the bay of Saint-Brelade

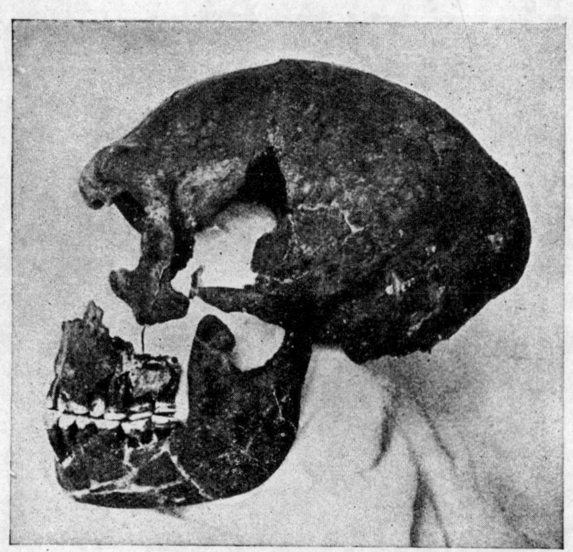

FIG. 117.—Skull from La Quina, seen in profile. One-third natural size. (Photograph by Henri Martin.)

to the south-west of the Island of Jersey,[3] some human molar teeth were collected along with bones of the Woolly Rhinoceros (*R. tichorhinus*) and the Reindeer, and also with Moustierian dressed flints.

Hillebrand has mentioned bones of children taken from the

[1] Martin, H., "Sur un squelette humain de l'époque moustérienne trouvé en Charente" (*C. Acad. Sci.*, 16th Oct. 1911); "Homme fossile de La Quina" (*Arch. Morph. gen. exper.*, 15, 1923).

[2] *L'Anthropologie*, xxxi, 1921, p. 331.

[3] Marett, R., "Pleistocene Man in Jersey" (*Archæologia*, lxii., 1911). Keith, A., and Knowles, F., "A Description of Teeth of Palæolithic Man from Jersey" (*Journal of Anatomy and Physiology*, xlvi., 1911).

Cave of Balla, near Repashuta in Hungary; but their stratigraphical position has not been definitely established.

In 1915, Messrs Pacheco and Obermaier described a lower jaw found, in 1887, in a Pleistocene calcareous tufa at Bañolas in Catalonia. They classified it as of the Neanderthal type.[1]

Summary. Of this score of discoveries, having credentials sufficiently reliable to allow of their being allocated to Mid-Pleistocene times, about half consist only of mere fragments. The remainder have yielded bone-remains in such condition as to render them suitable for thorough morphological study — Neanderthal, Gibraltar, La Naulette, Spy, Malarnaud, Krapina, La Chapelle-aux-Saints, Le Moustier, La Ferrassie, and La Quina. So that to-day our knowledge of the Neanderthal type of human fossil has been gained from well preserved, easily studied specimens, numbering at least some fifteen individuals.

What, then, of the ideas of Pruner-Bey, Gratiolet, Virchow, Hartmann, and others, regarding the exceptional, quite aberrant or pathological nature of the skulls of the Neanderthal type? It is hardly necessary to point out that they are now only of historic interest. The opposition of German scientists, who have since bought the Moustierian skeleton for its weight in gold, was still at its height in 1892, when the "Neanderthal Race" was looked upon as "imaginary," as a "creation of fancy," and was "laid to rest" by Virchow, von Hölder, Haas, Kolmann, and others.[2]

To-day, then, we are in possession of a comprehensive collection of materials, relating to the same homogeneous human type, which differs greatly from all living types, and indeed is better known than many of them. This human type, which exhibits many characters of inferiority, must be known (for reasons we shall consider later) by the name of *Homo neanderthalensis*. I shall give here a brief but as complete a descrip-

[1] Pacheco, E. Hernandez, and Obermaier, H., "La mandibula neandertaloide de Bañolas" (*Comision de Investigaciones paleontologicas y prehistoricas*, No. 6, Madrid, 1915).

[2] The late J. Fraipont published at this time an eloquent paper in vindication entitled, "La race imaginaire de Cannstadt ou de Néanderthal" (*Bull. de la Soc. d'Anthrop. de Bruxelles*, xiv., 1895).

tion as possible of it, taking as a basis the skeleton from La Chapelle-aux-Saints, on which I published a monograph,[1] and the still unpublished descriptions of skeletons from La Ferrassie.

Description. The Skull.

The skull of the man from La Chapelle-aux-Saints looks strange even to the eyes of an observer only slightly acquainted with anatomy (Fig. 118). First its great size strikes us, especially regarded in relation to the small stature of the individual to whom it belonged, and who, as we shall see later, was less than 1.60 metre in height. Next we are impressed by its bestial appearance, or rather by the general effect of its simian characters. The brain-box, elongated in form, is much depressed; the orbital arches are enormous; the forehead is very receding; the occipital region very projecting and much depressed; the face is long and projects forwards; the orbits are enormous; the nose, separated from the forehead by a deep depression, is short and very broad; by a prolongation of the malar bones, the upper jaw forms a kind of muzzle; the lower jaw is strong and thick; the chin rudimentary.

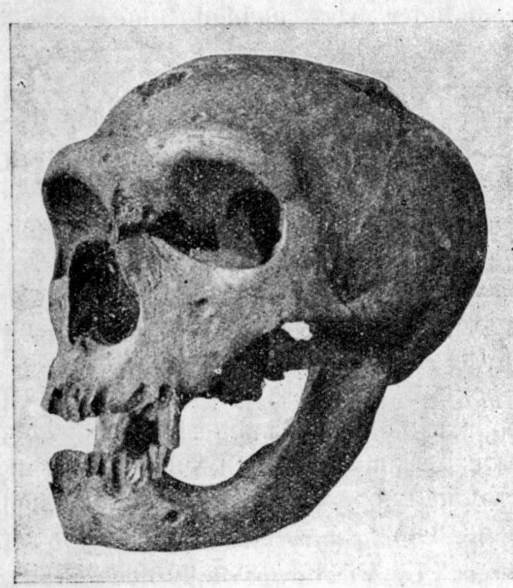

FIG. 118. — The Skull from La Chapelle-aux-Saints. Three-quarters view. About one-third natural size. Palæontological Gallery, French National Museum of Natural History, Paris.

[1] Boule, M., "L'Homme fossile de La Chapelle-aux-Saints" (*Annales de Paléontologie*, 1911-13). Numerous bibliographical references will be found there to works already published on the morphology of *Homo neanderthalensis* or useful for consultation in studying it, but these I cannot repeat in this volume.

General Morphology of the Skull.

Beginning with the general morphology of the skull, we find at first glance that the face is highly developed in comparison with the brain-box, and here, as we have seen (p. 72) lies a zoological character of the highest importance. It may be strikingly depicted in diagrammatic form by superimposing the profiles of the skulls of this fossil, of a Chimpanzee, and of a modern Frenchman upon a common base line closely following the boundary between the facial and the cerebral portions, after

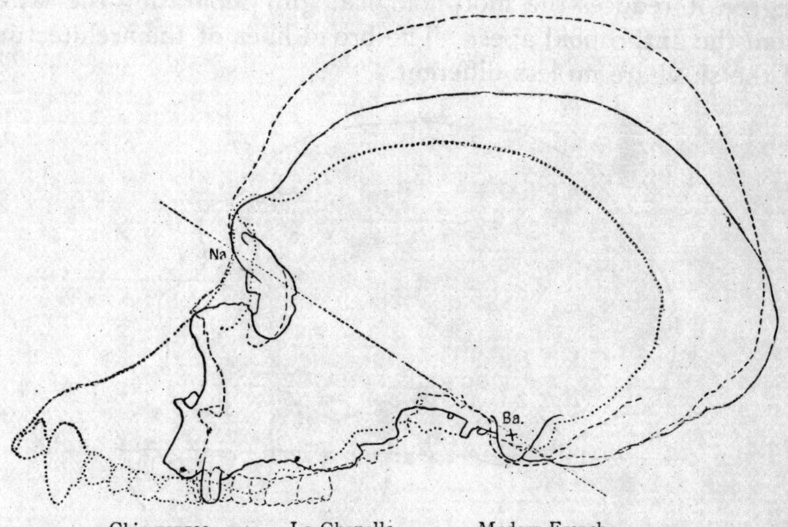

FIG. 119.—Profiles of the Skulls of a Chimpanzee, of the Man from La Chapelle-aux-Saints, and of a Modern Frenchman, superimposed along the basio-nasal line, which has been made the same length in each. Ba, Basion; Na, Nasion.

having, it must be clearly understood, made the three basi-axial lines of the same length, Na-Ba, and reproduced the skulls in due proportion (Fig. 119). In the Chimpanzee, the facial region is almost as great as the cerebral region. In the Frenchman, on the contrary, the face is much reduced, and the brain-box greatly developed. Between these two extremes lies the skull from La Chapelle-aux-Saints. While it is more akin to that of its congener, Modern Man, in regard to the face, in regard to the brain-box it is almost intermediate between Man and the Chimpanzee.

No less striking is Fig. 120, in which are shown, superimposed, the profiles of the skulls of our fossil Man and of the famous American palæontologist, Cope. They are almost equal in cranial capacity, but we see how great is the difference between the size of the face in one of the most intelligent of men and in our savage of Quaternary times.

In the relative development of its facial and cerebral regions, the skull from La Chapelle-aux-Saints differs widely from the skulls of living men even of the lower races, yet to a certain degree it reduces the morphological gulf separating the latter from the anthropoid apes. The broad lines of the architecture of the skull are no less different.

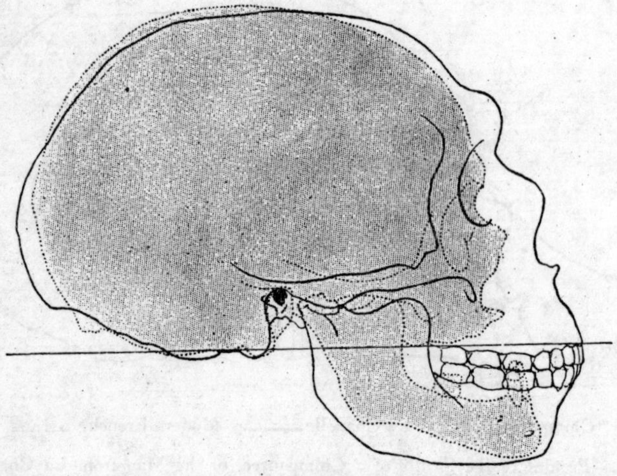

FIG. 120.—Superimposed Profiles of the Skulls of the famous American naturalist, Cope (shaded), and of the Fossil Man from La Chapelle-aux-Saints (unbroken line). One-third natural size.

All mammalian skulls are composed of the same parts, but the relative development and correlated arrangement of these component parts vary according to the nature of the skulls. The essential portion, constituting the keystone of the cranial edifice, is the base of the skull, formed, from behind forwards, by the basi-occipital, the basi-sphenoid, and the presphenoid. Around this *basicranial axis*, as Huxley named it, the principal modifications of the brain-box occur, under the influence of the development of the various parts of the brain. This axis itself is straight in the majority of Mammals, but it becomes bent, curved, and more or less broken, as we consider ascending types of skull from the lower monkeys and anthropoid apes up to Man.[1]

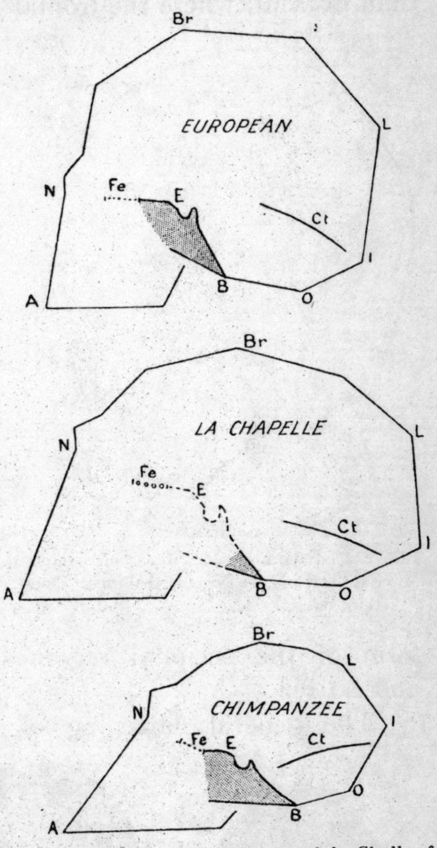

FIG. 121.—Geometrical Diagrams of the Skulls of a Chimpanzee, of the Fossil Man from La Chapelle-aux-Saints, and of a Modern Frenchman. About one-fourth natural size.

A, prosthion or incisive alveolar point; N, nasion (root of nose); Br, bregma (point of contact of the frontal and two parietal bones); L, lambda (point of contact of the two parietals and of the occipital); I, inion (base of the external occipital protuberance); O, opisthion (posterior margin of the foramen magnum); E, ephippion (anterior margin of the *sella turcica*); Fe, ethmoidal fossa; Ct, upper level of cerebellum.

On these facts we have constructed geometrical diagrams of median longitudinal sections of the skulls of a Chimpanzee, of a European and of the Man from La Chapelle-aux-Saints (Fig. 121). They clearly show the intermediate position of the last. The same result follows from a study of the relations of the various *planes* and *lines of orientation* used by anthropologists.

The Regions of the Skull. After this general view, we proceed to the examination of morphological details. Seen from above, (*norma verticalis*) all the skulls of Neanderthal Man appear remarkably uniform (Fig. 122). All are dolichocephalic; their cephalic index,[2] ranging from seventy to seventy-six, corresponds well with the average human index as it occurs in an

[1] Huxley, T., *Evidence as to Man's Place in Nature*, London, 1863. Topinard, P., "La transformation du crâne animal en crâne humain" (*L'Anthropologie*, ii., 1891).

[2] For definition of this term, see footnote, p. 99.

archaic type. The skulls are much more enlarged behind than in front, where the frontal bone recedes greatly above the great swelling ridges of the orbital arches.

FIG. 122.—Superimposed Profiles of different Skull-caps of the Neanderthal type. One-third natural size.

An interesting comparison of the fossil skull with that of a modern Chimpanzee and of a modern Man, photographed in the same position, is shown in Fig. 123.

In the lateral aspect of the skull (*norma lateralis*) the chief characters of the cerebral region are displayed. We can here observe the importance of the face, the overhanging projection of the supra-orbital ridges, the very receding profile of the forehead, the general flattening of the vault, the form of the occipital region like a flattened protuberance, and so on.

The general flattening of the skull (*platycephalia*) is a character of the greatest importance which the whole series of Moustierian skulls possesses in common (Fig. 124), and in which,

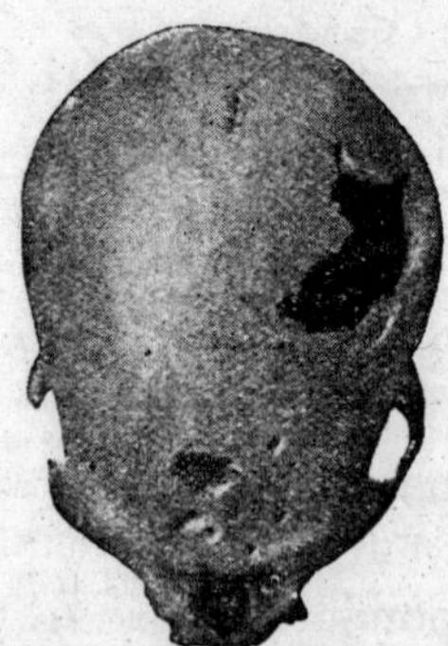

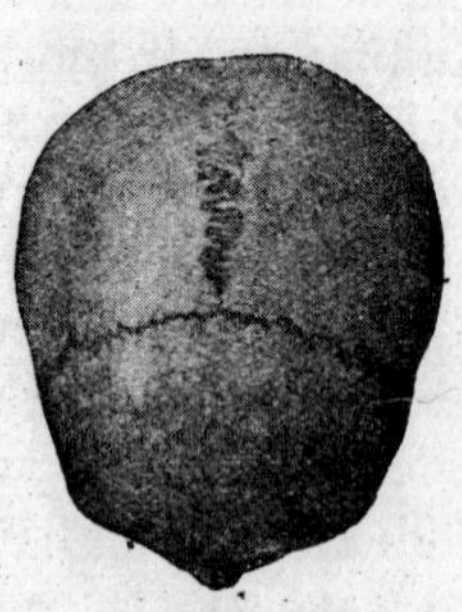

FIG. 123.—Photographs showing a comparison of the upper aspect of the Skulls of a Chimpanzee, of the Fossil Man from La Chapelle-aux-Saints, and of a Modern Frenchman.

at first glance, it seems to differ widely from modern skulls and to resemble those of the anthropoid apes without sagittal crests, such as the Gibbons and the Chimpanzees, even to such a degree as to warrant its inclusion in this group.

In its receding forehead, or, one might almost say, its absence of forehead, the Neanderthal type still occupies the lowest rung of the human ladder and strikingly resembles the anthropoid apes, as is shown by the various methods invented by anthropologists for measuring the development of the forehead.

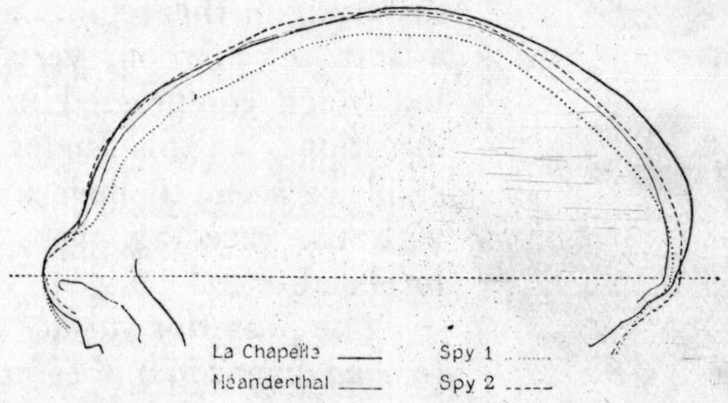

FIG. 124.—Superimposed lateral Profiles of different Skulls of the Neanderthal type, oriented on the glabello-iniac line. One-third natural size.

The parietal bosses are projecting and situated far back. The temporal lines are not more marked than in modern Man.

The line of suture of the temporal and parietal bones is not much arched; it is known that while this line is rather straighter in apes and new-born children, it is more or less arched in the various human races, where, in the case of higher types, it finally forms a well-rounded curve. The zygomatic apophyses are massive, more inclined forward than in modern Man; the post-glenoid apophyses, relics of an ancestral condition, are more developed. The supra-mastoid crest is also very projecting. The mastoid apophyses are extremely small. The resemblance of the petrous portion, with its great extent, its disposition as a large, oblique surface, forming with the neighbouring parietal and occipital surfaces a single plane which inclines inwards and backwards, is much more akin to the same region in the skull of the Chimpanzee than to that of modern human skulls.

The occipital region has a somewhat strange profile. Huxley had already described it: "To the eye of an anatomist, the posterior portion of the Neanderthal skull is even more remarkable than the front portion." The new specimens from La Chapelle and from La Ferrassie, complete and well preserved, are admirably fitted for the study of this region, which forms a sort of chignon, very prominent but much compressed in a vertical direction. In this hinder part of the skull we seem to have a correlation with the receding character of the forehead.

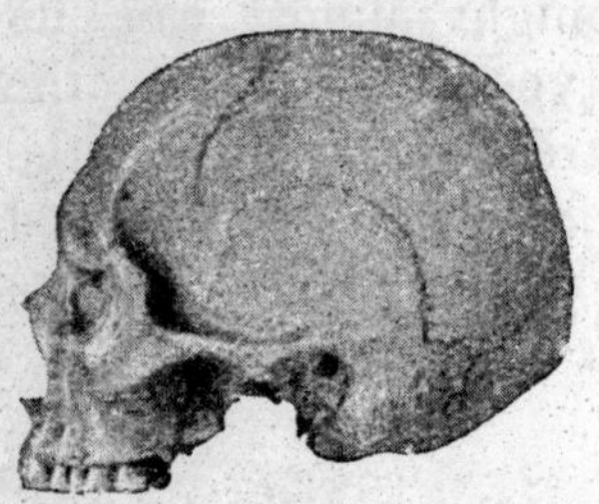

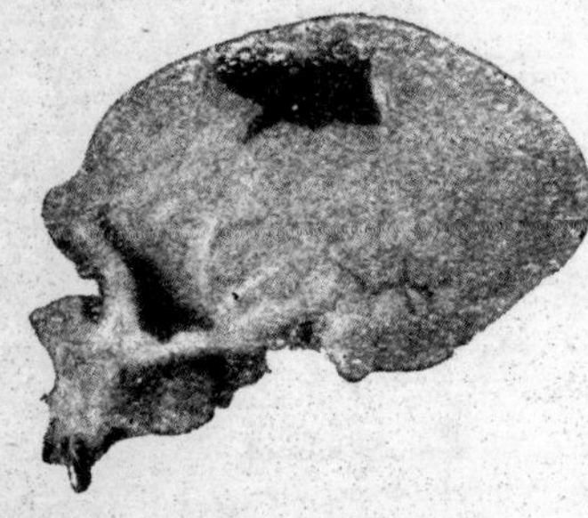

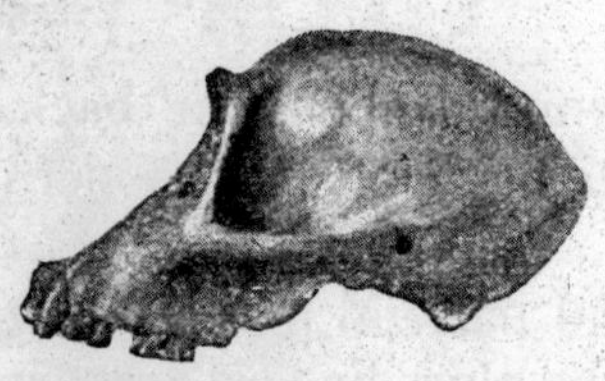

FIG. 125.—Photographs comparing the lateral aspects of the Skulls of a Chimpanzee, of the Fossil Man from La Chapelle-aux-Saints, and of a Modern Frenchman.

The posterior surface of the skull (*norma occipitalis*) does not show the usual pentagonal form, but has an almost circular contour, which is in keeping with the general platycephalia (Fig. 126). A marked transverse ridge, arched in shape, the *torus occipitalis transversus*, divides the upper from the lower part of the skull. This ridge is much more developed in apes, where it connects the temporal and the supramastoidean crests, forming together with them one continuous ridge. This uninterrupted protuberance is present, as we have said, in *Pithecanthropus*, but in lesser degree. In Neanderthal Man, no junction takes place between the occipital protuberance and the temporal crest (Fig. 58, p. 101).

There is no external occipital protuberance. In its place we find, on the transverse ridge, a sort of cup or cupule, which is present on every one of the known examples. The central point of the cup does not correspond with the division between the cerebral hemispheres and the cerebellum; it is situated higher up. The cerebellar protuberances are much reduced.

The whole surface of this region is very uneven; the projections and depressions representing the muscle-imprints

FIG. 126.—Photographs comparing the occipital aspects of the Skulls of a Chimpanzee, of the Fossil Man from La Chapelle-aux-Saints, and of a Frenchman.

are sharply defined, and this denotes an exceptional development of the muscles of the nape of the neck, in keeping with the enormous size of the head.

The lower surface (*norma basilaris*) of the skull of Neanderthal Man reveals a whole series of important characters,

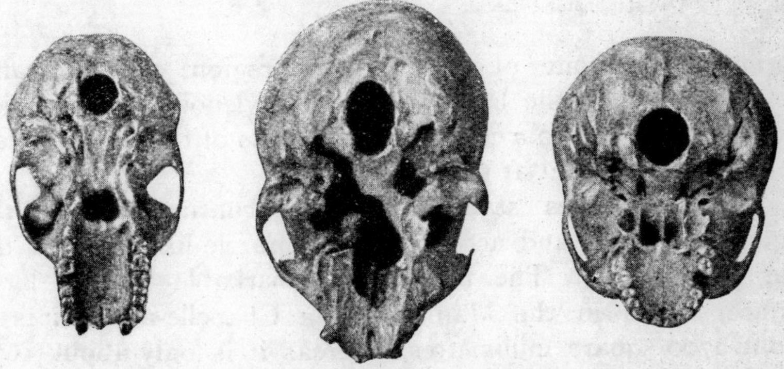

FIG. 127.—Photographs comparing the lower surface of the Skulls of a Chimpanzee, of the Man from La Chapelle-aux-Saints, and of a Modern Frenchman.

although, even in the best examples, its state of preservation leaves much to be desired, because of the fragile nature of the bones of which it is composed (Fig. 127).

The foramen magnum is placed in a relatively backward position. Though still widely removed from the apes in respect of this character, the Man from La Chapelle nevertheless resembles them rather more than does Modern Man. The direction of the plane of the foramen magnum is likewise intermediate.

Seen from below, the temporal bones are remarkable for their general flatness, that is to say, for the slightness of their prominences and hollows. This appearance differs from that of the temporal bones of modern Man, and resembles that found in apes, where, however, it occurs in accentuated degree. I have already pointed out the slimness of the mastoid apophyses, hardly more developed than those of Orangs and

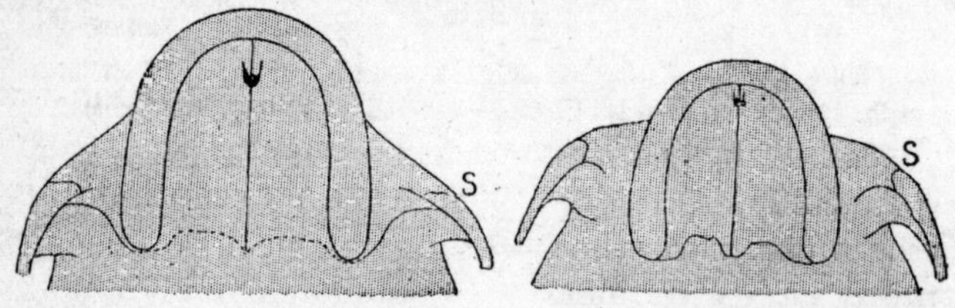

FIG. 128.—Front portions of Skulls from La Chapelle-aux-Saints on the left, and of a Modern Frenchman on the right, seen from below to show the difference in the profiles of the region of the Upper Jaw and Cheeks. One-third natural size.

Gorillas. In front of the mastoid region, the tympanic resembles that of the large apes. The glenoid cavities, wide and shallow, recall to a certain degree those of the Chimpanzee, which are almost flat at the bottom.

The maxillaries stand out as a continuation of the zygomatic arches, and accentuate the muzzle-like form of the face (Fig. 128). The palate is remarkable for its large surface, which in the Man from La Chapelle-aux-Saints is about 2700 square millimetres, whereas it is only about 1670 square millimetres in the skull of a normal Frenchman.

The Face.

There remains to be examined the anterior surface of the skull (*norma facialis*), which corresponds to the face in the living subject (Fig. 129). From the La Chapelle specimen, we obtained our first accurate and complete knowledge of the physiognomy of Neanderthal Man.

In its dimensions, the face exceeds the largest known human faces. If it does not appear very high in relation to

FIG. 129.—Photographs contrasting the Faces of the Skulls of a Chimpanzee, of the Man from La Chapelle-aux-Saints, and of a Modern Frenchman.

its breadth, this is due to the flattening of the brain-box and consequent reduction of the forehead (Fig. 130).

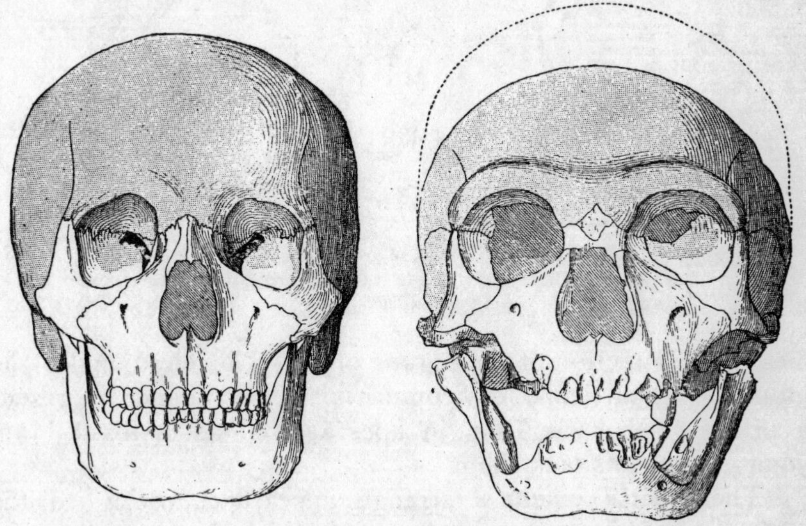

FIG. 130.—Skulls of a Frenchman and of the Man from La Chapelle-aux-Saints, seen full face, in order to show the difference of the relative development of cerebral and facial portions. One-third natural size.

The dotted line indicates the supposed outline of the cranial vault of the fossil Man, had it been of the same proportions as in the French specimen.

An immense development of the orbital arches characterizes all known skulls of Neanderthal Man. They unite in a projecting unbroken ridge, absolutely resembling the bony visor of the skulls of Chimpanzees, of Gorillas, and also of *Pithecanthropus* (see p. 99). Opinion is undecided as to the physiological part played by these protuberances. The fact that they are more developed in males might lead to their being looked upon as a character favouring sexual selection. Turner suggested that perhaps they heightened an appearance of ferocity which was of some value in the struggle for life. In the opinion of many anatomists their development is directly correlated with that of the jaws and the apparatus concerned in mastication. For my part, I regard them as a

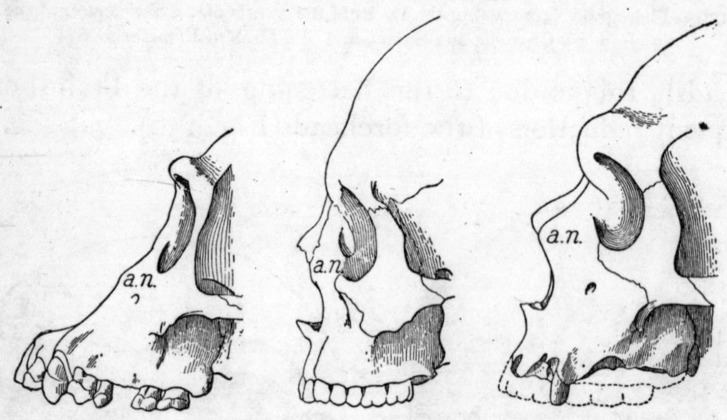

FIG. 131.—Facial profiles of Skulls of a Chimpanzee, of a Frenchman, and of the Man from La Chapelle-aux-Saints. One-third natural size. *a.n.*, nasal apophysis of the maxillary.

means of protection of the organs of sight, acquired under the influence of particular environments. They do not represent a primitive character, for, in apes as well as in fossil Man, young specimens lack them.

The orbits present a strange appearance owing to the projection of the superciliary ridges, which, by extending the roof of the orbits, form over them a kind of awning. They are widely separated from each other, and are rounded in form. Their size is relatively very great, half as great again as in a modern Man of the same brain capacity.

The base of the nose (nasion), marked by the point of junction with the fronto-nasal sutures, is situated in a deep depression, as in native Australians. This arrangement is the opposite of that found in apes, even in the apes with strong orbital ridges. The nose was very large (Platyrrhinian), as in the majority of modern men of the black races. The nasal bones, preserved in the Gibraltar skull, are essentially human and fundamentally different from those of monkeys. In other characters also, such as the development of the nasal apophysis of the maxillary and the area of the nostrils, the nose of Neanderthal Man, far from resembling that of anthropoid apes, differs from it much more than does that of living Man. This fossil man, which in so many of his characters approaches the apes more than any other man, is nevertheless so widely divergent from them as regards his nasal region, that instead of being simian in this respect, he might rather be looked upon as *ultra-human* (Fig. 131).

The arrangement of the malar or jugular bones indicates flat or receding cheeks.

The maxillaries are strong and massive, and project prominently forwards. Their anterior surface, instead of being concave and exhibiting *canine fossæ*, as in the case of all modern men without exception, is almost flat. This flat surface exactly continues the plane of the external surface of the malar bones, and so gives rise to the muzzle-like appearance which I have already described (Fig. 128), and which is one of the most characteristic traits of our Moustierian Man. It has the effect of adding greatly to the bestial appearance of the face of this man. I do not think, however, that we can regard these structures as simian, for, although the Chimpanzee has no canine fossæ, they are very deep in the Gorilla and Orang.

The sub-nasal or incisive area is very large. It continues the general direction of the profile of the face without projecting forwards. There is no alveolar prognathism.[1]

The Lower Jaw.

At the present day the lower jaw of Neanderthal Man is known from a large number of fragments which have all the appearance of a family resemblance. These jaws are massive and very strong, and their size corresponds to that of the skulls.

[1] For an illustrated account of the recently discovered child's skull from La Quina, see Appendix, p. 472.

They have no chin, or only a rudimentary and receding one, a fact first observed in the jaw from La Naulette, the discovery of which caused a great sensation. Regarding it, Broca expressed the following opinion: "I have no hesitation in saying that the jaw from La Naulette is the first evidence to provide the Darwinists with an anatomical argument. It is the first link in the chain which, according to them, ought to lead from Man to monkeys."[1]

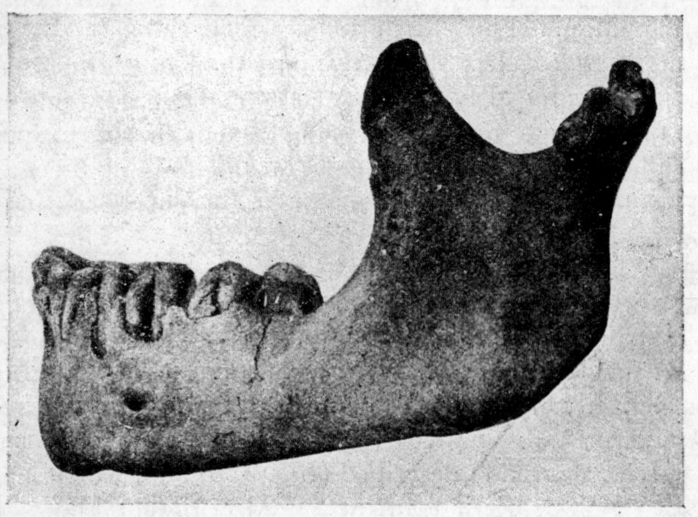

FIG. 132.—Lower Jaw, seen in profile, of the Man from La Ferrassie. Three-fourths natural size. Palæontological Gallery of the French National Museum of Natural History.

Assuredly we are here dealing with a most remarkable character, of the significance of which, from the point of view of its morphological rank in the Primate group, there does not seem to be any question. Anatomists regard the projection of the chin as an essentially human characteristic. Now, as regards the retreating character of chin, the jaws of Neanderthal Man are exactly interposed between the jaws of anthropoid apes and those of groups of modern human beings; and this is so even if we chose for comparison the races lowest in type in this respect. One of the most receding, if not the most

[1] In Topinard, P., "Les caractères simiens de la mâchoire de La Naulette," *loc. cit.*

receding chin in the Anthropological Gallery of the French National Museum of Natural History, is that of the famous Hottentot Venus. Yet in it the symphyseal angle (the angle formed by the anterior and inferior margins of the jaw) does not exceed 94°, whereas in the Man from La Chapelle-aux-Saints it reaches 104°. It must be added that certain fossil jaws exhibit a chin in process of formation, such as that of the skeleton from La Ferrassie, in which there is really an indication of a chin triangle (Fig. 132). While the Heidelberg Man (Mauer jaw) from the Lower Pleistocene, possessed a jaw just as much lacking in chin as that of the apes, a study of various fossil jaws of Mid Pleistocene origin from various districts in France shows the gradual formation of a chin, the special attribute of *Homo sapiens.*[1] This is shown in the series of diagrams in Fig. 133.

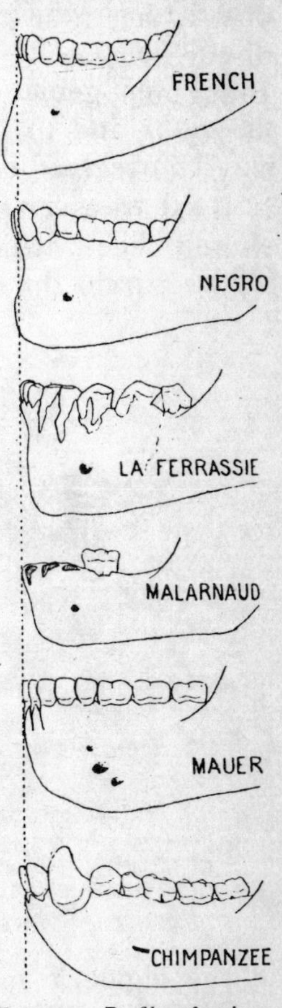

FIG. 133.—Profiles of various types of Lower Jaws showing a gradual series from chinless forms (Chimpanzee and Mauer Jaw) to forms with very pronounced chin (Frenchman), passing through intermediate forms (Malarnaud and La Ferrassie). Two-fifths natural size.

The inner or buccal surface of the symphyseal region, or area of the chin, inclines in general direction from front to back; that is, the hinder as well as the anterior surface of the chin recedes, so that when we place a lower jaw of Neanderthal Man on a table and look at it from above, we see, not the anterior surface of the body of the bone, but the posterior surface, whereas in the case of the jaw of Modern Man we see exactly the reverse.

The morphology of this inner surface shows certain interesting peculiarities

[1] Important researches have been made of late years with regard to the development and significance of the chin, by the anatomists Mies, Walkhoff, Toldt, and others. Bibliographical references will be found in my monograph, *L'Homme de La Chapelle-aux-Saints.*

(Fig. 134). We find here, but rather less developed, the genial fossa which we observed in the Mauer jaw, and which is still deeper in apes. But there are well-developed upper and lower genial apophyses, just as in modern Men. To the right and to the left of the lower genial apophysis, there may be observed two rounded, elliptical or depressed tubercles, situated transversely, and united to the mylo-hyoid or oblique internal lines, which are very marked. These protuberances (*b.t.*) separate the sublingual fossa (*f.l.*) from the impressions

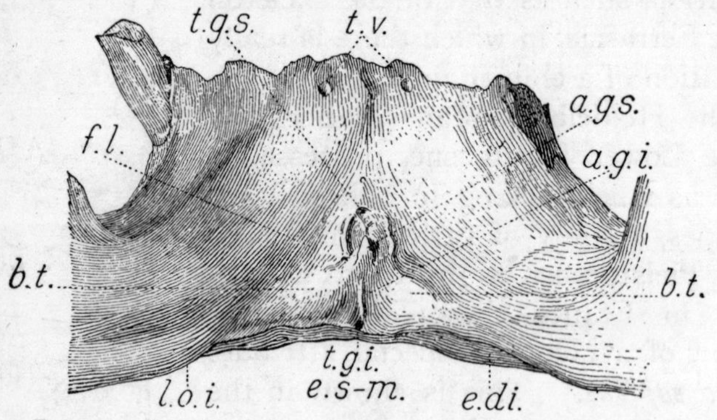

FIG. 134.—Posterior surface of the body of the Jaw of the Man from La Chapelle-aux-Saints. Natural size.

t.v., openings for blood vessels; *t.g.s.*, upper genial foramen; *a.g.s.*, upper genial apophysis; *a.g.i.*, inferior genial apophysis; *t.g.i.*, lower genial foramen; *l.o.i.*, internal oblique line; *b.t.*, transverse prominence; *f.l.*, sublingual fossa; *e.di.*, impression of digastric muscle; *e.s.m.*, submental notch.

of the digastric muscles (*e.di.*), and may be taken as simian characters. But, on the whole, this arrangement of the genial area is quite human, although a little different from the usual arrangement.

The development of the genial apophyses appears to have no connection with the development of articulate language, contrary to what has been believed and taught only too long. All that can be said of Neanderthal Man in this respect is that he had marked projections for the insertions of the genio-glossal and genio-hyoid muscles, whose functions are specially related to the movements of mastication and of swallowing, which are as necessary in animals as in Man.

The lower part of the body of the jaw in the symphyseal region is arranged in two facets, bearing impressions of the insertion of the digastric muscles. Thus we find here the arrangement already noted in the Mauer jaw, but in this case the impressions are shorter and differ less from those on the jaw of modern Man, where a true edge is developed, and where the marks of the attachment of the digastric muscles, smaller in size, are situated on the inner surface of the bone. So there exists a series of intermediate morphological stages which are brought out clearly in Fig. 135.

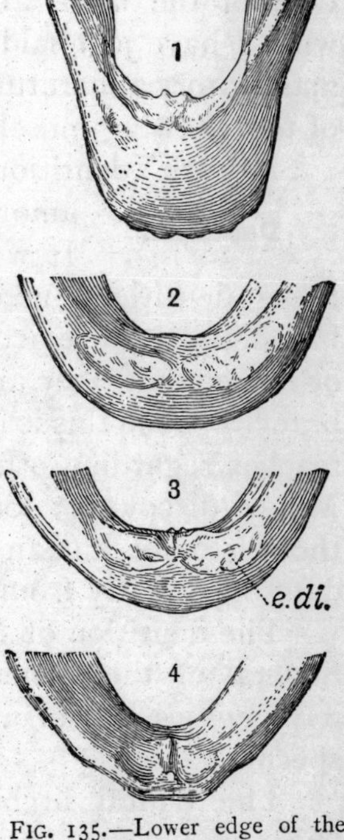

FIG. 135.—Lower edge of the Jaw and impressions of the Digastric Muscles, in various Jaws. About one-half natural size.

1. Chimpanzee; 2. Mauer; 3. La Chapelle; 4. French. *e.di.*, impressions of digastric muscles.

The vertical branches are very wide, somewhat narrower, indeed, than in the Mauer Man, but wider than in modern Man. On the inner surface the rugosities for the insertion of the interior pterygoid muscle are very marked. The angle of the jaw, that is to say, the junction of the lower margin of the horizontal branch and of the posterior margin of the vertical branch, is as if it had been truncated and replaced by an oblique line corresponding to the whole area of attachment of the internal pterygoid muscle. This arrangement occurs in the majority of the anthropoid apes, particularly in the Gorilla.

The region of the angle is very thin, and at the same time deviates greatly from the vertical. But the deviation takes place inside the angle, somewhat as in the case of Marsupials, instead of outside the angle, as in the majority of human jaws; and this inversion is another indication of the great power of the pterygoid muscles. It really seems that the muscles of mastication, intended for the lateral movements of the jaw, were relatively more developed than the elevator or biting muscles (the masseter and temporalis muscles).

The coronoid apophyses are low, broad, and blunt; the sigmoid notch, shallow. The proportions of the condyles are really simian; their structure merely forms the counterpart to that of the large and shallow glenoid fossæ: they confirm what I have just said with regard to the strength of the whole masticatory apparatus and of the relatively greater development of the chewing muscles than of the biting muscles.

Dentition.

Human dentitions preserve a much greater similarity amongst themselves than do human skulls. It would seem that the general characters of these dentitions must be very ancient, and that the human branch must have acquired them, so to speak, from its origin; or rather that they are associated with the very origin of this branch. And this is in accord, not only with what palæontology teaches regarding other groups of mammals, but also with the Mauer discovery; for it revealed a dental morphology which, though quite human in character, was associated with a jaw differing greatly from other known human jaws.

The dentition of Neanderthal Man is also definitely human. Several of the somewhat special secondary characters which it possesses are of a primitive nature, but they can have no specific value.

The dental arches are very wide. The upper curve (Fig. 136) corresponds almost exactly to the architectural form known as a semicircular or Romanesque arch. On the arched portion are placed the incisors, canines, and premolars; the relatively short pillars of the arch are formed by the series of the three molars. This is rather different in plan from the parabolic, hyperbolic, or elliptical forms found in the majority of human jaws; it is, indeed, rather the U-form of the anthropoid apes, but in this case the vertical branches of the U are shorter and diverge slightly, thus showing an approach towards a parabolic form.

The form of the lower arch (Fig. 137) is more akin to that of the upper arch than is usual in modern Man. It follows that the two arches stand to each other in a somewhat different

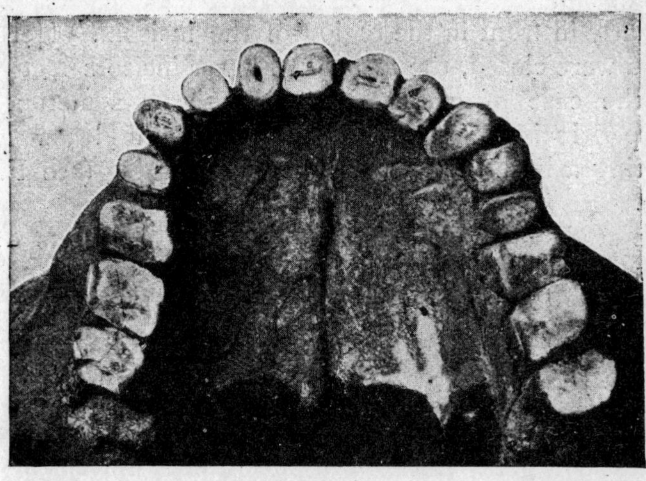

FIG. 136.—Palate and upper Dentition of the Skull of a Man from La Ferrassie. Natural size. Palæontological Gallery, French National Museum of Natural History.

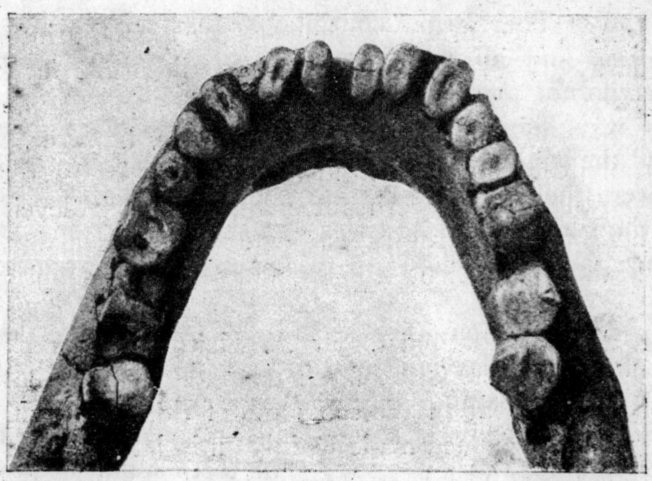

FIG. 137.—Lower Jaw and Dentition, seen from above, of the Skull of a Man from La Ferrassie. Natural size. Palæontological Gallery, French National Museum of Natural History.

relationship from what is the rule in modern Man, especially in civilized races, where the upper teeth overlap the lower, particularly in front in the region of the incisors. Here, when the jaws were closed, the upper and lower incisors met exactly, as in the case of modern Australians; or sometimes even the upper incisors were slightly behind the lower incisors, suggesting a little the condition in the bull-dog (see Fig. 116, p. 191).

The study of the worn surfaces of complete sets of teeth, such as that of the skull from La Ferrassie (Figs. 136, 137), shows that mastication must have been performed partly by a forward movement of the lower jaw against the upper jaw—a movement which furthered the great surface development and slight depth of the glenoid cavities, as well as the barely projecting form of the temporal condyles. This structure indicates also great freedom in the lateral movements of the jaws, an indication confirmed by the strength of the pterygoid muscles; all of which, in short, points to a dentition more employed for chewing than for biting, and implies a vegetarian rather than a carnivorous diet.

The teeth are strong, like the jaws which bear them (Fig. 138). They form a continuous series without intervals or diastemas, and all the crowns are at the same level. The canines do not project beyond the neighbouring teeth, and are in no way simian. The alveolar border, however, extends beyond the last molar in both upper and lower jaws; so that not only were the wisdom teeth not cramped in their development, as in the jaws of civilized races, but there even remained space beyond them sufficient to accommodate a supplementary molar.

An English anatomist, Sir A. Keith, regards as a distinguishing characteristic of the teeth of Neanderthal Man a special development of the roots and of the pulp cavity. His observations have been made mainly on isolated teeth found in a cave in Jersey (see p. 192). The teeth found in position in the skull from La Ferrassie do not exhibit the character recorded by Sir A. Keith, a point on which I satisfied myself by means of X-ray photographs.

The teeth of Neanderthal Man became very rapidly worn, because of the coarseness of the food, mingled with earth, which they had to grind; so that study of the crowns can only be carried out in specimens of young individuals. Some specimens from Krapina are particularly helpful in this respect (Fig. 138).

No important characteristic distinguishes the incisors, the canines, or the premolars from the corresponding teeth in modern Man. The true molars were almost equally developed. They present the primitive human form so well described by Albert Gaudry (see p. 155). The upper molars had all four cusps, well developed. All the lower molars had five cusps, as in the jaw of Heidelberg Man and in the jaws of many modern savage races.

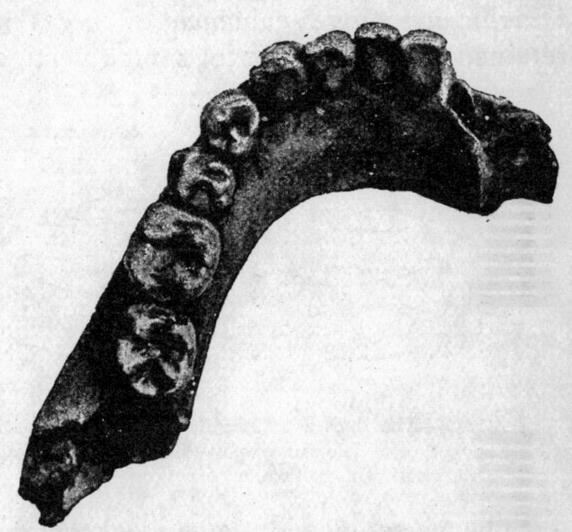

FIG. 138.—Portion of a Lower Jaw from Krapina, remarkable for the size and good preservation of the teeth. Natural size. From a cast by Gorjanovic-Kramberger.

To sum up, the dentition of Neanderthal Man does not differ in any important character from that of the men of to-day; but, as was to be expected, the points of resemblance are with savage, and not with civilized races. This similarity in character extends even to their pathology. The jaws of the Man from La Chapelle-aux-Saints bear numerous signs of diseases, which must have caused serious suffering to their owner, due to inflammation of the gums (simple pyorrhœa) or to pathological growths on the socket-rims of the teeth (general pyorrhœa). But in not a single specimen have I observed a trace of caries.

The Trunk and Limbs.

Until recent years, the study of other parts of the skeleton was somewhat neglected. The skeletons from La Chapelle-aux-Saints and La Ferrassie, which were almost complete and well-preserved, have enabled the study to be taken up afresh and completed.

In general these skeletons, composed of relatively short thick bones, with large articular heads and powerful muscle attachments, give evidence of great strength. They have retained a large number of simian vestiges.

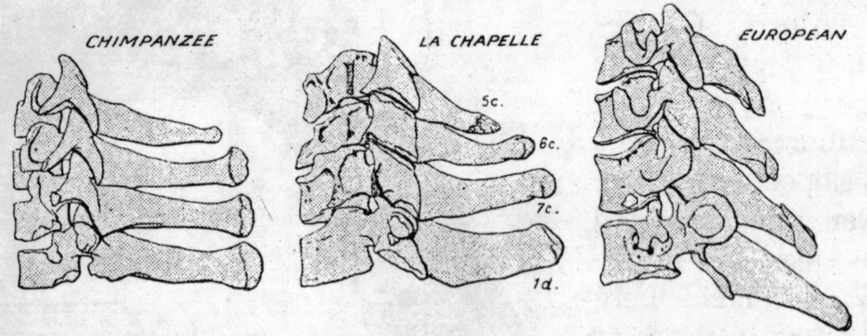

FIG. 139.—The three last Cervical Vertebræ, and the first Dorsal Vertebra, seen in profile, of a Chimpanzee, of the fossil Man from La Chapelle-aux-Saints, and of a European. One-half natural size.

Vertebral Column.

The vertebral column was short and massive. The first vertebræ are much more like those of a Chimpanzee than those of a Man (Fig. 139). Their spinous processes are long, lying at right angles to the axis of the vertebral column, instead of inclining backwards, and they are not bifurcated at the end. The articular apophyses are remarkable for the slight obliquity of their facets. These peculiarities seem to indicate in the cervical region of the vertebral column either a complete absence of curves, or a slight curve, in a direction opposite to that in modern Man, prolonging the dorsal curve, as occurs in anthropoid apes, such as the Chimpanzee. If the reader bears in mind what has been said regarding the differences of the curves of the vertebral column in the Primate series (see p. 75), he will appreciate the interest of these observations.

The vertebræ of the other regions are not so well preserved. It would seem as if the lumbar curve were less pronounced than in the majority of modern men. The portions of the sacrum which we possess point to narrow sacral vertebræ, slightly arched, and deeply set between the iliac bones. These are simian characters.

The ribs are extraordinarily strong, denoting a large thorax with very powerful intercostal muscles.

Girdles and Limbs.

The clavicles are long, slender, and much arched, like those of the Chimpanzee. The shoulder-blades (scapulæ) show, at their anterior or axillary margin, a special formation which I have not found in the skeleton of any modern Man.[1]

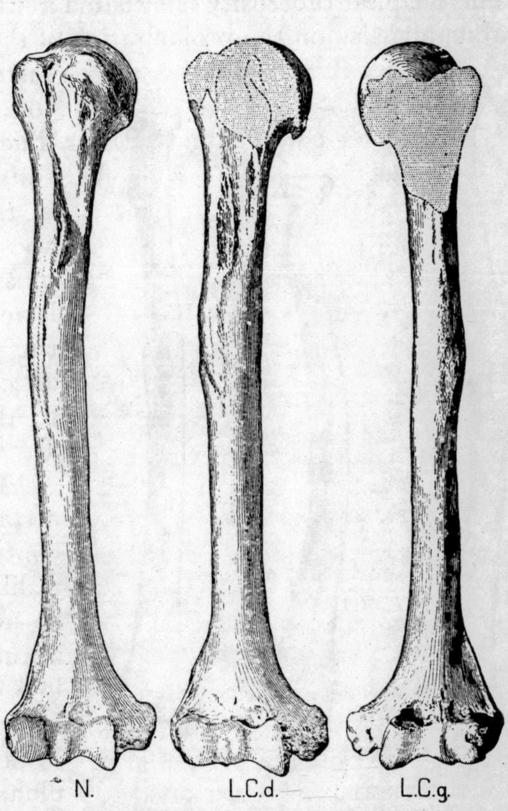

FIG. 140.—Humerus seen full face. N., from Neanderthal; L.C.d., L.C.g., right and left Humeri from La Chapelle-aux-Saints. One-third natural size.

The upper-arm bones (humeri), which are short, strong, and possess large articular heads, do not differ in any essential character from the humeri of modern Man. The right humerus is always a little stronger than the left: Moustierian Man was already right-handed (Fig. 140). Such observations are not surprising, as the forelimb provides one of the chief characteristics of the genus *Homo*, acquired at an early stage in the evolution of our most distant ancestors.

[1] See my monograph, p. 121.

The bones of the forearm are equally strong. All the known specimens of radius, instead of being almost straight as in modern Man, show a very pronounced curve. Hence the interosseous spaces for the muscles of the forearm, were greater. The bicipital tuberosity is situated a little lower than in modern Man, and lies on the prolongation of the inner crest of the bone, instead of being diverted to the anterior surface. These are altogether simian traits, which are reproduced in the radius of anthropoid apes (Fig. 141). But we must hasten to add that the lower end of the bone and the surface of the carpal articulations, corresponding to the region of the wrist, are very human.

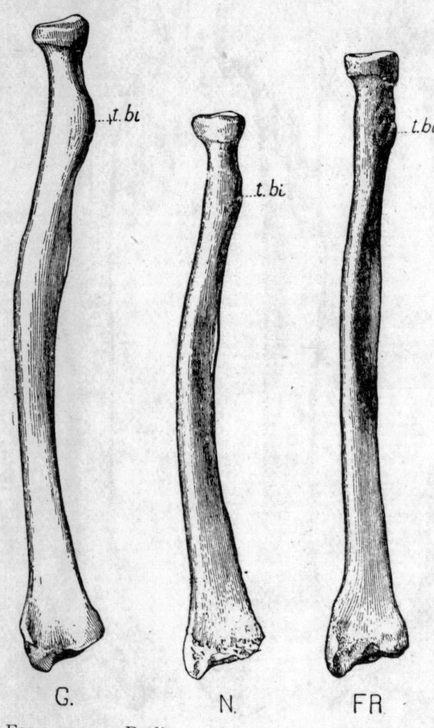

FIG. 141. — Radius, seen from inner surface. G., Gorilla ; N., Neanderthal Man ; Fr., Frenchman. One-third natural size. *t.bi.*, Bicipital tuberosity.

The ulna also shows certain peculiarities. But contrary to the case of the radius, it does not exhibit so much resemblance to the anthropoid apes as to the lower apes ; that is to say, I consider that Neanderthal Man has retained in his ulnar bones, as in the majority of the other bones of his skeleton, some relics of a primitive state far distant from the point at which the various branches of the larger Primates split off.

When extended, the upper arm and forearm lay almost in line with each other, whilst in our case the forearm forms an angle with the upper arm scarcely ever exceeding 170°. On the other hand, the strong development of the olecranon process of the ulna must have impeded the full extension of the forearm, as in the lower monkeys, whereas in the anthropoid species, complete extension is as easy as possible.

The hand is already very human in character. The carpal, however, is relatively small, as in the great apes. The metacarpals are large and thick, that of the thumb being relatively shorter than is the rule in modern Man. The articulations of the metacarpals permitted of easy movement. All the fingers were comparatively short (Fig. 142).

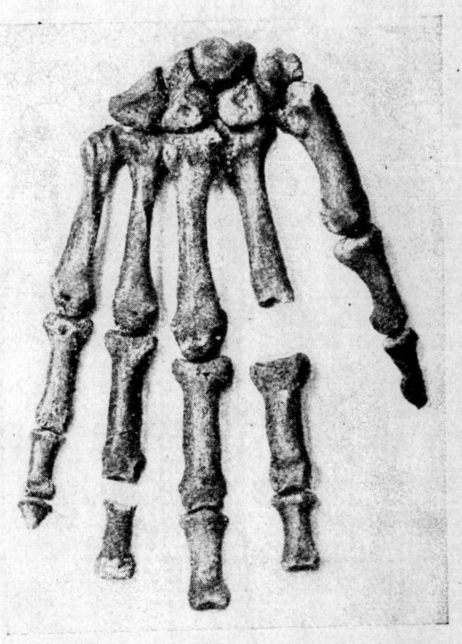

FIG. 142.—Right Hand of the Skeleton of a Woman from La Ferrassie. Half natural size. Palæontological Gallery, French National Museum of Natural History.

We already know that the pelvis of Man differs from that of all other Primates, because of its arrangement in correlation with the upright position and erect gait (see pp. 76, 77). We possess only portions of the pelvis of Neanderthal Man, but these portions are sufficient to allow a complete reconstruction to be made and to enable us to ascertain the persistence of certain vestiges of simian structure—great length in relation to breadth, flattened haunch (iliac) bones, much reduced sciatic spine, more pronounced ischial tuberosity, etc.

A priori, the osteology of the lower limbs must of necessity be interesting, since it is so closely correlated with the more or less upright carriage of the body.

The femora have massive shafts and large heads, and are much bent (Fig. 143). They thus resemble the femora of Gorillas and Chimpanzees, whereas modern Man has almost straight femora, like the Orangs and Gibbons. In the cylindrical form of their shafts they differ from the femora of anthropoid types and resemble those of Macaques, Dog-faced Monkeys, etc. There is often a *third trochanter* and *sub-*

trochanteric fossa, osteological peculiarities related, it is said, to the development of the gluteal muscles and to muscular

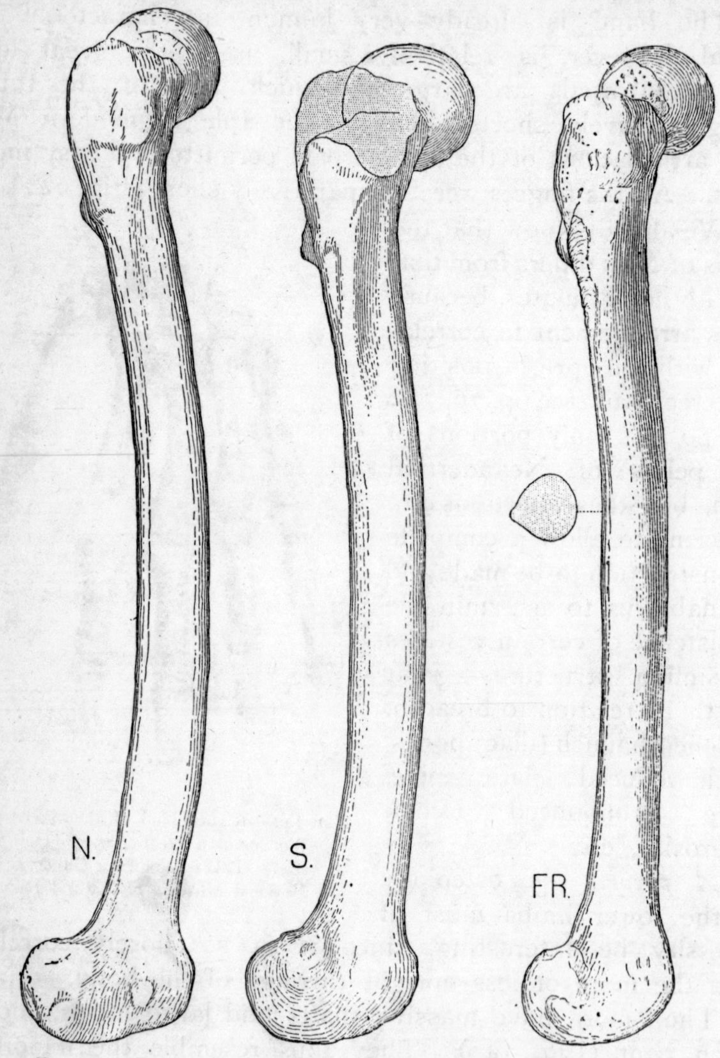

FIG. 143.—Femora seen in profile, exterior surface. N., of Neanderthal Man; S., of Spy Man; FR., of a Frenchman. One-third natural size.

activity adapted to movements in mountainous country. Certain frictional surfaces seem to indicate that the owners of these femora habitually maintained a bent posture.

The tibiæ were short but very strong (Fig. 144). The upper head is bent back, that is to say, curved towards the rear, which causes the plane of articulation with the femur to slope downwards and backwards. In this character, first

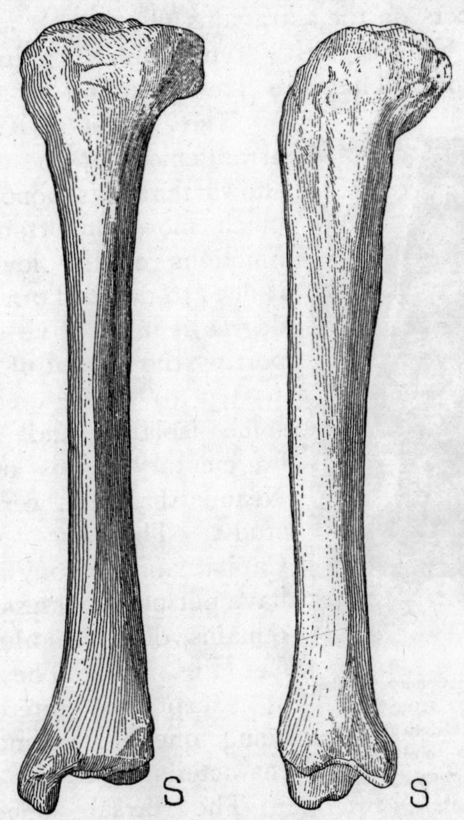

FIG. 144.—Tibia of Spy Skeleton, seen full face and in profile, external surface. One-third natural size.

described by Dr Collignon and further observed by Fraipont and Manouvrier,[1] the tibia of Neanderthal Man resembles that of the apes; only rarely is it present, even in much less pronounced form, in Man to-day. Presently we shall discuss

[1] Collignon, Dr, "Description des ossements fossiles humains trouvés à Bollviller" (*Revue d'Anthrop.*, 1880). Fraipont and Lohest in their memoir on Spy, *loc. cit.* Manouvrier, L., "Etude sur la retroversion de la tête du tibia" (*Mém. de la Soc. d'Anthrop.*, 2nd series, vol. iv., 1893).

the interpretation of this fact. At the same time, the general slope of the articular plane, from without inwards, is much more pronounced, as in the Gorilla. The lower head of the tibia bears one or two supplementary facets corresponding with analogous facets on the astragalus (Fig. 145), such as may be met with in certain native types which habitually maintain a bent position. They are likewise found in many apes.

FIG. 145.—Lower extremity of the Tibia and the Astragalus of a Woman's Skeleton from La Ferrassie. Two-thirds natural size. *fa.*, supplementary facets.

The fibula is strong. The arrangement of the articular facets shows that this bone played then a much more important part in the functions of the lower limb than it does to-day. To a much greater degree it must have shared in supporting the weight of the body.

Up to a few years ago, only some isolated and more or less fragmentary bones of the foot of Neanderthal Man were available for study. The discoveries of MM. Capitan and Peyrony at La Ferrassie have put us in possession of skeletal remains of three almost complete feet (Fig. 146). These remains are of exceptional interest, the foot being one of the most distinctive characteristics of the genus *Homo.*

The tarsal bones are particularly instructive, because their characters and variations depend on physiological conditions relatively easy to determine.

The astragalus or ankle-bone is short, high, and broad. The head is much bent, denoting that the great toe was widely separated from its neighbours. The articular surface for the scaphoid points to a much depressed instep. The malleolar facets for the tibia and fibula show a development comparable to that observed in apes (Fig. 147). In its extent, the facet for the fibula recalls that of anthropoid types.

These arrangements and some others, into the details of which I need not enter here, teach us that the foot must have rested chiefly on its outside edge; and we understand how, in order thus to support the weight of the body, the fibula required to be of stronger structure. To sum up, the astragalus of our fossil Man is the astragalus of a walking

FIG. 146.—Photograph of Right Foot of the Female Skeleton from La Ferrassie, still partially in its matrix. Half natural size.

mammal, which, however, has retained many relics of a former climbing state. The example from the small skeleton of La Ferrassie is very remarkable for the prominence of these vestigeal simian characters, now effaced in the white races, but found in a transitory form in the newly born, a fact which confirms their phylogenetic significance.

The heel-bone (calcaneum), no less robust, likewise recalls

in its structure the heel-bone of the newly born European child. It is remarkable for the extraordinary development of

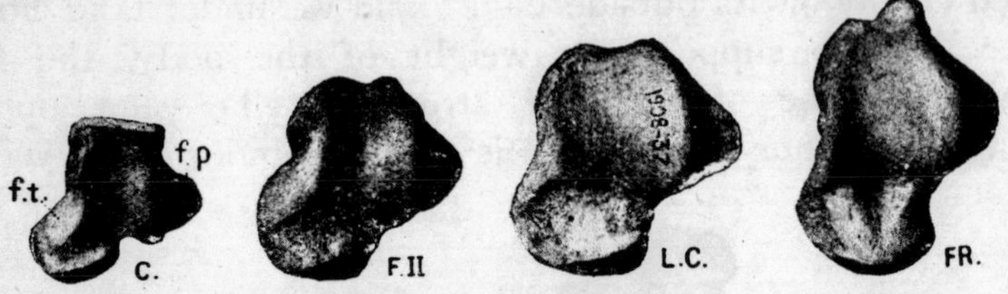

FIG. 147.—Photographs of the Astragalus or Ankle-bone seen from the upper surface. Half natural size.

C., of a Chimpanzee; F.II., of the Female Skeleton from La Ferrassie; L.C., of the Man from La Chapelle-aux-Saints; FR., of a Modern Frenchman; *f.t.*, tibial facet; *f.p.*, fibular facet.

the small apophysis (Fig. 148), a development comparable to that present in the great apes, especially the Chimpanzee and

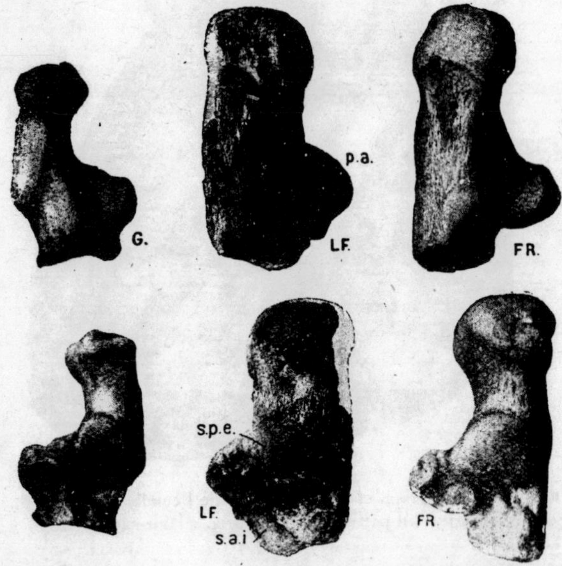

FIG. 148.—Photographs of the Heel-bone (Calcaneum) seen from below (upper row) and seen from above (lower row). Half natural size.

G., Gorilla; LF., Man from La Ferrassie; FR., Frenchman; *p.a.*, small apophysis; *s.a.i.*, anterior inner surface; *s.p.e.*, posterior outer surface.

Gorilla. This kind of support, this *sustentaculum tali*, must have borne a large proportion of the weight of the body by way of the astragalus and tibia.

Another peculiarity of the heel-bone relates to what is termed its torsion. In apes, whose feet rest on the ground, mainly on the outer edge, the axis of the posterior surface of the heel-bone is much inclined from within outwards. In men of the civilized races, whose feet rest equally upon the whole sole surface, this axis oscillates on an almost vertical plane, the torsion being rather from without inwards. In our fossil,

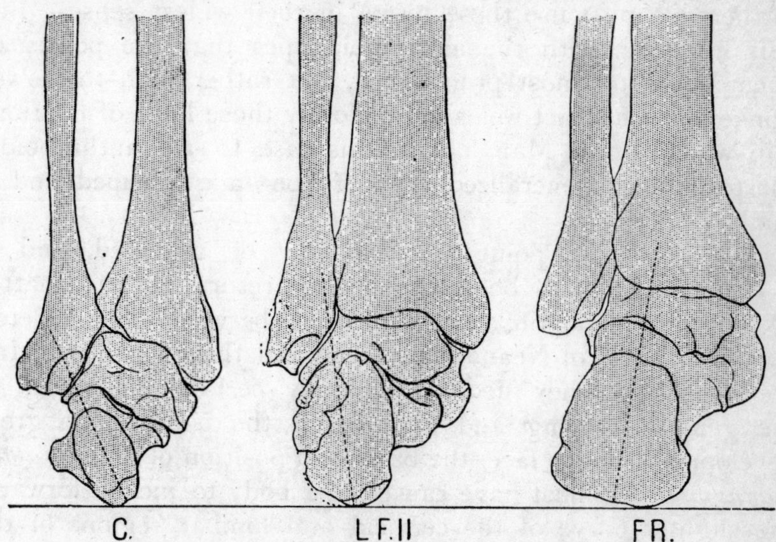

FIG. 149.—Hind view of a part of the Leg and Foot; C., of a Chimpanzee; LF. II, of the Woman's Skeleton from La Ferassie; FR., of a Frenchman. One-half natural size. The dotted lines represent the axis of the hind face of the heel-bone.

then, as also in modern Veddahs, the torsion is intermediate (Fig. 149).

The other bones of the tarsus and metatarsus do not exhibit any important peculiarity. But when their various components are brought together and articulated, we find a considerable separation of the first toe, a separation which the deviation in the neck of the astragalus had led us to expect. This arrangement has been described in the Negritos and Veddahs, whose feet, having become prehensile, readily lend themselves to the action of climbing. The forms of the articular facets of the majority of the bones also point to greater mobility of the other toes.

The lower limb of Neanderthal Man was, then, not altogether like that of modern Man. Yet it differed much less in possessing new characters than in presenting the conjunction of certain morphological characters which, indeed, were already known, but were scattered amongst several living races still leading a life of savagery. The majority of these characters could be readily described as simian or pithecoid, but here we must use these terms in their widest sense. For it is not even with the anthropoid apes that the points of resemblance are most pronounced, but rather with the lower monkeys. This fact welds more closely those links of relationship which unite Man, not in this case to the anthropoids, but to a more generalized type of ape—a quadruped and a climber.

Attitude and Proportions of the Body.

However this point of view may be regarded, and it will be necessary to return to the matter again, the differences between the skeleton of Neanderthal Man and that of modern Man are such that they necessarily imply certain differences in the general bearing and attitude of the body. The great development of the face, the backward position of the *foramen magnum* which must have caused the body to incline forward, the slighter curve of the cervical and lumbar regions of the vertebral column, and the distinctly simian arrangement of the spinous processes of the cervical vertebræ, all testify to this fact. With regard to the lower limb, it is clear that if the formation of the pelvis and the great development of the gluteal muscles indicate that a biped attitude had already been attained, the anatomical characters of femur and tibia, seen in profile in the upright position, show that the leg and thigh, when extended, could not have been in a precisely straight line with each other; that the femur must have sloped downwards and forwards, and that the tibia, sloping in a contrary direction, must have formed behind a wide angle with the femur. So that, without being mechanically impossible, the total extension of the knee could not have been normal, and the habitual attitude must have been one of semi-flexion (Fig. 150).

The fibula, stronger in character, had a most important part to play as a support. The general appearance of the articulations of the foot indicates a greater degree of mobility and freedom. The foot, still only slightly arched, must have rested on the ground on its outer edge, and must have assumed naturally an in-toed position; the wide separation of the great toe shows that it may have played the part of a prehensile organ.

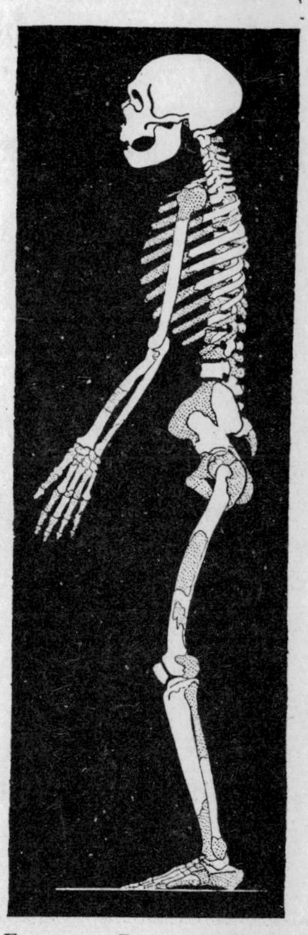

FIG. 150.—Reconstruction of Skeleton of the Man from La Chapelle-aux-Saints, side view. About one-fifteenth natural size.

In general, the ordinary, normal carriage of Neanderthal Man must then have differed in some degree from our own. This fossil Man often exhibits an infantile morphology, that is to say, a morphology the most striking and surprising traits of which are found in either the newly born or unborn infant of Europeans. It has long since been noted that the child is not born with the faculty of walking upright: he first of all tries to walk "on all fours," quite in the manner of a quadruped ape, and when he learns to walk in the biped position he places his foot on the ground on its outer edge. The crouching position, habitual to fossil Man and savage peoples, is likewise an ancestral survival. Here we have a case of individual evolution recalling and repeating the evolution of the race. Neanderthal Man represents a stage in this evolution certainly already far removed from the starting point, a stage closely resembling the modern state of Man, but still quite distinct from it.

Now that we have studied separately each component part of the skeleton of Neanderthal Man, let us try to imagine, as precisely as possible, the general appearance of his whole body.

For long, anthropologists have sought to determine the stature of a being by means of the size of the long bones. The formulæ used to this end are not altogether beyond criticism, especially in their application to fossil Man. While quite appreciating the results attained through these formulæ, I have made use of a different and more direct method. Having drawn, in profile and life-size, each part or each bone of the skeleton of the Man from La Chapelle-aux-Saints, I cut out these drawings and fixed them upon a panel, one by one, according to their anatomical relationships and as exactly as possible. I thus obtained a kind of working-plan of the skeleton in orthogonal projection and of natural size (Fig. 150). The total height of this drawing, that is to say the height of the skeleton, is 1.52 metres [4 ft. 11.8 ins.]. As we must add 20 millimetres to obtain the height of the living subject, this must have been from 1.54 to 1.55 metres [5 ft. 1 in.]. The skeletons from La Ferrassie give somewhat different figures the male skeleton, 1.60 metres [5 ft. 3 ins.]; the female, barely 1.45 metres [4 ft. 9 ins.]. We may therefore look upon the figure 1.55 metres [5 ft. 1 in.] as an approximate average.

The stature of Neanderthal Man was thus much lower than the average stature of modern Man—1.65 metres [5 ft. 4.96 ins.] according to Topinard. From Deniker's tables[1] we can compare in this respect the men of the Neanderthal type with the following ethnographical groups: the Ostiaks, the Veddahs, the Samoyeds, the Japanese of the lower classes, the Annamites of Cochin-China, and other Asiatic tribes; the Caribs of the three Guianas and Venezuela, the Labrador Esquimaux, the Fuegians in America; the Laps and the Voguls of Siberia among Europeans.

We know that the genus *Homo* clearly differs from the whole group of the anthropoid apes with respect to the relative proportions of the upper and lower limbs. In our fossil specimens, these proportions are quite human, but the forearm is very short in relation to the upper arm, and the leg is extremely short in relation to the thigh, more so than in any living race.

[1] Deniker, J., *Races et peuples de la Terre*, p. 659.

Recon-structions. The reconstruction of the whole skeleton of the Man from La Chapelle-aux-Saints, carried out according to the method I have just described, is reproduced in Fig. 150. The general aspect here shown may be described as follows: a large head upon a short and massive trunk; short, thickset, and very strong limbs; and a peculiar attitude, due to the somewhat different curve of the vertebral column and to the slightly bent lower limbs.

FIG. 151.—Reconstruction of the Muscles of the Head and Neck of the Neanderthal Man from La Chapelle-aux-Saints. About one-fourth natural size.

Can we go still further and restore the plastic form? Can we present a portrait of Neanderthal Man as he was in life? The artist is at full liberty to attempt to produce works of imagination, original in character and striking in appearance; but men of science—and of conscience—know too well the difficulties of such attempts to regard them as anything but pastimes and recreations. Certain accredited experts have published portraits in the flesh, not only of Neanderthal Man, whose skeleton is now sufficiently well known, but also of Piltdown Man, whose remains are so fragmentary; of Heidelberg Man, of whom we have only the lower jaw; and of *Pithecanthropus*, of whom we only possess a fragment of skull and two teeth. At the best, such productions might find a place in works aiming at extreme popularity, but they singularly mar the books, estimable in other respects, in which they have been introduced.

In only one attempt did I feel justified in indulging, and it was as follows: I put into the hands of a young sculptor, M. Joanny-Durand, an enthusiast in anatomical studies, a cast of the skull of the Man from La Chapelle-aux-Saints. I requested him to model, in plasticine, the most important muscles, and to superimpose them one by one on this plaster cast, proceeding from the deep to the superficial layers, and carefully marking their points of insertion, the strength of which helps us to judge to a certain extent of the power of the muscles attached to them. The bust thus obtained, shown of course without skin, is represented in Fig. 151. Far from emphasizing in his work a simian or bestial tendency, an easy thing to have done, the artist has clung as far as possible to the human feeling. Apart from the shape of the ears and of the extremity of the nose, for which we have no data, our reconstruction cannot differ greatly from the real appearance of the skinless head of our fossil Man. I leave the reader to study this physiognomy, to distinguish in it the basal structure of the skull, to compare it with the face of modern Man, and to enquire how the expression of this countenance might be modified by a covering of skin and hair, to say nothing of the more or less dramatic play of muscles represented here in a state of rest.

The Brain.

The osteological studies I have just summarized have shown us very satisfactorily the physical attributes and various aspects of the physical life of Neanderthal Man. We shall now attempt to imagine his psychic attributes, his mental life. First of all we may appeal to archæological evidence. All the Moustierian handiwork known to us is very primitive and rude in character, and does not argue in favour of the superiority of the brain which conceived and created it. A second kind of evidence is furnished by the study of the brain itself, so far as it can be carried out under conditions which I shall presently describe, when I have said a few words regarding the brain capacity, that is to say, the volume of the brain.

Brain Capacity.

Several anthropologists, including some of the most distinguished, Schaaffhausen, Huxley, Virchow, and Schwalbe, attempted to calculate the capacity of the Neanderthal skull, assuming that it was complete. The resultant figures varied from 1100-1300 cubic centimetres, very much lower than that of the average human being.

Measurements of the complete skull from La Chapelle-aux-Saints, obtained either by the use of the orthodox formulæ or by direct cubic measurements, lead to a very different result. The capacity of this skull may be fixed at about 1600 cubic centimetres.

The relative size of this figure may well astonish those who regard cranial capacity as a zoological and anthropological character of the first importance, directly related to the development of the intellectual faculties. Our fossil Man shows many signs of morphological inferiority, which somewhat lessen the gulf separating the human group from its nearest relatives, the anthropoid apes, and yet he clearly belongs to the human series. He even occupies one of the highest stages in it, as the following table shows :—

Anthropoid apes (maximum)	621 c.c.
Pithecanthropus (estimated)	855 ,,
Andamans (average of men)	1300 ,,
Piltdown Man (estimated)	1300 ,,
Australians (average of men)	1340 ,,
African negroes	1477 ,,
Parisians	1550 ,,
Auvergnese	1585 ,,
Esquimaux	1646 ,,

This result is all the more surprising in that it differs so greatly from the estimates suggested for the Neanderthal skull. Now, the dimensions of the skull-cap of the latter are so near those of the skull-cap of the skull from La Chapelle that the result obtained by direct cubing of the latter must cast the most serious doubt upon those earlier estimations. This we shall verify presently.

However that may be, the Man from La Chapelle had a brain as large as that of the most civilized of modern races.

But it must be remembered that we are dealing with the absolute value of this capacity. We must ask if the relative value is of the same order, and to discover this we must try to estimate the latter by taking into account the size of the head.

Now it is easy to calculate, according to formulæ in current usage, what would be the capacity of a normal skull, the horizontal diameters of which would equal those of the skull from La Chapelle-aux-Saints, and whose vertical diameter would be in the same relation to the former as in a modern skull. We thus obtain a figure exceeding 2000 cubic centimetres. Certain German skulls almost fulfil these conditions, such as that of Bismarck, whose cranial capacity was estimated at 1965 cubic centimetres.

We have seen (p. 195, Fig. 119) that by superimposing the profiles of the skulls of a Chimpanzee, of the Man from La Chapelle-aux-Saints, and of a white Man, and taking care to make the axes of the three skulls uniform in length, a figure is obtained plainly showing the differences between the *relative* capacities of the brain-box.

Thus there disappears, or is greatly lessened, the paradox seemingly indicated by the magnitude of the absolute volume of the La Chapelle skull, when due account is taken of the numerous signs of its structural inferiority. In reality, other things being equal, the brain is relatively rather less than the brains of modern large-headed men.

Further, it would appear that the example of Neanderthal Man from La Chapelle-aux-Saints was an individual who was specially outstanding in this particular. We now possess casts of the brain-cavity of other examples, notably specimens from Neanderthal, Gibraltar, and La Quina; and comparative measurements of these casts have enabled me to determine their capacity and to obtain the following figures:—

Neanderthal	1408 c.c.
La Quina	1367 „
Gibraltar	1300 „

The Man from La Chapelle, with his brain capacity of 1600 c.c., therefore stands far above the others in this small series. We are justified in supposing that he approaches the maximum figure, while the Gibraltar skull perhaps represents the minimum of the same type. The difference between the maximum and the minimum is by no means abnormal, being only about 300 c.c., while in modern human races, savage or civilized, it varies from 400 to 500 c.c.

The average brain-capacity of Neanderthal Man, calculated from the four individuals whose intracranial casts we possess, is about 1400 c.c., almost the same as that of the so-called inferior races of modern men. (See Appendix, p. 473.)

What do these figures signify? Do they give us the measure of the intellectual or psychic faculties of our fossil Man? Nothing could be more uncertain.

We know, of course, that the brain-capacity may vary enormously in a series of recent men selected for their eminence, from 1320 c.c. (the anatomist Meckel) or 1420 c.c. (Raphael) to 1950 c.c. (La Fontaine), that is to say in the relation of 1 to 1½. Considered by itself, then, this capacity cannot be taken as a criterion of the intellectual standard of a human being. There is an old saying: "A small chronometer is better than a big alarm clock," and the large heads are not always the best heads. As *Homo neanderthalensis* belongs to a type very different from modern types, it would appear that one of his characteristics was the great size of his head rather than the volume of his brain, whatever may have been the *quality* of the latter.

Further it must be enquired whether the organization, or more simply, the divisions of the brain substance do not also show certain differences. In some measure the study of the cast of the endocranial surface will answer this question.

Study of the Brain.

We cannot by means of such evidence claim to be able to penetrate all the secrets of the structure of the brain of any being. The results of such studies as can be carried out on casts of the cranial cavity may be likened to our idea of the form of a statue the covering of which we have been forbidden to raise, the covering being represented in the present case by the meninges which in the living body separate the brain substance from the bony surface of the brain cavity. Nevertheless, such casts, when made with the necessary skill and accuracy, enable us, on comparing them with similar casts made from skulls of apes or modern men, to arrive at some interesting conclusions. We now possess casts of the brain cavities of five individuals of Neanderthal Man: that from Neanderthal itself, from Gibraltar, from La Chapelle-aux-Saints, from La Quina, and from Le Moustier. Their general characters are remarkably uniform. The following description applies specially to the cast of the La Chapelle skull, which is the most complete and the most skilfully executed.[1]

The brain, the general form of which is reproduced in this cast, is, like the skull, long, broad and flattened. It presents a somewhat marked asymmetry; the left hemisphere of the brain was slightly more developed than the right, and this is in accord with the observation made on the inequality of the right and left upper arm-bones (humeri).

One of the most remarkable characters of the cerebrum, as the casts show, is the simplicity and coarse appearance of the convolutions. Casts of the brain cavity of modern men generally show traces not only of more numerous but of much more complicated convolutions. In respect of this character the brain of Neanderthal Man more resembles the brains of the great anthropoid apes or of microcephalic man.

The Sylvian fissure, which, in the brains of mammals, separates the frontal from the temporal lobe (Figs. 152 and 153), is shown here gaping in front, a disposition which seems to indicate a certain degree of exposure of the striated body called the *Island of Reil*. This arrangement is seen in modern Man in the course of embryonic development; in the adult, the Sylvian fissure is closed by the encroaching of the cerebral matter on all sides.

A study of the lobes of the brain reveals some important facts, first of all with regard to their relative development.

To-day we are agreed in admitting that the lobes, within the orthodox limits ascribed to them by anatomists, are far from

[1] For further details, see Boule, M., and Anthony, R., "L'encéphale de l'Homme fossile de La Chapelle-aux-Saints" (*L'Anthropologie*, xxii., 1911).

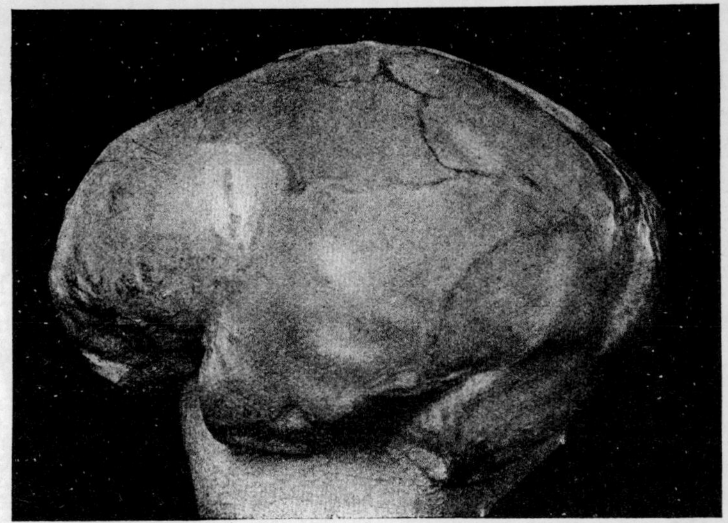

FIG. 152.—Photograph of cast of Brain Cavity of Fossil Man from La Chapelle-aux-Saints. Left side.

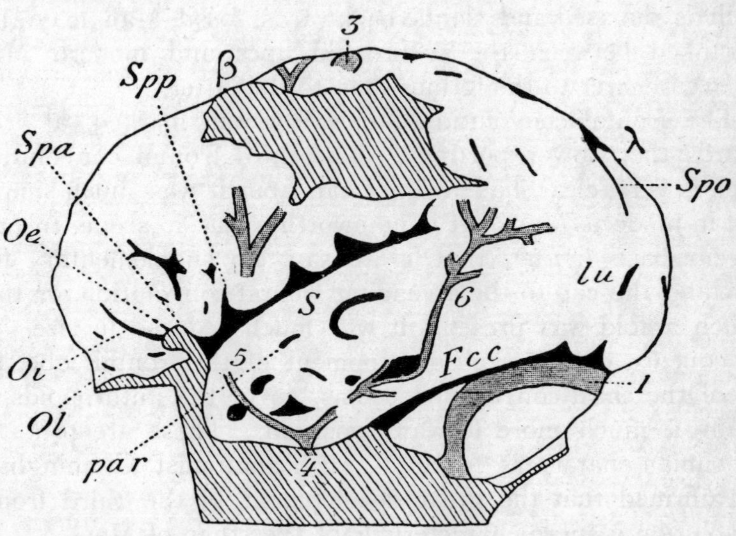

FIG. 153.—Topography of Surface of Left Side of Brain. Half natural size.

β, bregma; *λ*, lambda; 1, lateral sinus; 3, Breschet's sinus; 4, 5, 6, medial meningeal vessels; *Fcc*, cerebro-cerebellar fissure; *S*, fissure of Sylvius, along which are the parietal fissures of Broca; *Spa*, anterior pre-Sylvian branch; *Spp*, posterior pre-Sylvian branch; *Spo*, parieto-occipital fissure; *Ol*, olefactory groove; *Oi*, internal orbital groove; *Oe*, external orbital groove; *par*, parallel groove; *lu*, *sulcus lunatus*.

corresponding exactly to physiological regions. It is interesting all the same to measure, not the volume of these lobes, which would be impossible in the present case, but simply their external surfaces in relation to the total external surface of the cerebral hemisphere, and especially to compare the results of these measurements with those furnished by similar casts of anthropoid apes and modern Man.

M. Anthony and I have thus obtained certain numerical data, discussion of which cannot be entered into here, but the significance of which may be summarized in the statement that the relative development of the lobes in our fossil Man was not that found in modern Man. The difference is especially marked in the frontal lobe. In anthropoid apes the outer surface of this lobe forms about 32 per cent. of the total surface of the corresponding cerebral hemisphere. In modern Man the proportion averages about 43 per cent. In the Man from La Chapelle-aux-Saints it is about 36 per cent. From the point of view of the relative development of his frontal lobe, which is debased and slants backwards, fossil Man may thus be ranked between the anthropoid apes and modern Man, and even nearer to the former than to the latter.

The frontal convolutions are fairly well marked, at least towards their lower portions. The third frontal convolution, which is very clear, had a large cap,[1] but it was much simpler than in modern Man. It is noteworthy that it seems to have had no base (*pied*), that is to say, no supplementary fold attaching the cap to the ascending frontal convolution, or that, if such a fold was present, it was much reduced in size. In consequence of the weak development of the frontal lobe, the cap of the third convolution seems, as in the anthropoids, to occupy a much more forward position. These are primitive and simian characters, for an English anatomist, Cunningham, has affirmed that the great development of the third frontal convolution is more characteristic of apes than of Men.

There is nothing of importance to describe relating to the temporal and parietal lobes. The occipital lobes, which are

[1] The *cap* is the part of the frontal lobe between the two anterior branches of the Sylvian fissure named the pre-Sylvian branches (*Spa* and *Spp* in Fig. 153).

relatively highly developed, exceed and overhang the cerebellum much more than is the case in modern Man. In the sum of their characters, and in particular because of the presence of a well-marked *sulcus lunatus* or "Ape fissure," they greatly resemble those of the anthropoid apes.

The cerebellum is remarkable for the slight projection of the lateral cerebellar lobes, separated by a large fissure, at the bottom of which the median lobe, or *vermis*, must have been exposed as in the apes. In the anterior region, the cerebellum was, on the contrary, relatively more developed and more projecting than in Man, here again resembling that of the apes.

The medulla oblongata must have been disposed in a much more oblique direction from front to back than in modern Man, but less oblique than in the apes, even the anthropoids. This character agrees with what we have learnt from a study of the skeleton, particularly from the direction of the cervical portion of the vertebral column.

Functional Significance of the Brain.

The anatomical facts here revealed enable us to some extent to obtain an idea of the brain functions of Neanderthal Man.

If the great volume of the brain argues in favour of the intelligence of this Man, the coarse appearance and general simplicity of the design of the convolutions indicate, on the contrary, rudimentary intellectual faculties. The relative development of the different parts of the grey matter testify to this fact.

Flechsig has shown that the different regions of the cortex of the brain may be divided, from a physiological point of view, into two groups. The one comprises the sensory-motor regions, correlated with the different peripheral organs of sensibility and movement; the others comprise the "zones of association," where sensations are condensed or diffused, and movement is regulated. Such would be the "intellectual centres and the true organs of thought."

In the lowest orders of Mammals, the centres of association are almost lacking. In the apes their importance is already considerable, and their development almost equals that of the sensory-motor centres. In Man, in whom intelligence reaches its highest point, they have come to occupy two-thirds of the cortex.[1]

On the external surface of the human brain, three principal centres of association, wedged in between four sensory-motor zones, have been approximately mapped out. Comparing these facts and those derived from the study of the lobes of the brain of Neanderthal Man, we find that the visual zone (occipital lobe) was relatively more developed, and that the first centre of association was very ill-conditioned, since this centre corresponds to the anterior region of the frontal lobe, here specially reduced.

Now, if there is one assured fact in the physiology of the brain, it is that the anterior portions of the frontal lobes are indispensable to the intellectual life. Lesions in these parts affect neither sensibility nor powers of movement, but they give rise to intellectual disturbances; according to Dr Toulouse,[2] bilateral atrophy of the frontal lobes always causes dementia or defective control known as gatism.

"In the development of the cerebral hemispheres throughout the geological periods," Dr Houzé[3] writes, "it is the frontal lobe, the seat of the most complicated associations and of the most perfectly adapted mental combinations, which has increased in size. In Man, it has acquired such pre-eminence that it has done away with the necessity of adaptations for defence (*Homo nudus et inermis*). The frontal lobe has become the most formidable weapon for attack and defence."

It is probable, therefore, that Neanderthal Man must have possessed only a rudimentary psychic nature, superior certainly to that of the anthropoïd apes, but markedly inferior to that of any modern race whatever. He had doubtless only the most rudimentary articulate language. On the whole, the brain of this fossil Man is already a human brain because of the amount of its cerebral matter; but this matter does not yet show the superior organization which characterizes Modern Men.

[1] See Gley's excellent *Traité de Physiologie*, 4th ed., Paris, 1919, p. 1080.

[2] Toulouse, Dr, *Le cerveau*, Paris, 1901, p. 123.

[3] Houzé, Dr, "Les étapes du lobe frontal" (*Institut Solvay, Sociologie*, 1910).

Conclusions.

The long description the reader has just perused may be summed up in a concise form which may serve as a diagnosis of the Neanderthal type of fossil Man.

Diagnosis of the Neanderthal Type.

Body of short stature, but very massive. Head very large, with facial region much developed in comparison with cerebral region. Cephalic index medium. Skull much flattened; orbital arches enormous, forming a continuous ridge; forehead very receding; occiput protuberant and compressed in a vertical direction.

Face long and projecting, with flat and receding malar bones, upper jaws lacking canine fossæ and forming a kind of muzzle. Orbits very large and round. Nose prominent and very large. Sub-nasal space extensive.

Lower jaw strong and chinless, with large ascending rami, and truncated in the region of the angle.

Detention massive, structure of back molars retaining certain primitive characters.

Vertebral column and limb bones showing numerous simian characters and indicating a less perfect bipedal or upright carriage than in modern Man. Legs very short.

Brain capacity averaging about 1450 cubic centimetres. (See Appendix, p. 473.) Brain formation presenting numerous primitive or simian characters, especially in the relatively great reduction of the frontal lobes and the general pattern of the convolutions.

It is important to note that the physical characters of the Neanderthal type are quite in agreement with what archæology teaches us as to his bodily capacity, his psychology, and his habits. As we have already pointed out, there is hardly a more rudimentary or degraded form of industry than that of our Moustierian Man. His use of one simple material only, stone (apart probably from wood and bone), the uniformity, simplicity, and rudeness of his stone implements, and the probable absence of all traces of any pre-occupation of an æsthetic or of a moral kind, are quite in agreement with the brutish appearance of this energetic and clumsy body, of the heavy-jawed skull, which itself still declares the predominance of functions of a purely vegetative or bestial kind over the functions of mind.

Comparison with Modern Types.

A comparative study of the morphology of various living human groups confirms the idea that we are here concerned with an altogether special type, very different not only from the so-called superior races, but also from the Esquimaux, the Fuegians, the Bushmen, the Pygmies, African or Asiatic, the Veddas, the Polynesians, the Melanesians, and even from the Australians, with whom attempts at comparison have often been made.[1]

The skeleton of the last-mentioned racial type is as dissimilar as possible from that of Neanderthal Man (Figs. 154 and 155). It can no longer be asserted that the Australians are descended from our Moustierians; indeed, the idea of this relationship would probably not have occurred to the mind of the early observers, if, in place of having only a mere skull-cap, they had had the opportunity of examining a complete skull with its facial portion. All that can be admitted in this respect, is that the Australian group of men, certainly one of the least developed groups of modern mankind, is less far removed than other races from the primitive forms, and that, in consequence, it ought to have certain characteristics in common with the Neanderthal type. Perchance our Moustierian led the same wandering life as the modern Australians.

His Place in the Human Series

Neanderthal Man thus represents a new type in the human family. Ought he to form a separate genus, or ought he to take a place in the genus *Homo*? It was claimed, even at the time when we were acquainted with only a portion of his skull, that the fossil Man of Neanderthal differed generically from modern Man. Later, Sergi created the genus *Palæoanthropus* for this type, Bonarelli suggested the name *Protanthropus*, and Ameghino, regarding Neanderthal Man as the ancestor or forerunner of the genus *Homo*, called him *Prothomo*.

In my opinion, the human being (Hominian) of the

[1] These comparisons are detailed in my Monograph, *loc. cit.*

Neanderthal type is certainly a *Homo*. After all we are discussing a creature who greatly resembles us, and so the opinion I have expressed is that generally adopted, even by naturalists who are inclined to multiply genera. It would

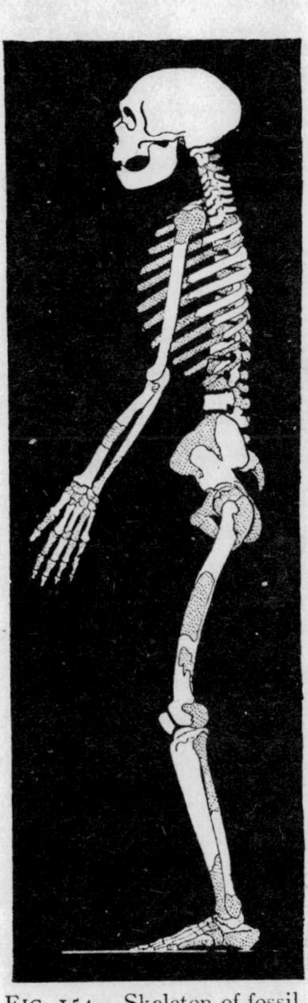

FIG. 154.—Skeleton of fossil Man from La Chapelle-aux-Saints, reconstructed, side view. About one-fifteenth natural size.

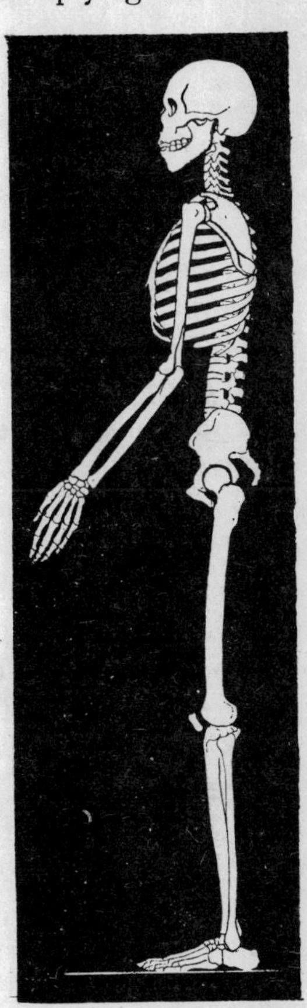

FIG. 155.—Skeleton, side view, of an Australian. About one-fifteenth natural size.

probably be a different matter if we were dealing with a feline, a ruminant, or a monkey!

It has still to be decided whether this fossil Man represents a distinct species, or race, or simply a variety of Man or of modern Men. The majority of anthropologists generally speak of the *Neanderthal Race*, but numerous specific names have also been created. One of them is of old standing, *Homo neanderthalensis*; others are more recent, *Homo primigenius, Homo europæus, Homo krapinensis, Homo mousteriensis, Homo antiquus*, etc.

If we argue from the multiple-genera point of view, it is at once clear that the Neanderthal type represents a new species, easily distinguished from all others. If, on the contrary, we believe in the specific unity of modern Man, this conclusion is not so evident; it may be deduced, however, from the following considerations.

A first point is that the fossil type differs much more from all the types existing to-day, than these types differ amongst themselves. Between the extremes of the series of modern skulls we can discover all the transition forms, but this series is clearly separated from the group of fossil skulls by a kind of hiatus, corresponding to a real morphological break.

A second point is that the osteological differences between our fossil Man and the mass of existing Men are much more important than the differences appealed to by mammalogists as a basis for separating the various species of mammalian genera. From this point of view, the Neanderthal type clearly represents a species different from *Homo sapiens* taken collectively, existing or fossil, because it presents a certain number of constant characters which are not met with, normally and in association, in any human race or species existing at the present day. The majority of these distinctive characteristics have a morphological value, and consequently a classificatory value, greater than the value of the characters made use of in mammalogy for classifying forms of the same genus. Each of them would be almost sufficient in itself to justify a specific distinction, were it a question of a mammal not belonging to the genus *Homo*.

We may consider the matter from yet a third point of view. Palæontologists, compelled to study beings simultaneously in space and in time, cannot have exactly the same conception of a species as zoologists who study them only in regard to space. Finding it impossible to designate by different names all the fine degrees of evolutionary development which are revealed by the never-ending modifications of forms of life, and which they trace out throughout the ages, palæontologists claim that new names should be reserved for those changes which mark a step, or, if I may be allowed the expression, a new rung in the evolutionary scale of these forms, or a quite distinct ring amongst the links of their chain.

Now the human fossil type which we have been studying corresponds ideally to the conception held by transformist palæontologists regarding species. No one can doubt that this type represents a degree in the human scale lower, morphologically speaking, than all the stages of modern humanity, and that it marks a step very distinctly separated from the step above it. The long series of primitive or simian traits which are imprinted on each constituent part of the skeleton, can only be interpreted as the marks of a more backward evolutionary stage than that reached by modern Mankind, and the difference is such that, according to the principles laid down, it fully justifies the distinction of a specific title.

What name should be applied to this species? A name much in use is that of *Homo primigenius*, which apparently we owe to Wilser,[1] and to which the fame of Schwalbe, the German anatomist, seems to have given some sort of sanction. It should be rejected. It has, to begin with, the very serious drawback of making an affirmation or suggestion which is certainly erroneous; for the Neanderthal type cannot be regarded as the absolutely primitive form of the genus *Homo*. It recalls a similar egregious confusion which has led to the application of the absurd name *Elephas primigenius* to the Mammoth, the last and most specialized of the Elephants.

It ought to be abandoned for yet another reason. The rules established by the International Congress of Zoology lay down that the proper name of a species shall be that which is first applied to it. Now, as early as 1864, an English scientist,

[1] Wilser, "Menschenrassen und Weltgeschichte" (*Naturw. Wochenschr.*, xiii., 1, 1898).

King,[1] created the expression *Homo neanderthalensis* for the skull-cap and other fossil bone-remains discovered at Neanderthal, and these remain the type specimens of the new species. In justice, then, this is the name which should be adopted. The terms *Homo antiquus*, etc., proposed by various authors, ought to be regarded as synonyms.

An Archaic and Extinct Species. That *Homo neanderthalensis* is a species with archaic characteristics is clearly evident from its general morphology. The numerous simian traits which it has retained are so many relics, still strongly in evidence, of an ancestral state.

Doubtless some of these characters are perhaps not altogether primitive traits, but rather the consequences of special physiological adaptations. The discrimination is often very difficult; but even if many of them may be regarded simply as the results of convergence, there is no doubt that the majority have a real genealogical value.

There is every reason to believe that *Homo neanderthalensis* is a species even more archaic than the geological age of the remains in our possession would indicate.

We know indeed that men of relatively higher organization, direct forebears of various forms of *Homo sapiens*, existed from very early times in Europe simultaneously with the Neanderthal type. We have seen, in retracing the history of the principal discoveries in human palæontology, that several finds of less primitive skulls or skeletons have been regarded as belonging to a very remote geological age: such as those at Olmo, Galley Hill, Denise, Clichy, Grenelle, Ipswich, etc. I have felt obliged to reject these discoveries, because they are not supported by satisfactory geological guarantees. But we have other proofs of the co-existence of *Homo neanderthalensis* with the ancestors of *Homo sapiens*. The early men of the Reindeer Age, the first Aurignacians, who abruptly followed the Moustierians in France, were men of the type known as *Cro-Magnon*, that is to say, men who were, as we shall see, very closely related to certain races of modern men, and who

[1] King, "The reputed Fossil Man of the Neanderthal" (*Quart. Journ. of Science*, 1864, p. 96).

differed from the Moustierians as much in their superior culture as in the superiority or diversity of their physical characters. Now these "*Cro-Magnons*," who seem suddenly to replace the Neanderthal people in France, must previously have existed somewhere, unless we can imagine a mutation so great and so sudden as to be altogether out of question.

The Grimaldi discoveries likewise prove that *Homo neanderthalensis* was not the sole form of humanity living on the earth about the middle of Pleistocene times, for the lower skeletons from the Grotte des Enfants could not differ greatly in geological age from the man from the Corrèze. Now, the Grimaldi Negroids already definitely belong to the group of *Homo sapiens*. Their existence in France at so distant a period, in close association with the much more primitive Neanderthal form, shows that *Homo neanderthalensis* could not be the ancestor of *Homo sapiens*, since representatives of both species were contemporary. The origin of *Homo sapiens* must be sought for in a much more remote past than *a priori* we could ever have supposed.

As to *Homo neanderthalensis*, all the evidence leads us to believe that he dates from a geological period more remote than the Mid Pleistocene; that, contemporaneous with the "Mammoth fauna," of which, in a way, he formed part, he occupied vast territories of western and southern Europe; but that, side by side with him, in other territories, and accompanied by probably a somewhat different fauna, there were already in existence human types, already in a higher stage of evolution, who represent the direct ancestors of modern *Homo sapiens*. In contrast with the latter, *Homo neanderthalensis* seems to us a belated form, a survival of ancestral prototypes.

What do we know of these ancestral forms?

Unfortunately, there are very few of the more ancient types with which comparison can be made. The only human fossils from a geological period unquestionably prior to that of *Homo neanderthalensis*, are the Mauer jaw and the Piltdown bone-remains.

The lower jaws of *Homo neanderthalensis* resemble the lower jaw of *Homo heidelbergensis* (*i.e.*, the Mauer jaw) in their general form, strength, and dimensions, so much so that if the Mauer jaw is articulated with the La Chapelle skull, the general aspect of the whole skull is but little altered. It is true that certain differences do exist, but the points of resemblance are so great that they lead us to believe in a close, if not direct, relationship between the ancient owners of these jaws. As the Neanderthal type can only be a survival, in some respects further evolved, of a still more primitive type, it is very possible that the Mauer jaw, from a much more remote geological period, may have belonged to a more primitive representative of this type, which would have been slowly modified in consequence of changes of environment, and perhaps, as Keith, Sera and Larger suggest, under pathological influences.

As to the Piltdown bone-remains, there are no grounds for attributing to them the least affinity with those just discussed; neither the jaw, which, as we have seen, is perhaps that of a Chimpanzee, nor the skull-cap, which from the generality of its characteristics—its prominent forehead, insignificant orbital ridges, and the structure of its temporal region—is more directly associated with the ancestors of *Homo sapiens* than with those of *Homo neanderthalensis*.

The latter species, whose origins are, from every aspect, extremely ancient, became extinct without leaving any posterity. It is fossil in a double sense; because it dates from a geological period prior to the present day, and because we are aware of no descendant, from the Upper Pleistocene onwards. In the Moustierian period it represented a belated type existing side by side with the direct ancestors of *Homo sapiens*; its relation to the latter was similar to that which exists at the present day between the races we call inferior and the superior races. Perhaps one might go so far as to say that it was a degenerate species.

In the Moustierian period this survival must have come to an end, for the Neanderthal race seems to disappear abruptly. The Aurignacians and the Magdalenians, who succeeded it in Europe, differ from it in their much higher organization, and I do not think, contrary to what has sometimes been said, that true transition forms have been found either in the Upper Palæolithic or in the present period.

Did there take place a simple substitution, or a migration, or even an extermination? I do not know. We have seen that it is impossible, among the modern ethnographical groups, to point to any single one which could be considered a descendant of the Neanderthal people.

The chief objection to this point of view is the existence, frequently insisted upon, of "Neanderthaloid" skulls found in prehistoric, historic, or modern burials in Europe. Numerous anthropologists have described and figured such specimens; there is to-day no important collection which does not contain at least one speçimen of this kind. Now the most "Neanderthaloid" of these skulls possesses only a very small number of the characters of the Neanderthal type, usually only a marked projection of the orbital ridges and a somewhat receding forehead. The face is always very different, the chin always well defined. In fact, all these "Neanderthaloid" skulls are only pseudo-Neanderthal types, that is to say, are really *Homo sapiens*, remarkable on account of the fortuitous presence of certain morphological traits, normally greatly developed in Neanderthal Man.

The appearance or reappearance, in sporadic fashion, of these characters is generally looked upon as atavistic. But this is not to say that *Homo sapiens* is descended in direct line from *Homo neanderthalensis*. It may be admitted that the characters in question are really primitive, that they formed part of a common inheritance from the remote ancestors of these two species. In *Homo neanderthalensis* they have been preserved at a stage much nearer to their origin; in the more highly developed *Homo sapiens* they no longer reappear except by accident.

Further, we cannot affirm that an infusion of Neanderthaloid blood, by way of hybridization, never entered into other human groups belonging to the branch, or to one of the branches, of *Homo sapiens.* But this infusion could only have been casual and had no great influence, as no modern human type can be considered as a direct descendant, even with modifications, of the Neanderthal type.

CHAPTER VIII

THE MEN OF THE REINDEER AGE

WE know that from the geological and palæontological points of view, it is difficult to make a sharp distinction between the Upper Pleistocene and the Mid Pleistocene (see p. 53). But from the anthropological standpoint the distinction is relatively clear and easy, at least in France.

The Reindeer Age.

From the prehistoric archæologist's point of view, the Upper Pleistocene corresponds to the *Upper Palæolithic*, the implements of which are very different from those of the Lower Palæolithic; it is also the period of the Reindeer or *Reindeer Age*. This latter term expresses an ethnographical rather than a palæontological fact, for the bones of the Reindeer are already numerous in Moustierian deposits. It means, first, that this animal has now become very abundant; it denotes, particularly, the age when the Reindeer played a great part in the life of man, for whom it provided food, clothing, and the raw materials for a large proportion of his industry.

Weapons and implements are more abundant, much more varied, and more perfect than in the Moustierian period. Thus archæologists, with more data at hand, have long sought to establish definite divisions of the Reindeer Age, based upon the different fashions of making use of this material. We already know that they now distinguish three successive stages: the *Aurignacian*, the *Solutrean*, and the *Magdalenian*, which was followed by the *Azilian*, a transition stage bordering on the *Neolithic* period, that is to say, on the *recent* times of geologists.

In spite of the differences which serve as a basis for these divisions, the Reindeer Age presents a combination of characters which give it great unity, and which denote considerable progress beyond the Moustierian world. So there exists a striking contrast between Men of the old and of the new Palæolithic periods.

Discoveries of human skeletons now bring us in touch with really superior types. The majority have a more graceful body, a finer head, and a large straight forehead. In the caves in which they lived, they have left so many evidences of their manual skill, of the resources of their inventive genius, of their artistic and religious pre-occupations, and of their powers of thought, that they truly merit the great name of *Homo sapiens*.

Each one of these types has, indeed, its own particular physiognomy, but the peculiarities are only of the degree exhibited in the various races of the present day, and in general appearance the types of the Reindeer Age do not differ more from modern men than these differ among themselves. We have reached a point after which the physical evolution of Mankind may be considered as finished; the problem of human origins loses its zoological character and becomes purely anthropological and ethnographical.

Nothing could be more engrossing than the study of these Men of the Reindeer Age, of the relics they have left us, and of the manners and customs which these relics reveal or lead us to surmise. It has already furnished material for many works and memoirs of the greatest interest, but a summary of these cannot find a place in the present volume, written, as it is, especially from the palæontological point of view. The essential facts must, however, be mentioned.[1]

Products of Industry.

The Men of the Reindeer Age lived by preference in caves or rock-shelters. They were mainly devoted to hunting and fishing. They had widespread commercial relations. Their dressed stones are

[1] The large bibliography is essentially French. It will be found partly in the work by S. Reinach, already quoted several times, *Antiquités nationales*: I., "Epoque des Cavernes"; in Déchelette, *Manuel d'archéologie préhistorique*, i. Consult also, Mortillet, G. and A., *Le Préhistorique* and *Musée préhistorique*. Cartailhac, E., *La France préhistorique*, and especially the *Matériaux* and *L'Anthropologie*. See also, further on in this work, some notes relating to the art of the Reindeer Age.

neither so heavy nor so uniform as those of the preceding periods. Their weapons, intended to be hafted, are fine, graceful, sometimes beautifully worked with a most skilful technique (Figs. 156-158). Their implements are no less slender, very varied in form, and adapted to manifold purposes : knives, scrapers, piercers, gravers, saws, etc., new types of which were constantly being developed. Dressed flints are found by hundreds and thousands, mixed with kitchen debris, in their ancient settlements in caves and rock-shelters.

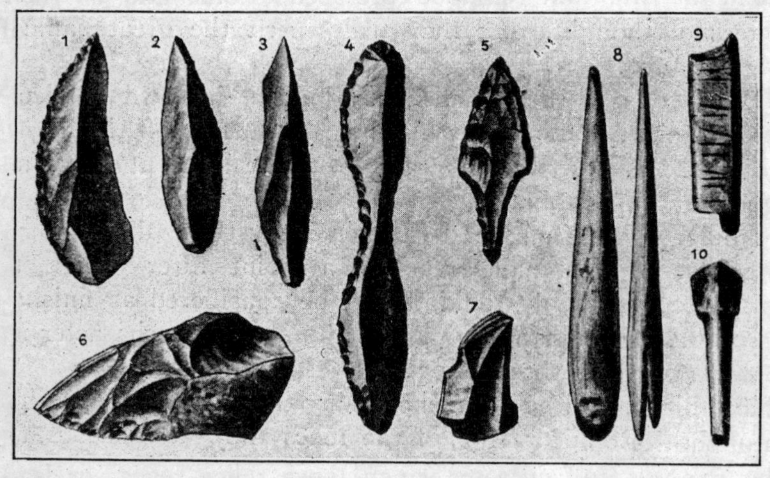

FIG. 156.—Characteristic Objects of the Aurignacian Period.

1, 2, Flint points with arched and retouched back, type known as "Châtelperron"; 3, flake of flint for cutting, formed by retouching, "La Gravette" type; 4, notched flake retouched along whole profile; 5, shaped flints retouched with a tang or tongue; 6, keeled scraper; 7, graver of arched type ("*busqué*"); 8, bone point of a dart, split at base; 9 and 10, bone objects. (After H. Breuil.)

Stone is no longer the only raw material used. Our industrious cave-dwellers knew how to work in varied fashion the ivory of Mammoth tusks, the antlers of the Reindeer, and the bones of all kinds of animals. From these they made daggers, javelins, or arrow-heads for the chase, harpoons for fishing, shafts for projecting their darts, etc. They also made utensils for domestic use : spatulæ, smoothing implements for preparing the skins of animals, needles pierced with an eye to sew their garments of fur, etc.

From such archæological materials, archæologists are enabled to establish their divisions.[1]

Subdivisions.

In the lower layers (*Aurignacian*), there still persist ancient Palæolithic forms, even amygdaloid flints. But new forms begin to appear. Elongated flakes become more frequent and are carefully retouched on their edges, which are often notched. There are gravers of a special

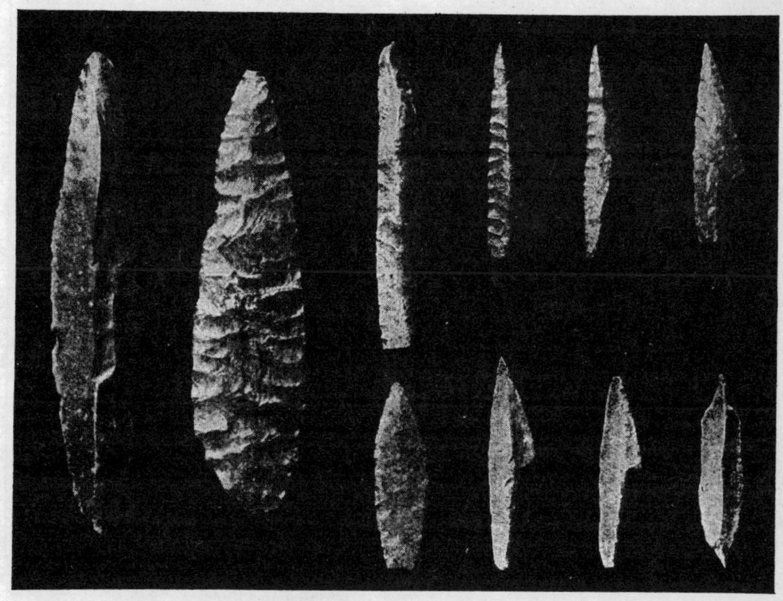

FIG. 157.—Solutrean flints, carefully dressed in shapes of laurel leaves and willow leaves, shouldered points (*pointes à cran*), and double piercers. One-half natural size. From Grotte du Placard in Charente. Palæontological Gallery, French National Museum of Natural History.

form known as "arched" (*busqué*), and thick massive scrapers known as "keeled" scrapers. Bone is now made use of to a great extent. Many shafts are to be found ornamented with striæ, or with "marks of the chasé," as well as flattened darts, often with split base, and various kinds of rude pins (Fig. 156).

The *Solutrean*, which succeeds the Aurignacian in some

[1] See particularly Breuil, H., "Les subdivisions du Paléolithique supérieur et leur signification" (*Congrès intern. de Génève*, 1912). Cartailhac and Breuil demonstrated the existence and the great importance of the Aurignacian period, also, it must be borne in mind, earlier foreseen by E. Lartet.

localities, is marked by extremely able and careful work in stone. The characteristic types are narrow points known as "willow leaves," larger points called "laurel leaves," and shouldered points (*pointes à cran*). These weapons are manufactured with much art, by means of skilled retouches, long and regular, which often cover both surfaces of the flints (Fig. 157). These are beautiful objects, precursors of the wonderful

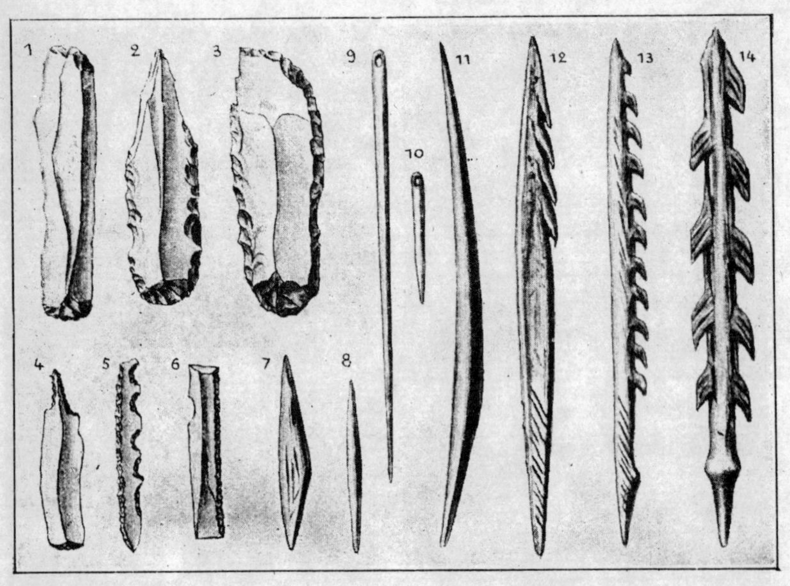

FIG. 158.—Characteristic Magdalenian Objects.

1, Flint flake, ending in double scraper; 2, flake dressed as straight graver at upper end, and as scraper at lower end; 3, scraper and graver of "parrot-beak" type; 4, small flint combining piercer and scraper; 5, flake of which one end has been trimmed by fine retouching, while its other edge bears a series of regular notches; 6, small flake trimmed for cutting; 7, javelin point with bevelled base; 8, bone piercer; 9 and 10, bone needles with eyes; 11, bone javelin point, bevelled and curved; 12, harpoon of reindeer antler, with barbs not prominent (primitive type); 13, harpoon with only one row of barbs; 14, harpoon with double row of barbs.

flints, lance-heads, arrow-heads, and daggers of the Scandinavian, Portuguese, and Egyptian Neolithic culture. The bone industry was further enriched by new forms of javelins with a pyramidal base and varied objects in ivory; the first needles with an eye appeared.

During the *Magdalenian* period, stone implements, far from attaining perfection, comprise countless flints generally small in size, and often hardly retouched at all, but varying greatly in form: flakes, scrapers, various gravers, piercers, saws, etc. Sometimes the same flint is adapted for different purposes; at one end it is a scraper, at the other a graver or piercer. The bone industry, on the other hand, is exceedingly diversified; javelins are numerous and varied, often decorated with designs; needles are abundant and also harpoons, which, to begin with, consist of simple notched bone sticks, and develop into barbed instruments, first with a single row, then with two rows of barbs (Fig. 158).

We must be careful not to accept the naïve idea of the prehistorians who are inclined to see in this sequence, established in these regions by means of data collated from numerous beds, the steady evolution of Upper Palæolithic industry. These subdivisions cannot have the general significance which they desire to attribute to them. On the contrary, all the evidence seems to show that the layers in each country, and even in each province, while possessing a common archæological basis, exhibit in addition certain special ethnographical characteristics which are only of local significance.

"It becomes more and more evident," says one of the writers most familiar with the Upper Palæolithic, "that what has been at first regarded as a continuous series due to the evolution of a single population on one spot, is, on the contrary, the result of the successive collaboration of numerous tribes reacting to some degree upon each other, either through the influence of industry or commerce alone, or by a gradual intermingling, or a sudden and warlike invasion of alien tribes."[1]

The future of the science depends principally on the establishment of this racial palæogeography, which, however, is certainly far more complicated than at present we can imagine. Already, following Breuil's guidance, we can discern two vast Upper Palæolithic regions, one of which we may call the *Mediterranean* and the other the *Atlantic*.

The First Artists.

The men of the Reindeer Age had also a taste for personal adornment; they used the teeth of wild animals killed in the chase (Fig. 171, p. 261), shells, pierced stones, etc., from which they made trophies, necklaces,

[1] Breuil, H., *Les subdivisions, etc.*, p. 169.

bracelets and amulets. Endowed with a sincere and deep feeling for nature, and great sensibility to beauty, they were true artists, *the first artists*, existing thousands of years before the Chaldeans, the Egyptians or the Ægeans.

FIG. 159.—Portrait of Edouard Piette.

It is in France, and by Frenchmen, that most of the wonderful evidences of this first blossoming of the plastic arts have been discovered. Next to the names of Edouard Lartet and Vibraye, that of Piette will always remain associated with the history of Quaternary art. After the death of Piette, the famous discoveries or labours of Rivière, Cartailhac, Capitan, Peyrony, Lalanne, Bégouen and particularly of Breuil, have enriched prehistoric archæology with a mass of relics of the greatest interest, still insufficiently known to the artistic world or to the enlightened public.[1]

FIG. 160.—Reindeer carved in Ivory, from Bruniquel, now in the British Museum. About one-half natural size. (After H. Breuil.)

All these productions, varying greatly in value, from indecipherable scrawls to true works of art, show that the men

[1] Apart from Manuals and Reviews, the most important works are: Lartet, E., and Chisty, H., *Reliquiæ aquitanicæ*, 1865-1875. Girod, P., and Massénat, E., *Les stations de l'âge du Renne dans les vallées de la Vézère et de la Corrèze*, 1888-1900. Piette, E., *L'art pendant l'âge du Renne*, 1907. The fine series of large monographs published by the Institut de Paléontologie humaine (founded by Prince Albert I. of Monaco), under the general title: *Peintures et gravures murales des Cavernes paléolithiques*, particularly: *Altamira* by Cartailhac, E., and Breuil, H.; *Font-de-Gaume*, by Capitan, L., Breuil, H., and Peyrony, D.; *Cavernes cantaliques* by Alcalde del Rio, Breuil and Lorenzo Sierra, etc. During the last few years, the Comisión de Investigaciones paleontologicas y prehistoricas de Madrid has issued fine memoirs by Cabré, Pacheco and Obermaier on the art of the Spanish caves. Finally, S. Reinach has published a very useful *Répertoire de l'art quaternaire* (Paris, 1913), with a complete bibliography of each production.

of the Reindeer Age must have had ample leisure, like all populations engaged almost entirely in hunting. During this leisure, they practised every kind of art work, bent on repro-

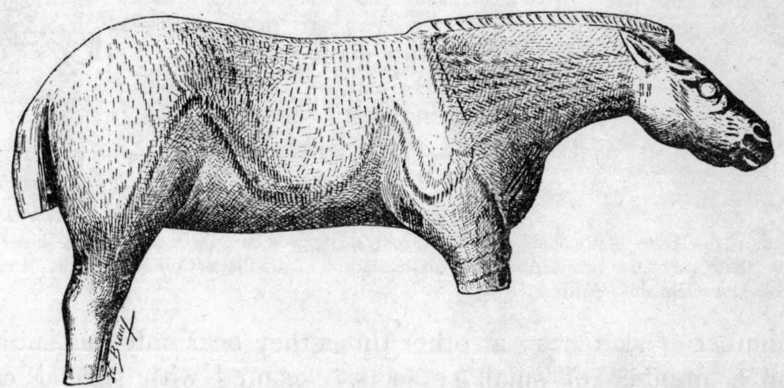

FIG. 161.—Statuette of a Horse in Ivory. Grotte de Lourdes. About 4/3 natural size. Drawing by H. Breuil.

ducing likenesses of the creatures living beside them, either by sculpture, by engraving, or by painting.

The ivory of mammoth tusks and soft stone served them

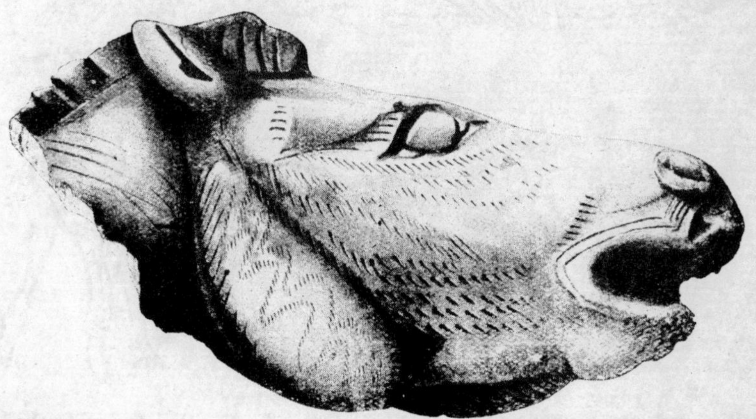

FIG. 162.—Head of Horse, carved in Reindeer antler, from the Cave at Mas d'Azil in Ariège. Piette Collection, in the Musée de Saint-Germain. About twice natural size. (After Piette.)

as the raw material for sculpture; they knew also how to model in clay. They engraved on any kind of bone, on reindeer antlers, on rounded pebbles from the river nearby.

Often these objects may be likened to regular leaves of a sketch-book, whereon are grouped, mingled, or superimposed any

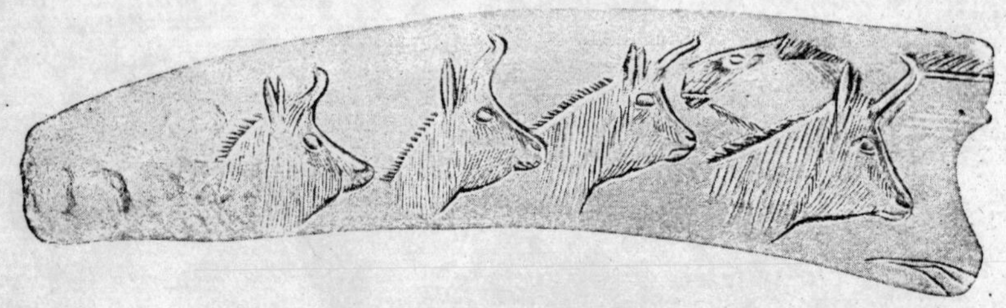

FIG. 163.—Heads of Chamois and of Badger (?), engraved on a piece of Reindeer antler from the Cave of Gourdan. About five-sixths natural size. Piette Collection in the Musée de Saint-Germain. (After Piette.)

number of sketches; at other times they bear only one subject, or a number of small subjects, executed with special care. Almost all domestic objects in daily use were decorated with

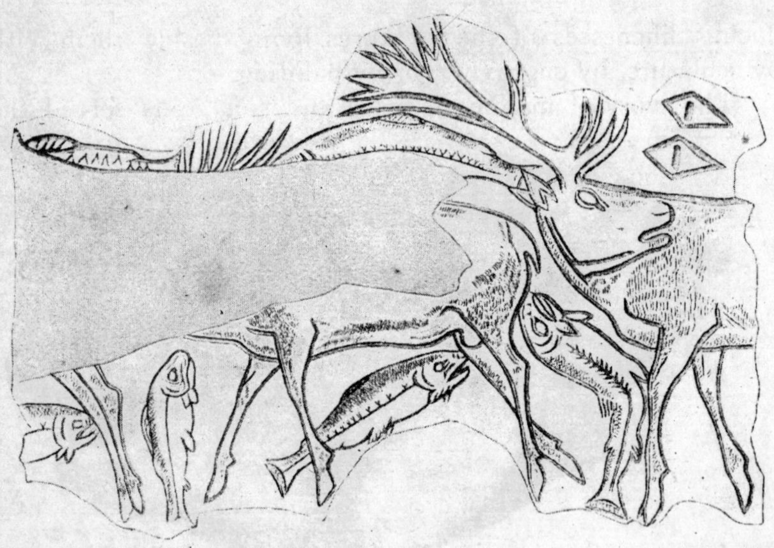

FIG. 164.—Red Deer and Salmon engraved round antler of a Stag from the Lorthet Cave in the Hautes Pyrénées. Piette Collection in Musée de Saint-Germain. Reduced by about one-fourth, and represented on the flat. (After Piette.)

Notice above to the right two lozenges divided by a small vertical line in the centre, perhaps representing the signature of the artist.

ornament or figures of living creatures. But coexistent with this mobile art, there was also a mural art. Our cave-dwellers sculptured, engraved, and painted large figures on the walls of dim caverns. The saucers and polychrome ochres of these

FIG. 165.—Reindeer engraved on stone from the Limeuil Cave in the Dordogne. Two-thirds natural size. (Photograph kindly lent by MM. Bouyssonie and Capitan.)

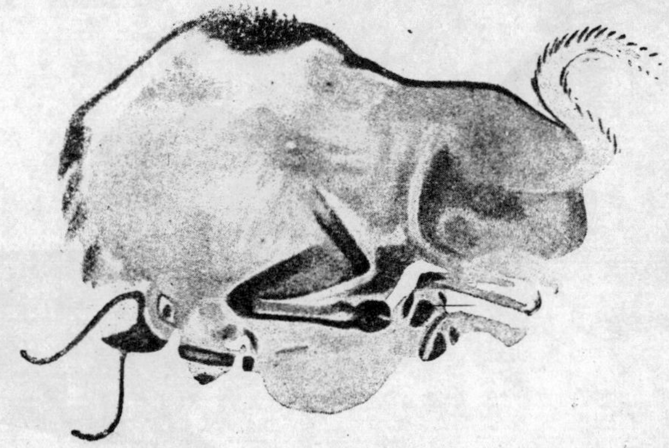

FIG. 166.—Bison in act of leaping. Polychrome painting on the ceiling of the Altamira Cave in Spain. Length of animal 1·55 m. (After Cartailhac and Breuil.)

primitive painters have been discovered, as well as primitive lamps by means of which they obtained light.

A great many of these productions bear strong evidence

FIG. 167.—Hind in polychrome, with small black Bison, from the Altamira Cave. Length of hind 2.20 m. (After Cartailhac and Breuil.)

FIG. 168.—Red and black Deer painted on a rock at Calapata, near Cretas, in Lower Aragon, Spain. About one-fifth natural size. (After Breuil and Cabré Aguila.)

of a real æsthetic sense, a masterly realism, skilled technique, and great strength of execution. They are frequently masterpieces full of life and movement, representations of Mammoths with long hair and bent tusks, Rhinoceroses of massive form, Horses of spirited bearing, bristly Boars, crowds of Ruminants of elegant build, Bears with bulging heads, Felines with supple body, and so on. The frieze of Horses carved life-size on the rocky wall of the Laussel Cave in the Dordogne, the Bisons modelled in clay at Tuc d'Audoubert in Ariège (*see Frontispiece*), and the polychrome paintings in the Altamira Cave in Spain—the "Salon Carré" of palæolithic painting—are only a few extraordinarily striking evidences of the intellectual superiority and masterly skill of these very ancient artists. I have here grouped together some masterpieces of different kinds; they will give an idea of the perfection of palæolithic art. (*Frontispiece* and Figs. 160-168).

Our cave dwellers, however, lived in the midst of unpromising surroundings. In some respects their material existence was very miserable: they knew nothing even of pottery. Brave in contact with Cave Bears and Lions, they were, nevertheless, peaceful and gentle in character; they possessed a sense of mystery, for their decorated caves seem to have served as sanctuaries, where they must have devoted themselves to practising the arts of magic[1]; they engaged in religious

[1] In agreement with M. Salomon Reinach, many prehistorians believe that the art of the Reindeer Age was by no means disinterested; that it had especially a practical purpose related to their magical practices. I do not entirely share this view.

The truthfulness of the drawings, the purity of the lines, the grace of the attitudes, cannot be accounted for by the simple practices of magic. Poor designs, childish drawings, and simple compositions, such as those of present day natives, would have sufficed for that. It is impossible that such figures, drawn with so great feeling for form, and really extraordinary certainty and delicacy in the manner of engraving, were not accomplished by the artist except for love's sake; he could not have attained to such heights of perfection or to such mastery except by disinterested study. No one, having sufficiently practised the art of drawing and having devoted some time to the study of palæolithic masterpieces, could think otherwise. The hypothetical suggestion of the practices of magic may satisfy scholars, but I do not think that it would be wholly accepted by artists. I do not hesitate to confess that I believe in the theory of art for art's sake, without at the same time refusing to allow a certain influence due to the practice of magic. But, in point of fact, diagrammatic and stereotyped art, better suited to this purpose, is more recent than true palæolithic art, which, to begin with, is realistic only. In its real, and much earlier beginnings, art is probably only a special manifestation of the general spirit of imitation, already highly developed in the apes.

practices, for their sculptures testify to funeral rites and to a true worship of the dead. We may therefore be proud of the spiritual and moral qualities of these distant ancestors of ours. We shall see that in their physical characters, also, they were as closely related to modern Mankind.

Historical.

To an English scientist falls the credit of having exhumed and placed in a museum the first human skeleton of the Reindeer Age. This skeleton, with bones coloured red, the famous *Red Lady of Paviland*, was excavated in the Paviland Cave in Wales by Buckland in 1823, and was placed in the Oxford Museum, where for long it remained forgotten. In recent years, Professor Sollas[1] has told us that the "Red Lady" is probably a man, whose skull-less skeleton dates from the Aurignacian period, and exhibits the characters of the race known as Cro-Magnon, as de Quatrefages and Hamy thought.

The skull discovered in 1833 by Schmerling, in the cave at Engis in Belgium, has been examined by the best anthropologists of last century, and is now generally attributed to Neolithic times.

It may be that among the skeletons discovered in 1852, in the famous cave at Aurignac in the Haute-Garonne, there were human remains contemporary with the settlements of the Reindeer Age, as Edouard Lartet believed. But this we have now no means of ascertaining.

The famous settlement at Solutré, in the Saône-et-Loire, discovered in 1866, has yielded investigators a great number of human bone-remains. Unfortunately, in this case, we have to deal with burials dating from different periods, and it is not easy to separate Quaternary bone-remains from more recent skeletons. De Quatrefages and Hamy estimated at about fifteen the number of human skulls at Solutré which might have been contemporary with the Mammoth and Reindeer, but only six of these are fit for detailed examination.[2] Their morphology is not very homogeneous, and while they certainly

[1] Sollas, W. J., "Paviland Cave" (*Journal of the Royal Anthrop. Institute*, vol. xliii., 1913).

[2] For bibliography of former discoveries, see *Crania ethnica* by these two scientists.

show some characters of Cro-Magnon type, they differ so much as to lead us, at the present day, more than ever to regard their age with certain reservations.[1] The Solutrean deposits require new and more methodical investigation.

But the following discoveries are of greater importance.

Cro-Magnon. Laugerie. Duruthy.

Cro-Magnon is a place-name in the Commune of Tayac, near Eyzies in the Dordogne. In 1868, during the construction of the railway from Périgueux to Agen, workmen found under a sheltering rock the remains of five human skeletons, placed upon floors of occupation containing numerous bone-remains of animals, dressed flints, and great quantities of sea-shells. Informed in time, an experienced geologist, Louis Lartet, son of Edouard Lartet, went to Eyzies to continue the excavations and to devote himself to the scientific investigation of the bed.[2]

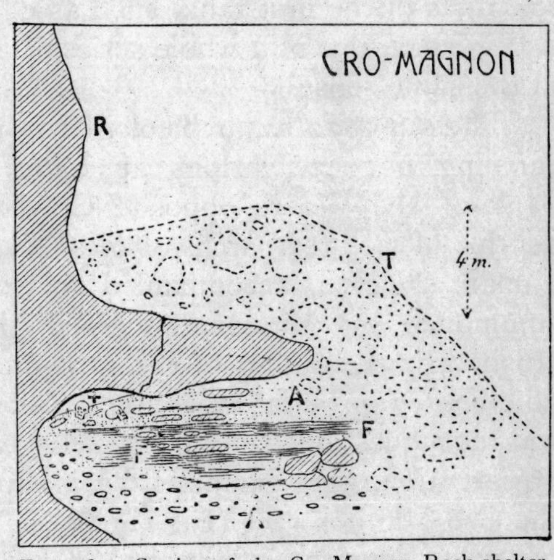

FIG. 169.—Section of the Cro-Magnon Rock-shelter. (After L. Lartet.) R, Cliffs of Cretaceous chalk forming shelter. T, Former talus; A, Cave deposits; F, Floors of occupation of Reindeer Age; + Bone-remains of the old man.

The human bones were grouped at the back of the shelter, in conditions suggestive of burial. Indeed, beside them there were collected numerous shells pierced with a hole for using as pendants, and other objects prepared for making up necklaces or other ornaments. The skeletons, studied first by Broca and Pruner-Bey, and later by de Quatrefages and Hamy, were considered by the latter anthropologists as prototypes of a new fossil race—the *race of Cro-Magnon*.[3]

[1] Arcelin, A., "Les nouvelles fouilles de Solutré" (*L'Anthropologie*, i., 1890).

[2] Lartet, L., "Une sépulture des troglodytes du Périgord" (*Bull. de la Soc. d'Anthrop. de Paris*, iii., 1868. *Annales des Sciences naturelles*, Ser. 5, vol. x.; *Matériaux*, 1869).

[3] *Crania ethnica.*

No doubt could be cast on the antiquity of this bed, which was definitely established by L. Lartet, for even a casual inspection of the section reveals the fact that the deposit of human bone-remains was necessarily prior to the formation of the enormous mass of fallen débris, covering, and so to speak sealing up, the prehistoric shelter (Fig. 169). Its age was, however, hotly disputed, especially by G. de Mortillet, who all along refused to believe that fossil men could have practised the veneration of the dead. And as his authority among prehistorians was considerable, almost all the latter rallied to his opinion, in spite of the protestations of a few scientists of the first rank, such as de Quatrefages and Hamy, and in spite also of a whole series of other discoveries testifying in the same sense.

Thus in 1872, an archæologist from Brive, named Massénat, carrying out excavations at Laugerie-Basse, on the farther bank of the Vézère opposite Cro-Magnon, discovered in the depths of a great archæological deposit of Reindeer Age a human skeleton, accompanied by sea-shells. M. Cartailhac, summoned by Massénat, agreed that the man had been crushed by the falling of the rock under which he lay.[1] Since no question of burial arose, the palæolithic age of the skeleton was not questioned. The bone-remains, which were in poor preservation, were examined by Hamy, who attributed them to the Cro-Magnon race (Fig. 170).

Some months afterwards, Louis Lartet and Chaplain-Duparc, exploring the Duruthy rock-shelter at Sorde in Landes, found at the bottom of the deposit, in palæolithic surroundings, a human skeleton, the elements of which were scattered and crushed, and were accompanied by ornaments.[2] On this occasion no shels were present, but about forty canine teeth of Bears and three canines of Lions were found, almost all pierced with a hole for suspension, ornamented with

[1] Massénat, E., Lalande, P., and Cartailhac, E., "Un squelette humain de l'âge du Renne à Laugerie-Basse" (*Matériaux*, vii., 1872). Hamy, E. T., "Description d'un squelette humain fossile de Laugerie-Basse" (*Bull. de la Soc. d'Anthrop. de Paris*, 1874).

[2] Lartet, L., and Chaplain-Duparc, "Une sépulture des anciens troglodytes des Pyrénées" (*Matériaux*, ix., 1874).

engravings (Fig. 171), and "divided into two groups lying at unequal distances from the skull, as if one had formed a necklace and the other a girdle."

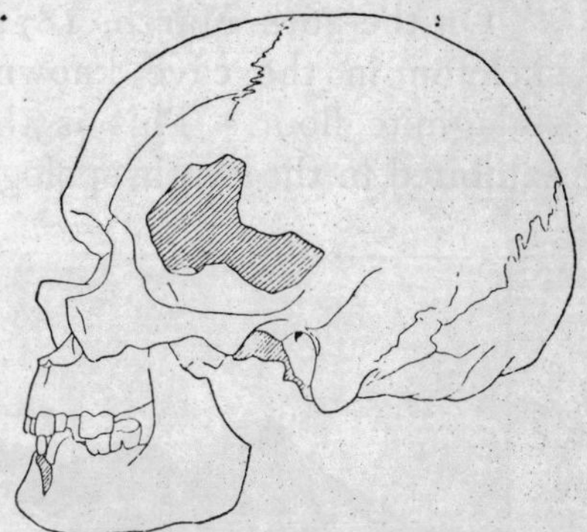

FIG. 170.—Profile of Skull from Laugerie-Basse. One-quarter natural size. (After Hamy.)

In spite of their poor state of preservation, Hamy was able to prove the resemblances of the human bones to those from Cro-Magnon.

The Grimaldi Caves.

From the same year, 1872, date the first discoveries of fossil men in the caves which are variously known as the Mentone, Baoussé Roussé, or Grimaldi Caves.[1]

These caves were nine in number, but one has been completely destroyed. Except the Prince's Cave, they are relatively shallow, and lie wide open towards the blue sea, in the midst of a charming landscape (Fig. 172). So they have been frequently visited at all times, and, before 1872, numbers of archæologists carried out there excavations of a superficial kind. At this period, M. E. Rivière, living at Mentone, undertook more thorough investigations. His observations and results were rendered easier by the work executed in the construction of the road from Marseilles to Genoa, which, skirting the foot of the cliffs,

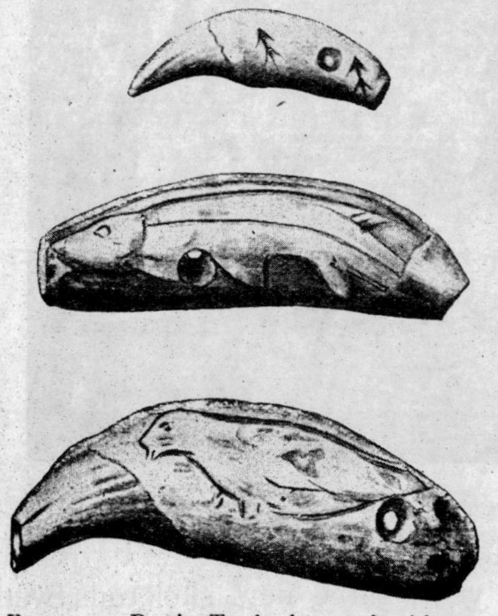

FIG. 171.—Bear's Teeth decorated with engravings (arrow-heads, a fish and a seal) and pierced by a hole for suspension. From the Duruthy Cave near Sorde (Landes). (After Louis Lartet.)

[1] The *Baoussé Roussé*, or *Red Rocks*, are indeed in the neighbourhood of Mentone, but they lie in Italy, in the territory of the Commune of Grimaldi, so that the most accurate name for the natural excavations into which the rocks have been hollowed, is the *Grimaldi Caves*.

cut into the talus of debris and pierced the Red Rocks by a tunnel.

On the 26th March, 1872, M. Rivière discovered a human skeleton in the cave known as the Cavillon Cave, below a stalagmite floor. This is the famous "Mentone Man," now exhibited in the Anthropological Gallery of the Paris Museum.

FIG. 172.—The Red Rocks, or *Baoussé Roussé*, at Grimaldi, near Mentone. General view of the Caves.

(Fig. 173). In the following year, three new skeletons were dug up in the cave known as the Cave of *Baousso da Torre*. In 1874 and 1875, M. Rivière extracted two skeletons of children from Cave No. 1, since called on this account the *Grotte des Enfants* (Fig. 174, p. 267).

All these human spoils were found in conditions resembling those already observed in the Dordogne. Here as there, the skeletons were accompanied by a whole series of objects, particularly by a quantity of shells which had been used for purposes of decoration or of dress. It is to Rivière's credit that he understood and affirmed that these were Palæolithic burials. His opinion was contested by the majority of prehistorians, headed by Gabriel de Mortillet. Later discoveries, which I shall presently discuss, have proved that Rivière was not mistaken.[1]

FIG. 173.—The "Mentone Man," found by M. Rivière. (Anthropological Gallery of the French Nat. Hist. Museum, Paris.)

Chancelade.

In 1888 two archæologists from Périgueux, Messrs Féaux and Hardy, excavated a rock shelter at Raymonden, a commune in Chancelade, near Périgueux, and found a human skeleton under the lower Magdalenian floors of occupation, that is to say, in the upper stages of the Reindeer Age. The doubled-up, unnatural posture of the skeleton pointed to deliberate burial. As at Mentone, the corpse must have been sprinkled with red ochre. The skeleton, exhumed with great care, was examined by Dr Testut,[2] who established it as the type specimen of a new race called the *Chancelade race*. It is now in the Museum at Périgueux.

Brünn. Predmost.

The Quaternary silts on which is situated the village of Brünn, in Moravia, have long been explored on account of their rich animal remains and dressed flints. In 1891 they yielded a human skeleton, lying $4\frac{1}{2}$ metres below the surface of the soil, along with some Mammoth and Rhinoceros remains. This skeleton, partly

[1] Rivière, E., *De l'Antiquité de l'Homme dans les Alpes-Maritimes*, Paris, 1887.

[2] Testut, L., "Recherches anthropologiques sur le squelette quaternaire de Chancelade" (*Bull. de la Soc. d'Anthrop. de Lyon*, viii., 1889).

destroyed at the time of its discovery, is said by Obermaier[1] to have been richly decked with ornaments; round about it there were collected more than six hundred pieces of the Tusk Shell (*Dentalium*) once strung into a necklace or a breast ornament, several discs of stone pierced or decorated, and a tiny human figure wrought in ivory. Some bones still preserved traces of deep red coloration.[2] This again, then, was also a case of true burial. The skull, badly preserved, is dolichocephalic; certain anthropologists refer it to the Cro-Magnon type, others regard it as distinct from that type.

The locality of Predmost near Prerau, which also lies in Moravia, possesses an important Palæolithic settlement enclosed in a covering of gravels and clays surrounding the "Hradisko" rock. Excavations carried out from time to time since the year 1880 have revealed a cold climate Pleistocene fauna, so rich that the Mammoth is represented by remains of 800 to 900 individuals. In addition, this settlement has yielded a collection of flint implements comprising more than 30,000 specimens, a whole series of products of an industry in ivory, in bone and in Reindeer horn, and works of art, in particular a curious statuette in ivory representing a Mammoth.

Some human remains have also been found at different times.[3] In 1894 Maschka discovered a large burial containing forty complete skeletons, as well as the remains of six other individuals.[4] The bodies were protected by a sort of stone rampart. The skeleton of a child bore a necklace formed of forty small oval beads of ivory. This burial must have been prior to the main archæological layer, which belonged to the Solutrean period. The skulls are dolichocephalic; the male skulls have marked orbital ridges; the long bones point to great height (1.80 m.—5 ft. 10.8 ins.) They show no neander-

[1] Obermaier, H., "Les restes humains quaternaires dans l'Europe centrale" (*L'Anthropologie*, xvi., 1905, and xvii., 1906). Numerous bibliographical references regarding Brünn, Predmost, and other similar localities.

[2] Makowsky, A., *Der Mensch der Diluvialzeit Mährens*, Brünn, 1899.

[3] Maschka, K., *Der diluviale Mensch in Mähren*, 1886. Kriz, M., *Beiträge zur Kenntnis der Quartärzeit in Mähren*, 1903.

[4] *L'Anthropologie*, xii., 1901, p. 147.

thaloid character, as has been asserted. The "Predmost race" represents only a variety of *Homo sapiens*; according to Szombathy,[1] it should be classified as belonging to the Cro-Magnon type.

The chronological order of the discoveries brings us back to France. In the Ain, at the mill of Les Hoteaux, near the village of Rossillon, there is a cave containing floors of occupation dating from the Reindeer Age. In 1894, Messrs Tournier and Guillon[2] brought to light a burial in the most ancient of these settlements. The bones of the skeleton were covered with red ochre. The articles accompanying the burial consisted of a Red Deer's tooth, pierced with a hole for suspension, dressed flints, and a "bâton de commandement" of Magdalenian Age. The skeleton is that of a youth of sixteen to eighteen years of age. The publication of these facts revived the discussions between the partisans and opponents of the theory of Palæolithic burials. Agreement could only be reached after further and conclusive discoveries; and such were not long in forthcoming.

Les Hoteaux.

The exploration of the Grimaldi and Mentone caves did not cease with the labours of M. Rivière. Different archæologists undertook investigations, particularly in the fifth cave, known as *Barma Grande*. In 1884, one of the most energetic of collectors, Julien, found there a human skeleton. Unfortunately this skeleton was almost completely destroyed, in consequence of a dispute between its discoverer and the proprietor of the cave. It was found possible to restore the skull, and it is now in the Mentone Museum. This Barma Cave seems to have been the richest and most interesting of the Grimaldi group. The proprietor of it, Abbo, the owner of the quarry, has partially destroyed it by exploiting it as a quarry for building stone. His work of excavation, unfortunately

The Grimaldi Caves. Further Explorations.

[1] Szombathy, J., "Un crâne de Cro-Magnon trouvé en Moravie" (*L'Anthropologie* xii. 1901, p. 150). This skull, from a cave near Lautsch, certainly belongs to the Cro-Magnon race. But its age has been disputed.

[2] Tournier (Abbé) and Guillon, C., *Les hommes préhistoriques dans l'Ain*, Bourg, 1895.

lacking in any scientific method, has produced several human skeletons, which have been examined by Dr Verneau.[1]

These finds of Abbo in the Barma Grande afforded material for fresh discussion, which would probably have continued to this day but for an intervention as generous as it was enlightened.

The late Prince Albert I. of Monaco, whose noble spirit of scientific enquiry was so many-sided, had already taken a great interest in the Grimaldi caves. From 1883 onwards he had himself worked with great method in the Barma Grande Cave. In 1895, anxious at least to advance, if not to decide, the solution of the important problems arising out of the Baoussé Roussé investigations, the Prince gave orders for the carrying out of a systematic exploration. The excavations, conducted with rare skill and patience by M. the Canon de Villeneuve assisted by M. Lorenzi, were first of all directed to the large cave known as the Prince's Cave, then almost untouched. They produced interesting results from the geological and palæontological points of view, thanks to which we now possess accurate information regarding the succession of events in Quaternary times in this part of the Côte d'Azur. Innumerable fossil bones of animals were recovered from about 4000 cubic metres of the deposits of this cave; but not a single human relic was met with. Herein lay disappointment; for everyone had counted on the Prince's Cave to decide definitely the ages of the various skeletons of fossil men discovered in the neighbouring caves, concerning which there had been so much dispute.

The Prince then decided to transfer his enterprise to other points. The Grotte des Enfants (Fig. 174) had only been imperfectly excavated. M. Rivière's work had stopped at a depth of 2.70 metres, whilst below this level cave-deposits, nearly 8 metres deep, still remained untouched. Here the investigations met with the greatest success from the anthropological point of view, for in the cave four human skeletons were discovered at three different levels. Since, at the same time, many bone-remains of animals were collected and the

[1] Verneau, R., "Nouvelle découverte de squelettes préhistoriques aux Baoussé Roussé, près de Menton" (*L'Anthropologie*, iii., 1892).

stratigraphy of the cave could be accurately determined, the geological age of the skeletons was readily and indisputably

FIG. 17.—View of the "Grotte des Enfants," in the Baoussé Roussé, at Grimaldi.

established. The Prince having confided to me the task of collaborating in his work, my geological and palæontological observations were able to furnish a firm chronological basis for my friends, Messrs Cartailhac and Verneau, to whom was assigned the archæological and anthropological study of the layers.[1]

The main result of these observations was that all the human skeletons were really Pleistocene and dated from the earliest part of the Reindeer Age. Contrary to what was previously asserted, the Reindeer formed part of the Pleistocene fauna of the Côte d'Azur; I discovered its bones associated with those of its usual companions of the cold climate fauna, superimposed, here as elsewhere, upon the more ancient warm climate fauna.

The new skeletons exhumed from the Grotte des Enfants had been the objects of true burial, the articles deposited with them resembling those accompanying the skeletons previously discovered—perforated shells, objects of decoration, bones coloured red, and so on. Dr Verneau recognized the fact that the two skeletons from the lowest level represented a special race, which he called the *Grimaldi Race.* The skeletons from the upper layers, like those discovered by Rivière, Julien, and Abbo, entirely agree with the Cro-Magnon type. These precious relics, along with all the results of the Prince of Monaco's excavations, have been collected and are now exhibited to the public in the Anthropological Museum of Monaco under the directorship of M. de Villeneuve.

Combe-Capelle. Obercassel.

In 1910, Hauser, the dealer in antiques, found a skeleton ornamented with sea-shells in a bed at Combe-Capelle in the Dordogne. This skeleton was acquired by the Berlin Museum for which Hauser acted in the capacity of agent. An attempt has been made to establish it as the type specimen of a particular species under the name *Homo aurignacensis*, and the German anthropologist, Klaatsch, has propounded the most extravagant hypothesis with regard to it.[2] As a matter of fact it also is nothing more than a variety

[1] *Les Grottes de Grimaldi* (*Baoussé Roussé*), *Historique et Description*, par L. de Villeneuve; *Géologie et Paléontologie*, by M. Boule; *Anthropologie*, by R. Verneau; *Archéologie*, by E. Cartailhac. 2 vols., Monaco, 1906-19.

[2] Klaatsch, H., and Hauser, O., *Homo aurignacensis Hauseri* (*Præhistorische Zeitschrift*, i., 1910). Klaatsch, H., "Die Aurignac-Rasse und ihre Stellung im Stambaum der Menschheit (*Zeitschrift für Ethnologie*, 1910).

of the Cro-Magnon Race, exhibiting, according to Giuffrida-Ruggeri, certain Ethiopian characters (Fig. 175).

In 1914, the German physiologist, Verworn, published an account of the discovery of a double burial, the skeletons of a man and a woman, from a settlement of the Reindeer Age at Obercassel, near Bonn. These skeletons, accompanied by some bones engraved in the Magdalenian manner, had been coloured red. They show the chief characteristics of the Cro-Magnon race, together with some features of the Chancelade skeleton.[1]

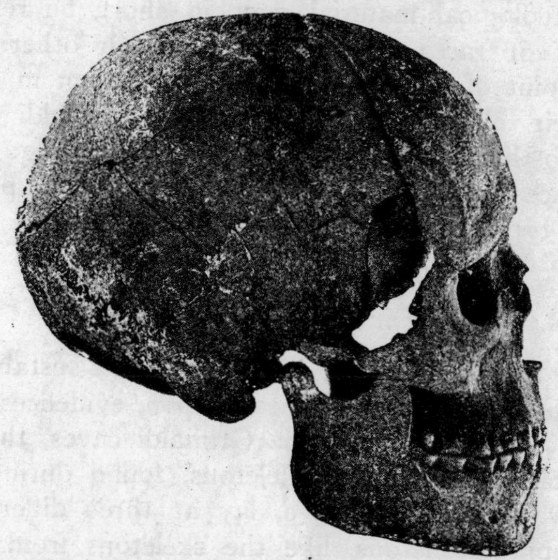

FIG. 175.—Skull from Combe-Capelle. (After Klaatsch.)

Such are the principal data we possess for the study of the fossil men of the Upper Pleistocene. I might describe certain other discoveries, but either they consist of bone fragments barely sufficient to provide any accurate information, or else some doubt exists as to the conditions of their deposition. We may mention those at Bruniquel (Tarn-et-Garonne), at La Madeleine (Dordogne), at Gourdan (Haute-Garonne), at Le Placard, near Vilhonneur (Charente), at Freudenthal and Kesslerloch, near Schaffhausen in Switzerland, and in the Prince John Cave, near Lautsch in Moravia.

[1] Verworn, M., Bonnet, R., and Steinmann, G., "Diluviale Menschenfunde in Obercassel bei Bonn," Wiesbaden, 1919.

To these osteological records we must now add evidences contributed by works of art depicting human beings, which we shall study separately.

With regard to the skeletons, which we shall consider first of all, the opinions of anthropologists are somewhat at variance. In all these Men of the Reindeer Age, some anthropologists distinguish varieties of only one race of *Homo sapiens.* Others are inclined to distinguish as many special types as there are specimens. I think that the truth lies between these two views. All this osteological material may, in short, be relegated to three types or races, closely akin to each other from the zoological point of view. We shall study them in succession and in their chronological order: the Grimaldi Negroids, belonging to the earliest part of the Reindeer Age; the Cro-Magnon Race and its varieties, to the Aurignacian period; the Chancelade type, to the Magdalenian period.

The Grimaldi Race.

Its Geological Age.

As its name indicates, this race was established by Professor Verneau upon evidences obtained from one of the Grimaldi caves, the "Grotte des Enfants." The human skeletons, found during this excavation by M. de Villeneuve, lay at three different levels. The two upper skeletons, like the skeletons from the other caves and like the famous "Mentone Man," belong to the Cro-Magnon race, which we shall presently study. The two skeletons of the Grimaldi race lay at a lower level. Discovered on 3rd June 1901, they were carefully exhumed and conveyed to the Monaco Museum.

It is important to study exactly the conditions of their embedding. A section of the cave-deposits of the "Grotte des Enfants" is shown in Fig. 176, which I was enabled to make from M. de Villeneuve's descriptions and my own observations. This cave-deposit consists of a whole series of layers, superimposed to a depth of 10 metres, and chiefly composed of materials brought thither by human beings, and of cinders mixed with rock fragments accruing from the disintegration of the rocky walls; the floors of occupation distinguished by the excavators are merely zones showing more charring than the neighbouring zones. The whole of these cave deposits are Pleistocene, for bone-remains of the Reindeer are found even in the uppermost layers. This first conclusion, determined by a study of the fauna, is one of the most important from the point of view of a solution of the much discussed problem of the age of the human skeletons found in all the Grimaldi caves.

The deepest layers, resting on the rocky floor of the cave and containing Merck's Rhinoceros, must, in my opinion, be considered as forming a transition from the Lower Pleistocene to Mid Pleistocene. The skeletons of Negroids, situated at a depth of 8½ metres on a floor of occupation directly above these deposits, must then belong to the Moustierian period, and this is the conclusion at which I first arrived. Now, we have here to deal with a human type which is quite distinct from that of Neanderthal and exhibits many characters common to certain African races of *Homo sapiens*, and this co-existence in Western Europe, at one and the same geological period, of two human forms so widely different would seem to be a fact of very great importance.

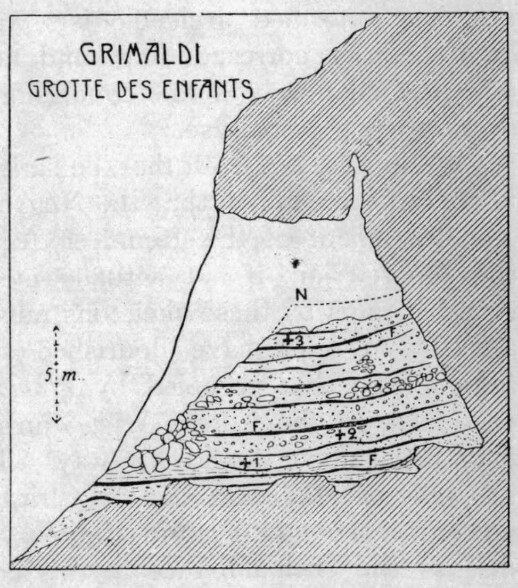

FIG. 176.—Section of the Grotte des Enfants.

N, original level of the soil; F, layers of occupation; +1, spot at which the Negroids lay; + 2, + 3, spots at which the skeletons of Cro-Magnon type lay.

But here we must make some reservations. M. de Villeneuve's observations, contained in his diary of the excavations, show indeed that the Negroids lay in a pit 0.75 metre in depth. They were, therefore, contemporary, not with the layer of occupation at the level of which they were encountered, and which I believe must be considered as dating from the Mid Pleistocene, but rather with a higher layer of occupation which marked the uppermost limit of the pit, and the products of which, according to M. Cartailhac, are already definitely Aurignacian in character. The Negroids are, therefore, Aurignacian, like all the Cro-Magnon Men. In addition to the somewhat abrupt change between the products of the Moustierian and the Aurignacian industries, now recognized by the best informed archæologists, there would seem also to have been a corresponding, and no less important, anthropological change, and this conclusion appears to be best fitted to meet the present case.

We must not, for all that, be satisfied with mere words. It is none the less a fact that the Negroid skeletons date from the commencement of the Reindeer Age, a period bordering on the Moustierian, if not actually merging with it. The race revealed to us by these skeletons must therefore be much more ancient; it must have flourished somewhere, in a different region from that occupied by *Homo neanderthalensis*, and in particular on the Côte d'Azur, whither *Homo neanderthalensis* may never have found his way. The Aurignacian culture, probably of Mediterranean or African origin, is perhaps more ancient in this spot than in France. And though we may well discuss the Grimaldi race in this chapter devoted to Upper Pleistocene Man, with whom it forms the great mass of *Homo sapiens fossilis*, we might just as well have discussed it in connection with the Mid Pleistocene.

The Skeletons.

We have just seen that the Negroid skeletons had been objects of burial. The first is that of an old woman, the second that of a young man of fifteen to seventeen years of age. They lay side by side, their bodies drawn up, their lower limbs much bent (Fig. 177), in the manner of Peruvian mummies. The skull of the young man was protected by a sort of cist, formed of undressed blocks of stone. There were also found the remains of a head-dress and bracelets of shell work made from Dog-Whelk (*Nassa*) shells.

According to Professor Verneau,[1] whose important work

[1] Verneau, R., *loc. cit.*, *Les Grottes de Grimaldi*, vol. i., part i., "Anthropologie."

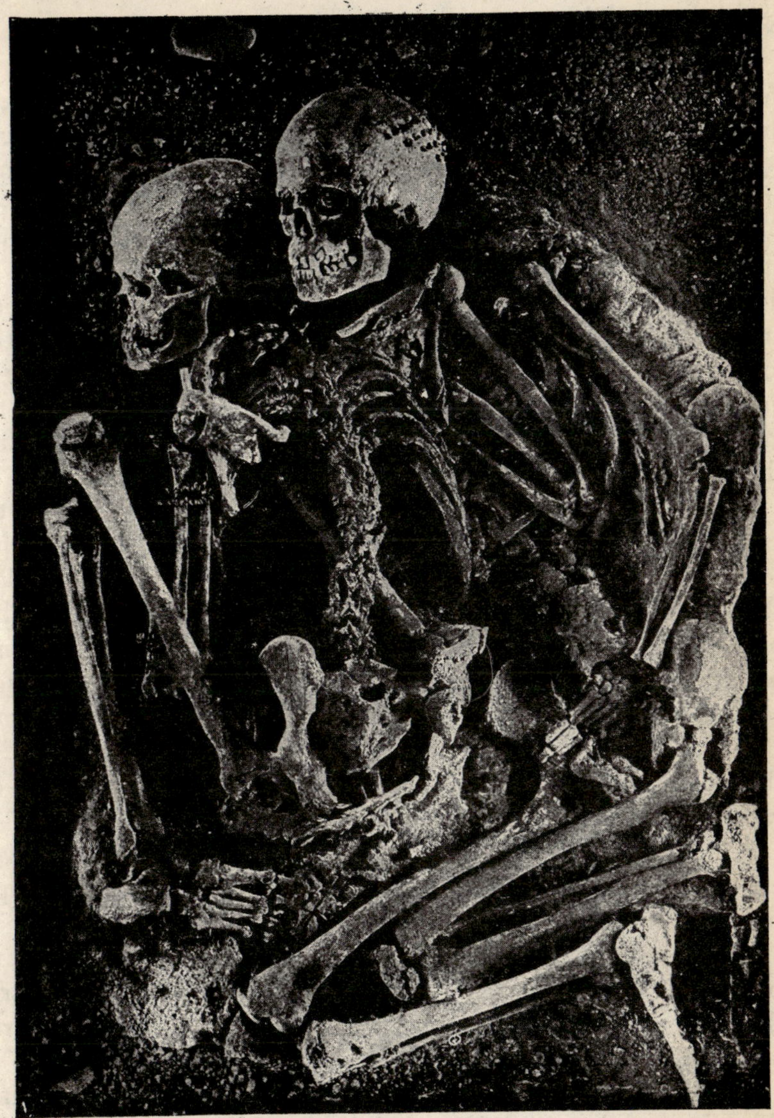

FIG. 177.—The two Skeletons of Negroids discovered by M. de Villeneuve in the Grotte des Enfants. Musée d'Anthropologie, Monaco. (After Verneau.)

I shall here summarize, "these two ancient human beings differ appreciably from those who succeeded them [in the same cave, that is to say those of the Cro-Magnon type], and they exhibit the most striking points of resemblance to each other." These two subjects, instead of attaining the great height of the other cave-dwellers of the Baoussé Roussé, scarcely exceed the average height of the French of our own day (1.56 metres [5 ft. 1.4 in.] in the case of the youth, 1.60 metres [5 ft. 3 ins.] in the case of the old woman).

When we compare the dimensions of the bones of their limbs, we see that the leg was very long in proportion to the thigh, the forearm very long in proportion to the whole arm; and that the lower limb was exceedingly long relative to the upper limb. Now these proportions reproduce, but in greatly exaggerated degree, the characters presented by the modern Negro. Here we have one chief reason for regarding these fossils as negroid, if not actually negro.

The Skulls.

The negroid affinities are likewise indicated by the characters of the skull. The heads are large; the skulls are very elongated, hyperdolichocephalic (indices 68 and 69), and, seen from above, they present a regular elliptically-shaped contour, with flattened parietal bosses (Fig. 178). The skulls are also very high, so that their capacity is at least equal to that of the average Parisian of our day: 1580 cubic centimetres in the case of the young man, 1375 cubic centimetres in the case of the old woman. The mastoid apophyses are small.

The face is large but not high, whilst the skull is excessively elongated from the front backwards; so that the head might be called unbalanced or disharmonic.[1]

The forehead is well developed and straight; the orbital ridges project only slightly. The orbits are large, deep and subrectangular; their lower border is everted towards the front.

The nose, depressed at the root, is very broad (Platyrrhinian). The floor of the nasal fossæ is joined to the anterior surface of the maxillary by a groove on each side of the nasal spine, as in Negroes, instead of being bordered by a sharp edge as in the white races. The canine fossæ are deep.

[1] In order to have balance or *harmony*, vertical lengthening of the face should accompany the lengthening of the skull.

The upper maxillary projects forwards in very marked fashion. This prognathism especially affects the subnasal or

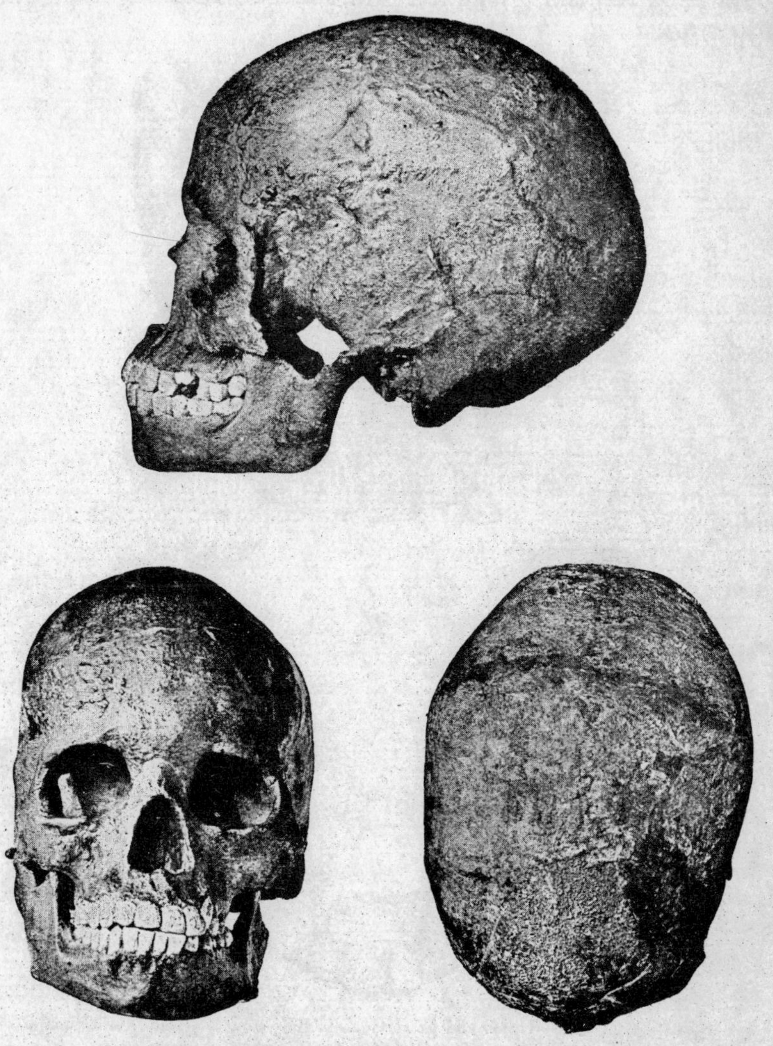

FIG. 178.—The Skull of the young Negroid from Grimaldi, seen in profile, in full face, and from above. One-third natural size. (After Verneau.)

alveolar region. The palatal arch, though only slightly developed in breadth, is very deep (Fig. 179).

The jaw is strong, its body very thick; the ascending

branches are broad and low. The chin is not greatly developed; a strongly marked alveolar prognathism, correlated with the upper prognathism, gives it a pronouncedly receding appearance (Fig. 180).

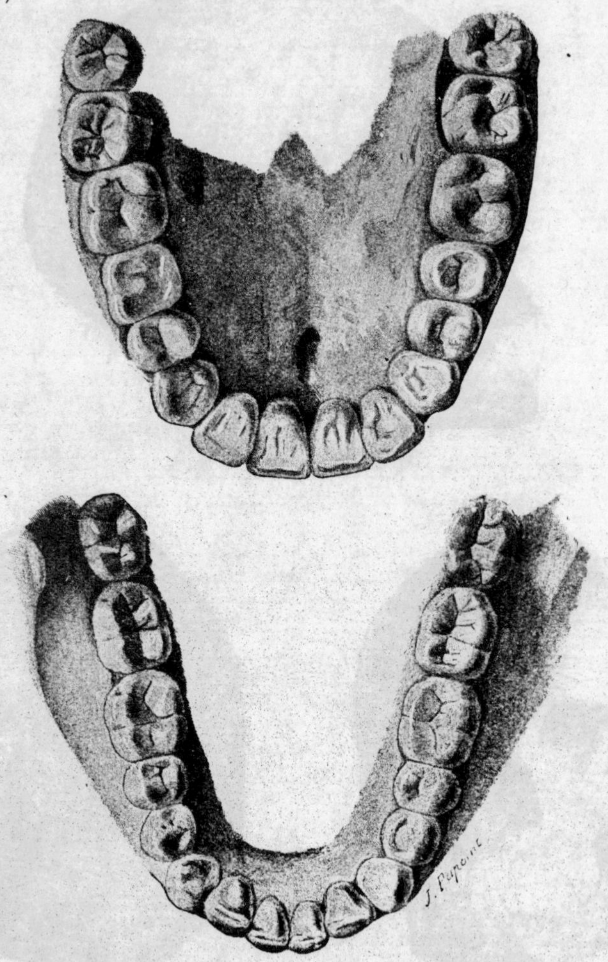

FIGS. 179 and 180.—Upper and Lower Jaws of the young Negroid from Grimaldi. Natural size. (After Albert Gaudry.)

The majority of these characters of the skull and face are, if not negro, at least negroid. Some of them may be regarded as indicating inferiority from the morphological point of view.

Dentition. The dentition of the young individual has been studied by Albert Gaudry: It strikes one at first sight by its unusual size. Its greatest resemblances are with the dentition of the Australian aborigines, which has retained many primitive characteristics. The dental arches are less widely divergent than in the superior races. The alveolar borders are more elongated; and the development of the teeth and of their cusps is in proportion to this prolongation.

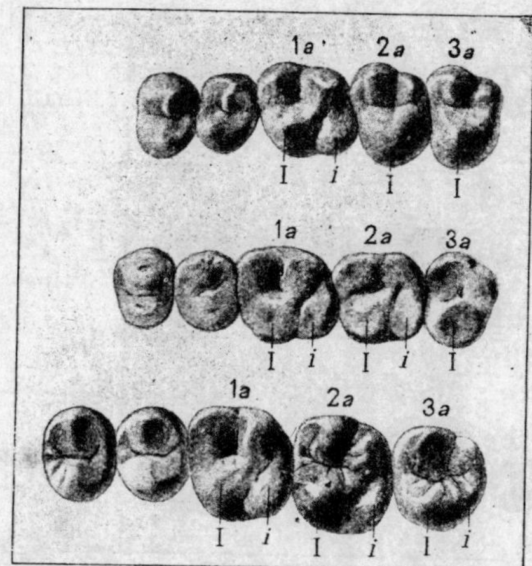

FIG. 181.—Comparison of the Upper Left Molars of the young Negroid, of an Australian, and of a Frenchman. Natural size. (After A. Gaudry.)

1*a*, 2*a*, 3*a*, first, second, and third true molars; I, inner cusp of the first lobe; *i*, inner cusp of second lobe, present in all the molars of the fossil.

So we find that the morphology of the molars retains several simian characters like those found in the fossil types we have already examined, as is the case in the modern primitive races with large dentitions. All the upper true molars have four well-developed cusps, even the furthest back tooth, which in civilized races has only three (Fig. 181). All the lower molars have five quite distinct cusps, even the second and third, which, in the white races, usually have only four (Fig. 182).

Albert Gaudry,[1] when examining this dentition, was struck with the narrowing of the fore part of lower jaw at the level of the

[1] Gaudry, A., "Contribution à l'Histoire des Hommes fossiles" (*L'Anthropologie*, xiv., 1903).

premolars and canines (Fig. 180). In this he sees an indication of inferiority in agreement with the marked prognathism of the lower jaw, the consequence of which was to leave less room for the development of the tongue and for its free forward movement.

Limb Bones.

We have seen that the proportions of the limbs and of their several segments are negroid proportions. M. Verneau recognized that "in the vertical direction of the haunch bones, in the curve of the iliac crest, in

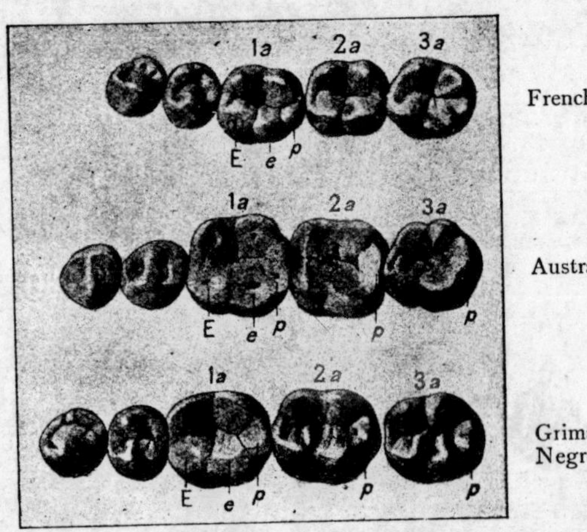

FIG. 182.—Comparison of the Lower Left Molars of the young Negroid, of an Australian, and of a Frenchman. Natural size. (After A. Gaudry.)

1*a*, 2*a*, 3*a*, first, second, and third true molars ; E, external cusp of first lobe ; *e*, external cusp of second lobe ; *p*, posterior cusp present in all the true molars of the Australian and of the fossil Man, and particularly well developed in the latter.

the reduced dimensions of the great sciatic notch, the pelvis of the old woman differs from the pelvis of the modern European female, and resembles, contrariwise, that of the Negress."

The shaft of the femur is rather strongly bent, and the tibiæ exhibit a certain degree of retroversion at the upper head. These characters, which we have already studied in Neanderthal Man, are here, however, much less marked.

Affinities and Survivals.

To sum up, in the most ancient skeletons from the Grotte des Enfants we have a human type which is readily comparable to modern types and especially to the negro or the negroid types. It would be interesting to go further than M. Verneau, with a prudence that we can readily appreciate, ventured, and to endeavour to define even more closely his comparisons with present-day types. For my part, I have been greatly struck by the resemblances these Grimaldi Negroids bear to the group of South African tribes, the Bushmen and the Hottentots. Comparisons which I have been able to make with the material at my disposal, in particular with the skeleton of the Hottentot Venus, have led me to note, for instance, the same dolichocephalic character, the same prognathism, the same flattening of the nose, the same development in breadth of the face, the same form of jaw, and the same great size of teeth. The only differences are to be found in the stature, and perhaps in the height of the skull. Professor Sollas of Oxford has also made similar observations.[1]

This comparison between two groups of human beings, so widely separated at the present day both in time and in space, seems to be confirmed, as we shall presently see, by an examination of the steatopygian statuettes of women, which some of the oldest deposits of the Reindeer Age have yielded.

M. Verneau has investigated the survivals of the Grimaldi race at different prehistoric periods. He has first of all compared this type with the Cro-Magnon, which succeeded it in place. "At first sight," he says, "the two races appear to differ greatly from each other; but on examining them in detail, we see that there is no reason why they should not have had some ties of kinship." M. Verneau even declares that the Grimaldi Negroids "may have been the ancestors of the hunters of the Reindeer Age."

My learned colleague and friend has devoted himself to a long and laborious enquiry for the purpose of discovering, in both prehistoric and modern races, any survivals or reappearances of the Grimaldi types.

"In Brittany, as well as in Switzerland and in the north of Italy, there lived in the Polished Stone period, in the Bronze Age and during the early Iron Age, a certain number of individuals who differed in certain characters from their

[1] Sollas, W. J., *Ancient Hunters*, 2nd ed., London, 1915.

contemporaries," in particular in the dolichocephalic character of their skull, in possessing a prognathism that was sometimes extreme, and a large grooved nose. This is a matter of partial atavism which in certain cases, as in the Neolithic Breton skull from Conguel, may attain to complete atavism. Two Neolithic individuals from Chamblandes in Switzerland are negroid not only as regards their skulls but also in the proportions of their limbs. Several Ligurian and Lombard tombs of the Metal Ages have also yielded evidences of a negroid element.

"All these facts," says M. Verneau, "show that our two Grimaldi Negroids are really representatives of a race which played an important part in Western Europe. Were it merely a matter of chance individuals who had accidentally expired in the Baoussé Roussé, we should not see their influence making itself felt, by atavism, during the Neolithic period and during the first Metal Ages, from Brittany to Switzerland and throughout the north of Italy."[1]

It is not only in prehistoric times that the Grimaldi race seems to have made its influence felt. M. Verneau has been able to see, now in modern skulls and now in living subjects, in the Italian areas of Piedmont, Lombardy, Emilia, Tuscany, and the Rhone Valley, numerous characters of the old fossil race. The persistence or reappearance of these characters, which anthropologists have long recognized without arriving at any understanding of them, are explained by the facts of atavism. "That we may still find at the present day so many traces of a racial type having characters recalling those which I have observed in the Grimaldi race," declares M. Verneau, "must of necessity have been due to the fact that this race was formerly represented in our country by a whole group." And he adds: "We must therefore admit that an almost negro element lived in south-western Europe towards the Mid Quaternary period, between the Spy race and the Cro-Magnon race."

[1] Since the publication of M. Verneau's work, other skeletons of Negroids have been described from the Neolithic of Illyria and the Balkans. The prehistoric statuettes from Sultan Selo in Bulgaria, dating from the Copper Age, seem to epresent figures of Negroids. (According to Zupanic in "Les premiers habitants des pays Yougo-Slaves," *Revue anthropologique*, 1919, p. 32.)

The Cro-Magnon Race.

This race derives its name from the locality where, in the circumstances recounted above, the remains of five human bodies were found—an old man, two adult men, a woman and an unborn infant. These remains, first investigated by Broca and Pruner-Bey,[1] represent the prototypes of that fossil human race, recognized, described and named by de Quatrefages and Hamy.[2]

We have seen how discoveries made in different regions have revealed the existence of this race throughout a great stretch of Europe, and in archæological conditions which indicate methods of burial and funeral rites of very great uniformity. Among these discoveries the most important is represented by the series of ten skeletons exhumed successively from the various caves of the Baoussé Roussé. The first few skeletons, discovered by Rivière, were studied by de Quatrefages and Hamy. The succeeding ones, discovered in the course of Abbo's work and the excavations of the Prince of Monaco, form the subject of a masterly monograph by M. Verneau.

It is now easy to define the Cro-Magnon race by certain general characters. Some readily observed differences accord with different beds or different regions, but such differences fall within the limits of those existing at the present day between individuals or, at most, between varieties of the same race.

The Cro-Magnon Skeletons.

To begin with, let us take as types the original specimens from Cro-Magnon, and in particular the skull of the old man, who exhibits the characters of the race in an especially clear, if not indeed even in a somewhat exaggerated degree (Fig. 183).

The skull is dolichocephalic (cephalic index—73.7) and very large (cranial capacity—1590 cubic centimetres). The cranial vault is high, *hypsicephalic*. Seen from above the brain-box exhibits a pentagonal contour, due especially to the marked projection of the parietal bosses. This is the so-called

[1] In the "Reliquiæ aquitanicæ," *Bull. de la Soc. d'Anthrop. de Paris*, 1868.
[2] *Crania ethnica.*

dolichopentagonal form of certain anthropologists. Seen in profile, we observe the forehead rising above relatively slight superciliary ridges; then the vault is developed in a

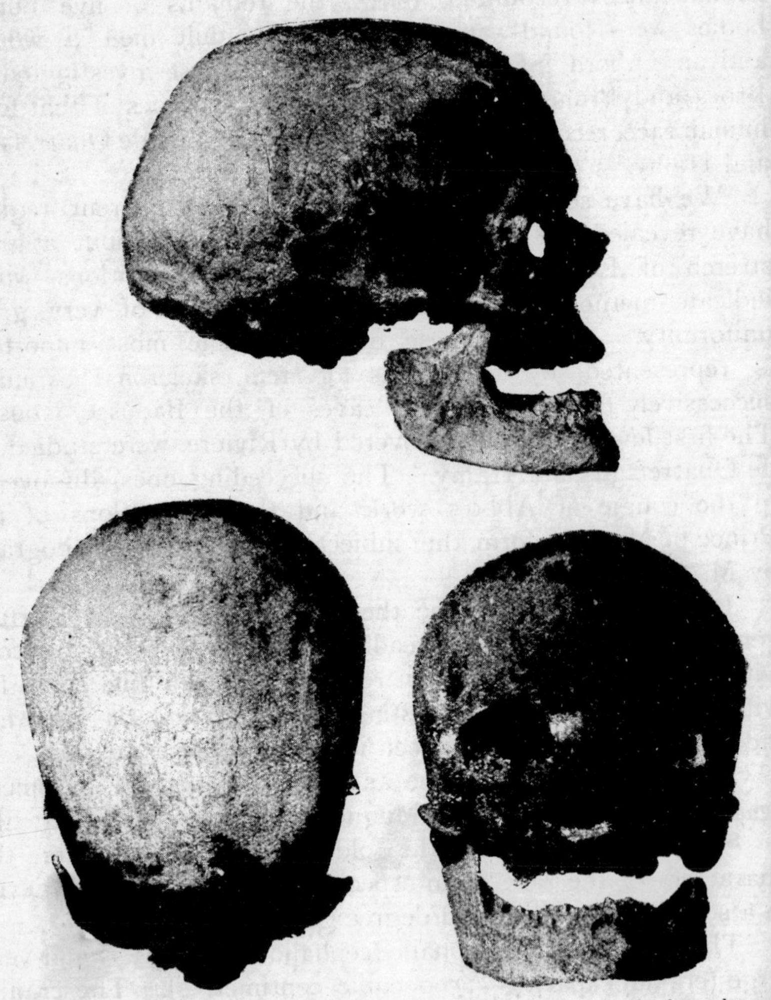

FIG. 183.—Skull of the "Old Man" from Cro-Magnon, seen in profile, from above, and full face. One-third natural size. Anthropological Gallery, French National Museum of Natural History, Paris.

regular curve in the anterior and middle regions, whilst t parieto-occipital region forms a broad flattened area succeed by the projection of the nape. "Thus," says de Quatrefag "in this savage contemporary of the Mammoth, the skull presents in a high degree all the characters regarded as indications of an intellectual development of the most advanced kind."[1]

The face, no less remarkable, is relatively flat and very broad, whilst the skull is narrow and long. The proportions of the head are disharmonious (see note, page 274). Below the wide high frontal bone, arched in the centre, open the orbits, likewise very wide, with almost rectilinear margins forming a quadrilateral contour. The cheek-bones are strong and prominent. On the other hand the nose is narrow, long and fine, *leptorrhinian;* the nasal bones project forwards. The upper maxillary, withdrawn to the level of the dental arches, exhibits a somewhat pronounced prognathism. The roof of the palate is relatively narrow, is shallow, and bears a median projection.

The long bones point to great height (1.82 metres [5 ft. 11.6 ins.]) and an athletic physique; their muscle imprints are strongly marked. In the femur the *linea aspera* is so developed as to form a kind of prominent column. The tibia is flattened like a sabre-blade, a *platycnemic* tibia. This characteristic, absent in Neanderthal Man, seems to be fairly general in the Cro-Magnon race.

The Grimaldi Skeletons.

M. Verneau's examinations of the fine series of skeletons from the Baoussé Roussé have put him in possession of new facts. He was able to study nine skulls, eight male and one female, all belonging to the Cro-Magnon type (Fig. 184). These skulls exhibit a certain degree of variation, but, according to M. Verneau, the differences do not exceed the individual variations to be met with in relatively homogeneous groups.

All the skulls are remarkable for their great size, due partly to the great stature of the individuals. There exists a striking lack of harmony between the skull, which is dolichocephalic, and the face, which is both flat and broad. The flattened parieto-occipital area, found in the Cro-Magnon skulls, is always present. In all the specimens, the face shows the

[1] Quatrefages, A. de, *Hommes fossiles et Hommes sauvages*, Paris, 1884, p. 65.

essential characteristics of the type: transverse enlargement due to the development of the cheek-bones and zygomatic arches, rectangular orbits, leptorrhinian nose, etc. The strong jaw has a prominent chin, triangular in shape.

But to these general characters there fall to be added certain peculiarities constituting a variant of the type. The parietal bosses are less pronounced, and, in consequence, the pentagonal form of the skull, seen from above, is somewhat

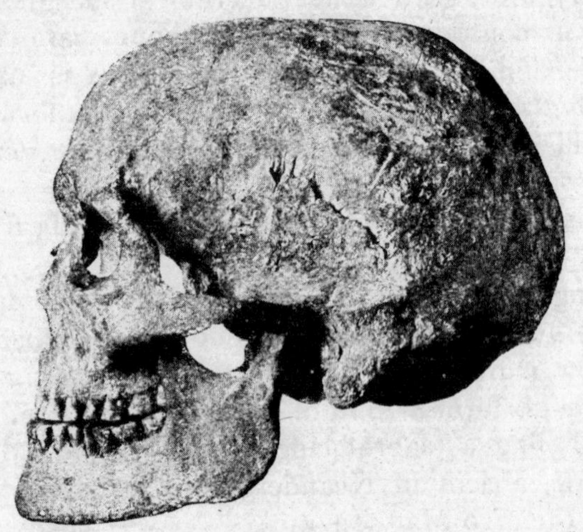

FIG. 184.—Skull of large Male Individual from La Grotte des Enfants, seen in profile. One-third natural size. (After Verneau.)

different; the occiput is less prominent, and the sub-nasal prognathism tends to disappear. "These peculiarities," says M. Verneau, "can in no way justify us in separating the man from the Baoussé-Roussé from the Vézère type. . . . The old man from Cro-Magnon exhibits this type in exaggerated form. Amongst the skulls which anthropologists agree in classifying in the same group, some are met with which present the weaker characters I have just described."

Thanks to the numerous specimens at his disposal, Dr Verneau was able to re-examine the problem of the stature of the men of Cro-Magnon type. Broca had estimated the stature of the old man at 1.80 metres [5 ft. 10.8 ins.]. M. Rivière had calculated the height of the three adults whose skeletons he had found, as varying from 1.85 to 2.05 metres [6 ft. 0.7 ins. to 6 ft. 8.7 ins.]. According to Verneau, these figures are too large. Those obtained by him vary from 1.79 to 1.94 metres [5 ft. 10.4 ins. to 6 ft. 4.4 ins.] for five male adults, giving an average of 1.87 metres [6 ft. 1.5 ins.]. The Grimaldi cave-dwellers must still be regarded, even after this correction, as men of very great height (Fig. 185).

FIG. 185.—Skeleton of the large Male Individual discovered by M. de Villeneuve in the Grotte des Enfants. Musée de Monaco. (After Verneau.)

My learned colleague has shown that they were also exceedingly strong; that they exhibited a very marked elongation of the forearm in relation to the whole arm, and particularly of the leg in relation to the thigh; that their upper limbs were very long in contrast with their lower limbs; that the trunk was of remarkable breadth at the level of the shoulder. "In the proportions of their limbs," writes M. Verneau, "as well as in the transverse development of the upper portion of their chest, the Grimaldi men differ from Europeans, and are more akin to the negro races."

On the other hand, the pelvis has no negro characteristic. "On the contrary, the well-developed iliæ and the harmony of its curves render it as fine in form as that of the white races in which it is most highly developed."

The femora, as in those from Cro-Magnon, here show signs

of remarkable strength. The ridge (*linea aspera*), broad and prominent, may really be described as a column, and is always accompanied by a *hypotrochanterian fossa*. This structural arrangement gives rise to a certain antero-posterior flattening of the shaft below the trochanters, that is to say, to a *platymeric* femur.

The tibiæ are equally strong: their shafts are more or less flattened transversely. This character of *platycnemia*, less

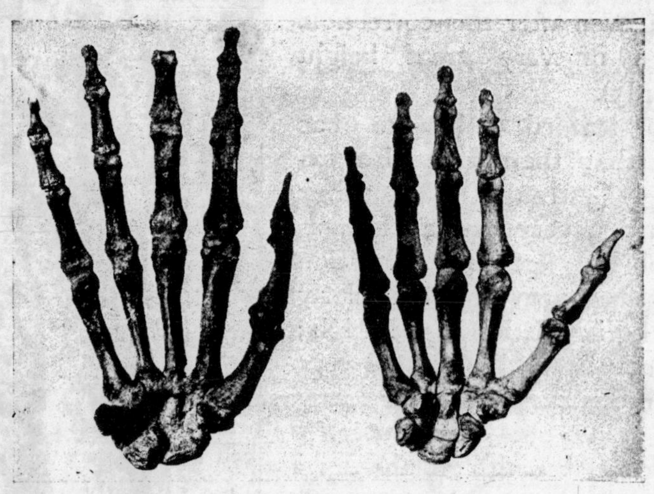

FIG. 186.—Hand of the large individual from the Grotte des Enfants, and of a modern individual 1.67 metres [5 ft. 7.5 ins.] in height, photographed to same scale. One-third natural size. (After Verneau.)

marked in some bones, is, in the case of others, as marked as in the old man from Cro-Magnon.

The hands are large, corresponding to the general build of the skeleton; the metacarpals are relatively longer and the fingers relatively shorter than in the case of a modern Frenchman (Fig. 186). These characteristics are exactly reproduced in the handprints in the Castillo cave, in the Cantabrian Pyrenees.[1] The feet are remarkable for the length of the heel.

[1] Sollas, W. J., "Cro-Magnon Man; Imprint of his Hand" (*Nature*, 7th May 1914).

Other Evidences. Varieties of the Type.

Bone-remains of the Cro-Magnon race have been collected in many parts of Western Europe. To this race de Quatrefages and Hamy attribute a large number of finds, apart from those at Cro-Magnon and Mentone; the skeleton from Paviland, in England; skulls from Engis and Engihoul, in Belgium; and more or less complete skulls from Aurignac, La Madeleine, Grenelle, Bruniquel, Laugerie-Basse, Solutré, and Gourdan, in France. To these must now be added a skull from the Grotte du Placard, at Vilhonneur in Charente, the skeleton from Les Hoteaux in the Ain, that from Combe-Capelle in the Dordogne, skeletons from Brünn and Predmost and the skull from Lautsch, in Moravia, as well as the skeletons from Obercassel, in Germany.

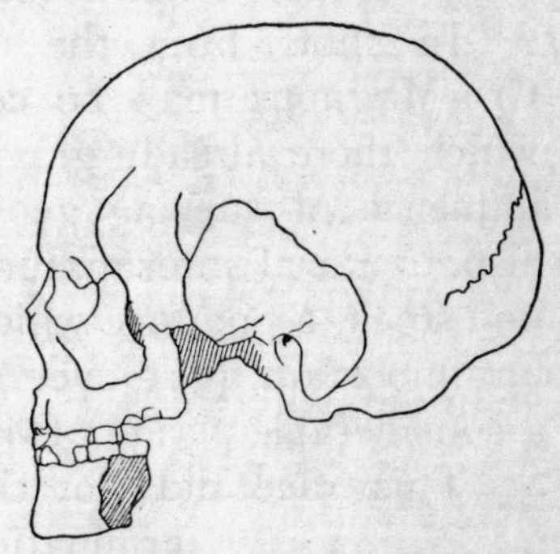

FIG. 187.—Profile of a Skull from the Grotte du Placard. One-fourth natural size. (After Hamy.)

This long list suggests several remarks. In the first place, the stratigraphical conditions of their beds and the true age of certain of these specimens are somewhat obscure, or have not been satisfactorily ascertained. In the second place, many of these finds consist simply of fragments of skulls or of lower jaws, and these, in my opinion, are insufficient to justify a classification within such narrow limits as those of a race. In the third place, the specimens of which the antiquity has been most clearly demonstrated and the state of preservation is most satisfactory often differ quite markedly from the Cro-Magnon type itself, and thus indicate a very marked degree of variation in this type. Such is the skull from Le Placard (Fig. 187), which differs from it in possessing a cephalic index that is almost sub-brachycephalic.[1] The Solutré skulls, regarded as authentic relics of the Reindeer Age, are even less homogeneous. The Obercassel skeletons are of smaller stature.

[1] Hamy, E. T., "Nouveaux matériaux pour servir à l'étude de la Paléontologie humaine" (*Congrès intern. d'Anthrop.*, Paris, 1889).

The differences may be even still more marked. On this account several anthropologists regard the skulls or skeletons from Brünn and Predmost as representing a special race, the Solutrean race, which, towards the east, impinged upon the western and southern race of Cro-Magnon. Its skulls are remarkable for their marked orbital ridges, a feature in which they bear a resemblance, though a misleading one, to *Homo neanderthalensis*, from whom in reality they differ greatly. According to Giuffrida-Ruggeri,[1] the skull from Combe-Capelle, which is more dolichocephalic, more prognathic, and more platyrrhinian or flat-nosed, shows certain Ethiopian affinities.

In short, from the osteological point of view, the true Cro-Magnons may be considered as a median type, around which there already gravitate variations, due probably to the influence of varying geographical environments, and perhaps also to racial intermixture. But as a whole they really form one stock, a fine race which, as de Quatrefages has said, played an important part over a considerable area and throughout a considerable period of time.

Survivals.

This race did not come to an end in France with the termination of Quaternary times. As we shall see later, it not only survived during the Neolithic age, but, even in our own day, it still appears sporadically in different parts of France, especially, according to Dr Collignon, in the Dordogne (Fig. 188). M. Verneau has been able to trace the Cro-Magnon race throughout Spain; it is found in burials, dating from more and more recent times the further one travels southwards.

Broca had noted the presence of structural affinities between the Basques, the Kabyles, and the Guanchos. De Quatrefages and Hamy proved that a fair number of craniometric characters present in the cave-dwellers of Périgord are also to be met with in the pure-blooded Kabyles, and these characters have been noted in remains collected in prehistoric tombs in Algeria.

But it is among the Guanchos of the Canary Islands that the Cro-Magnon type has been best preserved. This state-

[1] Giuffrida-Ruggieri, "Quatro crani preistorici dell' Italia meridionale e l'origine del Mediterranei" (*Arch. per l'Anthr. e la Etnol.*, vol. xlv., 1916).

ment, which we owe to de Quatrefages and Hamy, has been confirmed by Verneau's researches in the Canary archipelago, where the foundations of the population are formed of the Guancho element; that this is derived from the Cro-Magnon

FIG. 188.—Cro-Magnon type, still persisting in the Dordogne. (From photographs by Dr Collignon.)

race is proved by resemblances in the skulls, which are sometimes identical. Verneau has even discovered among the modern islanders implements such as were formerly used by our ancient Dordogne hunters.

The Chancelade Race.

The discovery at Raymonden, near Chancelade, to which we have already referred, was made and carried through under excellent scientific auspices. The bed proved to be very rich in archæological and artistic objects. The fauna, identified by A. Gaudry, comprises, among other interesting species, the Greenland Seal. The human skeleton was examined by Professor Testut, of the University of Lyons. All this valuable material is preserved in the Museum at Périgueux, where it has been well arranged by M. Féaux, one of the fortunate investigators at Chancelade.[1]

[1] Hardy, M., "La station quaternaire de Raymonden" (*Bull. de la Soc. histor. et archéol. du Périgord*, 1891). Testut, L., *loc. cit.* Féaux, M., *Catalogue du Musée du Périgord*, Périgueux, 1905.

The geological arrangement of the bed is extremely simple, consisting of three Magdalenian layers of occupation, alternating with gravels and silts, and forming a total thickness of 1.60 metres in depth. The lower floor of occupation lies directly upon the rock-surface, and the human skeleton rested in the lower part of it (Fig. 189).

The skeleton, found on the 1st October 1888, lay on its left side. The arms were raised; the left hand was placed under the head, the right hand under the left side of the lower jaw; the lower limbs were bent; the feet were, in consequence, drawn up towards the lower part of the pelvis, and the knees were just touching the jaw. Such a constrained position recalls that of certain Peruvian mummies, and M. Testut has advanced the opinion that the Chancelade Man "might also have undergone similar treatment, and have been firmly tied up with cords or flexible lianæ, and perhaps even been sewn in some kind of sack made of an animal's skin, all in order to reduce the body to the smallest size so that it should occupy the smallest possible space. A similar method of burial is found among a great many ancient and modern peoples, particularly in the case of modern Esquimaux."

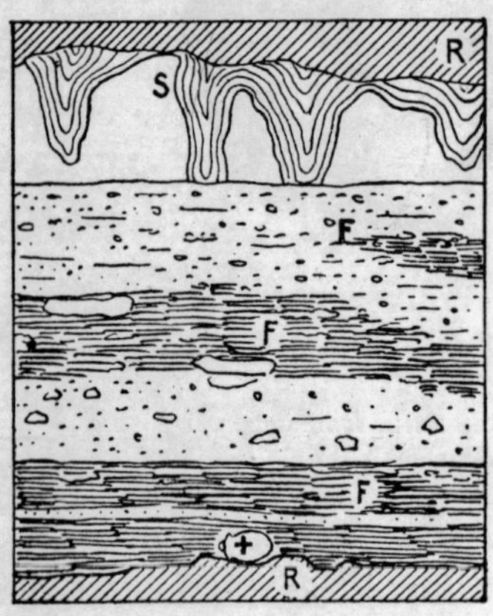

FIG. 189.—Section of the Rock-Shelter at Chancelade. (After Féaux.)

R, rock, floor and roof of shelter; FF, floors of occupation; +, human skull; S, stalactites.

As in the Mentone burials, the body seems to have been powdered over with hæmatite iron-ore, which had coloured red not only the bones but also the surrounding earth.

Let us glance at the chief results of Dr Testut's anthropological research.

The Skeleton.

The skeleton is that of a man who had died between the ages of 55 and 65 years. This man was of small stature, 1.50 metres according to Testut, 1.59 metres according to Rahon, or, let us say, about 1.55

metres [5 ft. 1 in.]. In this character, to begin with, he differs considerably from the Cro-Magnon type.

The state of preservation of the skull leaves much to be

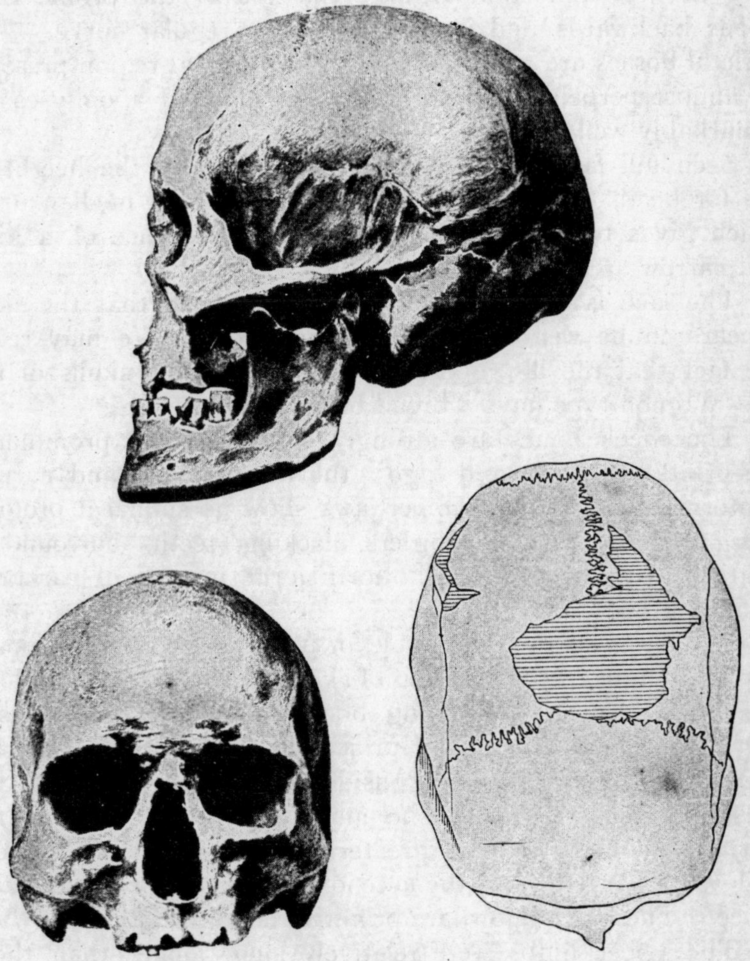

FIG. 190.—The Chancelade Skull, seen in profile, full face, and from above. One-third natural size. (After Dr Testut.)

desired, but it has been carefully reconstructed (Fig. 190). It is markedly dolichocephalic (cephalic index, 72), and remarkably high. Its capacity, estimated at 1710 cubic centimetres, greatly exceeds the average of present-day skulls, even of Europeans.

"Seen in profile, the Chancelade skull exhibits all the characters belonging to the higher races." Above the slight superciliary ridges, the broad, somewhat bulging forehead rises at first almost vertically; the line of the profile then slopes backwards and is continued in a regular curve. The parietal bosses are very marked. The occipital region presents an almost perpendicular surface. The mastoid apophyses are remarkably well developed.

Seen full face, the skull is characterized by the height of the forehead, and by the unusual height of its median area, which gives to the general aspect the appearance of a high and narrow arch.

The face is very broad and very high, so that the skull appears to be well-balanced or harmonious. We may recall the fact that the ill-proportioned or disharmonic skulls of the Cro-Magnon type have a broad but squat face.

The cheek bones are strongly developed and prominent; the orbits are large and high; the nose is long and narrow (leptorrhinian), and the upper jaws show no subnasal prognathism. The alveolar borders, lacking teeth, surround a relatively narrow palate, elliptic in form, instead of parabolic as in the Cro-Magnon type.

The lower jaw is remarkable for its narrowness, corresponding to the general elongation of skull, for its strength, and for the breadth of its ascending branches. The chin forms a very broad and prominent projection. The whole external appearance testifies to a considerable development of the muscles of mastication. The molar teeth were strong; the last, a wisdom tooth, was greater in size than those in front and was separated from the ascending ramus by a considerable space. These, we know, are primitive characteristics.

The upper limbs were relatively long, longer than those of modern Europeans, longer even than those of Negroes. Their bones, massive and thick-set, like the skeleton as a whole, indicate a vigorous frame and strong muscles; so that, to judge from their insertions, the supraspinatus, infraspinatus, deltoid, great pectoral, great dorsal, and great round muscles, all of which are attached to the humerus and play an important part in the act of climbing, were specially developed. The same applies to the lower limbs, to the gluteus maximus, the posterior muscles of the thigh, and, as a whole, to all the posterior muscles of the leg, which are the active factors in the erect position and in walking.

The femora are a little more bent than in modern Europeans. Like those of Cro-Magnon, they have the linea aspera formed like a column or rod, and a hypotrochanterian fossa. The upper ends of the tibiæ are markedly bent backwards, in consequence of which "in the upright position the

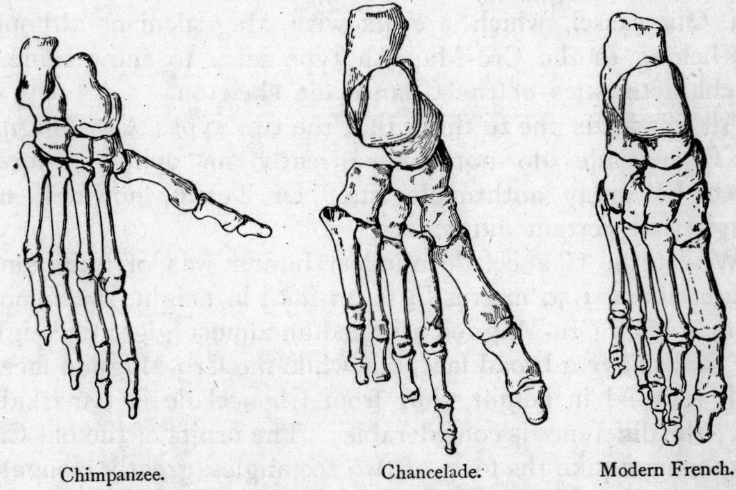

FIG. 191.—Skeleton of Foot of Chimpanzee, of Chancelade Man, and of a Modern Man, from Lyons. About one-fourth natural size. (After Testut.)

knees must have projected more prominently forwards than in modern races." These tibiæ have the shaft flattened in a transverse direction: they are slightly platycnemic.

The Chancelade Man had large feet, which in the usual position were turned inwards. The first metatarsal, corresponding to the great toe, was distinctly separated from the second toe, a little like that of the apes and exactly like that of Neanderthal Man. The foot of our fossil man, as in the case of some native races, could probably play the part of a grasper, able to grip objects between the first two toes (Fig. 191).

It would seem as though we were very poor in representatives of the fossil Chancelade race.

Comparisons.

Hervé has shown that we may attribute to it the skulls from Laugerie-Basse, classified by de Quatrefages and Hamy as Cro-Magnon in type. The skeleton of the cave-dweller from Sorde also exhibits these points of resemblance. Even the female skull from Le Placard, in spite of its sub-brachycephalic character and its close relationships with the female skulls of the Cro-Magnon race, must, according to M. Hervé, be added to the small Chancelade group and must confirm the homogeneity of the Magdalenians.[1] The skeletons from Obercassel, which are likewise Magdalenian, although they belong to the Cro-Magnon type seem to show some of the characteristics of the Chancelade skeleton.

All this leads one to think that the two types, Cro-Magnon and Chancelade, do not differ greatly, an opinion, indeed, shared by many anthropologists. Dr Testut, however, has brought out certain differences.

Whilst the Chancelade reindeer hunter was of quite small stature, about 1.50 metres [4 ft. 11 ins.] in height, the famous old man from Cro-Magnon attained an almost gigantic height. The skulls have a broad face, but while the Cro-Magnon face is much reduced in height, that from Chancelade is remarkably high; the difference is considerable. The orbits of the old Cro-Magnon man take the form of two rectangles, greatly elongated in a transverse direction; those of the Chancelade man have a quadrilateral shape, the height differing little from the breadth. It is evident that the general appearance of these two cave-dwellers of the Upper Palæolithic must have been very different.

Nevertheless it is also a fact that among the three types we have just described, the Grimaldi Negroids, the Cro-Magnon Men, and the Chancelade Man, there is a certain number of common and fundamental characters, which bear witness, as I have already said, to the unity and at the same time to the variability of *Homo sapiens fossilis*.

[1] Hervé, G., "La race des troglodytes magdaléniens" (*Revue de l'Ecole d'Anthrop.*, iii., 1893).

Dr Testut has clearly shown the resemblance of the Chancelade skeleton to the skeletons of the Eastern Esquimaux, who still live in a wild state amid the snows of Labrador and Greenland, and who, in all respects, represent a very ancient race. "Like the Chancelade Man, the Esquimaux are decidedly dolichocephalic; like him, they have a high skull, prominent sagittal ridge, a face at once very wide and very high, and orbits almost round in form. We also know that the Esquimaux have a large head and are of small stature. . . ."

This comparison, drawn from anatomical anthropology, confirms that which the prehistorians, Hamy, Gervais, and Dupont, made long since on the strength of archæological and ethnographical evidences, and of the resemblances between the physical environments. As long ago as 1870, Hamy said of the modern Arctic peoples that they "seem to be akin to the Quaternary cave-dwellers of our countries. They carry on into our own times, in polar regions, the *Reindeer Age* of France, of Belgium, and of Switzerland, with the same zoological and ethnographical characters, etc."[1]

This theory, shared by Pruner-Bey, Boyd-Dawkins, and Hervé, has recently been revived by Professor Sollas of Oxford. According to this scholar, a study of the modern Esquimaux, of their habits, their implements, and their artistic efforts, confronts one with an array of facts in favour of the existence of an actual relationship, which is so admirably confirmed by the Chancelade skeleton. The Cro-Magnon race may formerly have had relations with the Chancelade race, similar to those to be observed at the present day between the Algonquins and the Esquimaux. The Palæolithic peoples seem to have gradually taken possession of the circumboreal regions by way of the Behring Straits and the Aleutian Islands.[2]

This is not the opinion of certain anthropologists in the United States, Boas and Chamberlain,[3] who regard the Esquimaux as a race of American origin. Far from having

[1] Hamy, E. T., *Précis de Paléontologie humaine*, p. 366.

[2] Sollas, W. J., *Ancient Hunters*, pp. 513-516.

[3] Chamberlain, A. F., "Quelques problèmes ethnographiques et ethnologiques de l'Amérique du Nord" (*L'Anthropologie*, xxiii., 1912).

come from northern Asia, they may have formed part of an ancient centre of distribution in the interior of Canada, whence they spread towards the maritime regions they occupy in our own day. In which case it would appear that America had peopled or repeopled, at a certain time, a portion of the Old World. This is quite possible, but it does not explain the resemblances found between certain peoples of our Reindeer Age and modern Esquimaux. It would certainly seem, in a final analysis, that all the New World peoples may have come from the Old World, and that before becoming Americans the Esquimaux may first of all have been Asiatics, and perhaps even northern Europeans as well. De Quatrefages was fond of saying that the Cro-Magnons must have had the closest kinship with the Red Skins.

Attempts have been made to explain such ethnographical resemblances by a similarity in the conditions of life and the use of the same materials, reindeer horn, bone and ivory. These similarities may indeed exhibit a simple case of convergence quite apart from all direct relationships. But how can we explain by the same theory physical and mental similarities?

I shall only mention, in passing, Girod's[1] opposite theory of "the Esquimaux invasion," but, to use a popular expression, that is a case of "putting the cart before the horse."

Human Figures.

Let us now examine the works of art contemporary with the men whose skeletons we have studied. Judging from the accuracy of many of their animal drawings, this new source of information should be of great assistance to us. The artists of the Reindeer Age did, indeed, sometimes carve or engrave portraits of their fellows; but, unfortunately, they were unskilful in this form of art, and this is particularly true of the engravers. The sum total of the anthropomorphic figures of Quaternary art already forms a series of considerable importance from the point of view of the number and variety

[1] Girod, P., *Les invasions paléolithiques dans l'Europe occidentale*, Paris, 1900.

of subjects dealt with; but in its evident inaccuracy or lack of skill in drawing, this series is in contrast with the collection of animal figures, which are not only very numerous but amongst which real works of art abound. The majority of the engraved or sculptured portraits of fossil Man are most often simply childish caricatures, and many of these human personages seem to be dressed in animal masks which conceal their true features. In the absence of any work of art of really outstanding quality, on which we might consequently have relied, we must be cautious in the interpretation of certain characters or contours, which, far from representing real morphological characters, can only be the result of unskilful execution.

With these reservations, let us examine the principal evidences.

The Reindeer Age was of long duration, and in the course of it art, like industry, underwent an evolution; this evolution has been studied by Piette and Breuil.[1] The former thought that the sculpture of statues in full relief preceded sculpture in low relief, and that the latter itself preceded engraving. But Breuil has shown that this idea is correct only in a general way and in regard to the art of movable objects, while, so far as mural art is concerned, it applies with still less exactness. None the less, sculptures are more numerous during the most ancient period of the Reindeer Age (Aurignacian), and engravings attain their maximum development, in number as well as in quality, during the last phases of this age (Magdalenian).

Aurignacian Sculptures.

However this may be, Aurignacian deposits of different countries, far apart from each other, have yielded a certain number of very realistic statuettes or bas-reliefs.

There are, first of all, objects in Mammoth ivory, taken by Piette[2] from the Brassempouy Cave in Landes, the most important of which are here reproduced (Fig. 192).

[1] Breuil, H., "L'évolution de l'art quaternaire et les travaux d'Edouard Piette" (*Revue archéol.*, 1909).

[2] Piette, E., "La station de Brassempouy et les statuettes humaines de la période glyptique" (*L'Anthropologie*, vi., 1895).

Here we see, along with the head of a young woman, known as the "hooded figure," some mutilated statuettes representing torsos and bodies of men or women. One of

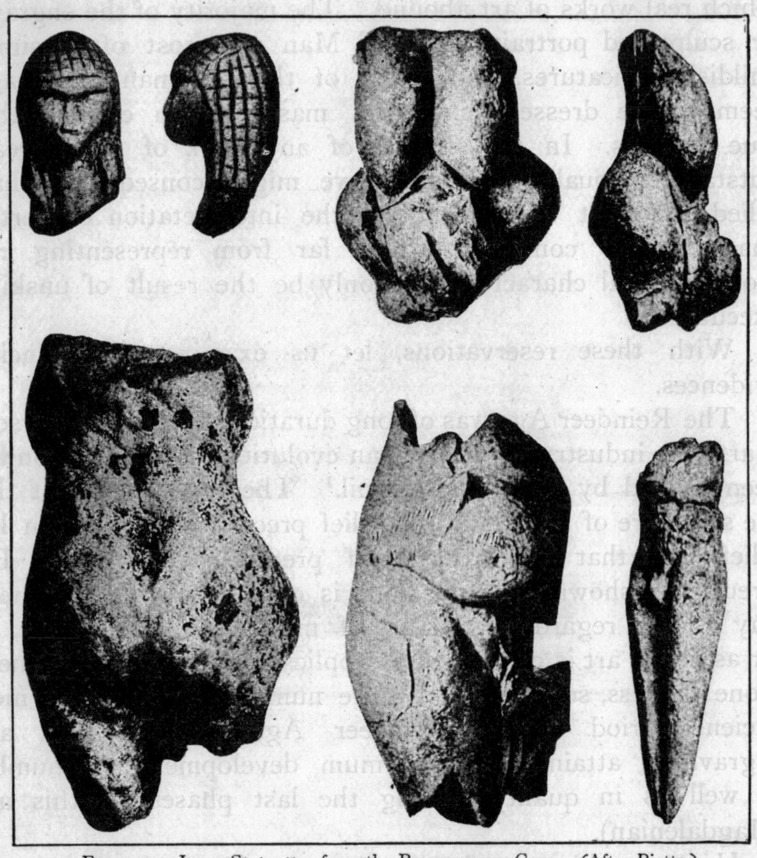

FIG. 192.—Ivory Statuettes from the Brassempouy Cave. (After Piette.)

Above, to the left, "hooded figure," full face and profile, slightly reduced. Above, to the right, damaged female body, seen full face and in profile, about four-fifths natural size. Below, to the left, female body, known as a "dagger haft," about four-fifths natural size. Below, in centre, the "Venus de Brassempouy," reduced by one-fourth. Below, to the right, "belted figure," reduced by one-fifth.

these statuettes, known as the *Venus de Brassempouy*, or *La Poire*, must have been a fine specimen, if we may judge from the delicacy of the modelling of the portions remaining intact. Another, regarded by Piette as the haft of a dagger, is a female body, remarkable for the exuberance of its form. A

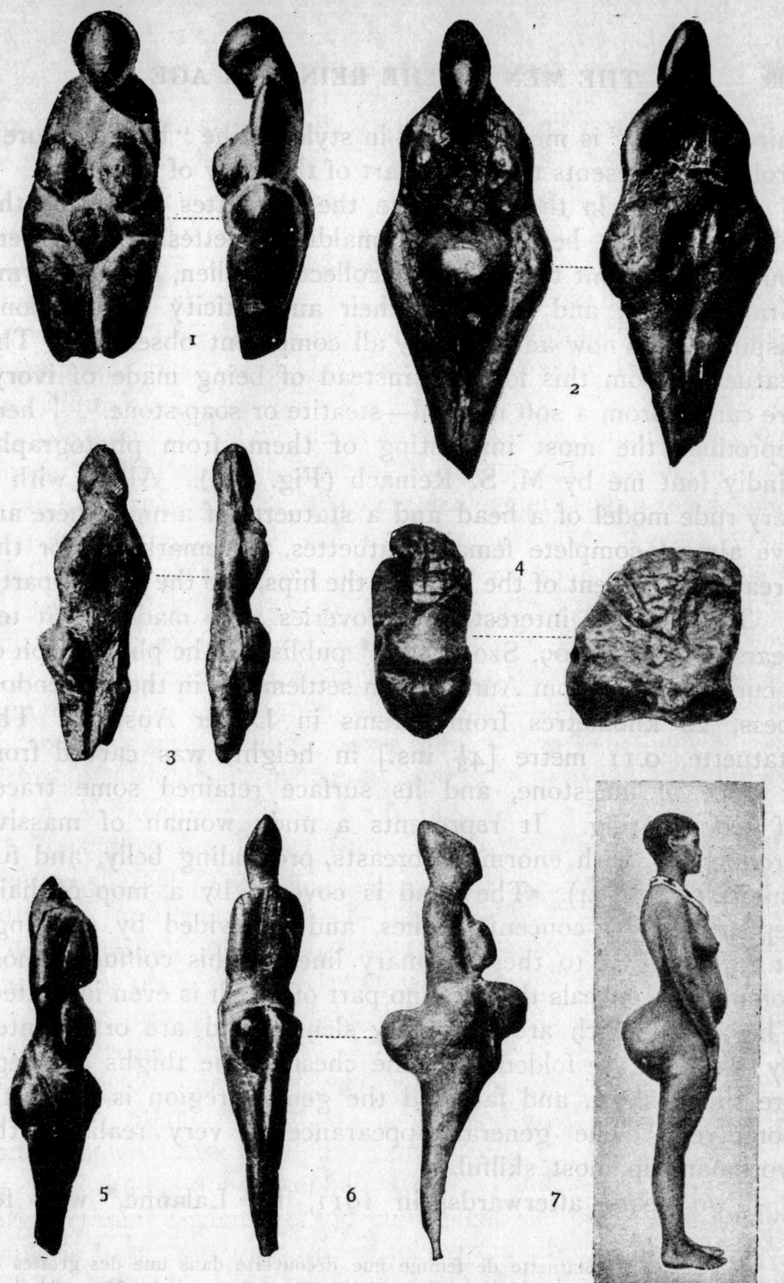

FIG. 193.—Statuettes in Steatite, from the Mentone or Grimaldi Caves. Natural size; from photographs. Musée de St Germain. Compare with photograph of Bushman Woman.

1, Female figure, seen full face and in profile; 2, female figure, front and back view; 3, female figure, back view and profile; 4, negroid head, full face and profile; 5, male figure, profile; 6, female figure, back view and profile; 7, modern Bushman woman, profile.

third "Venus" is more elegant in style. The "belted figure" probably represents the lower part of the body of a man.

There are, in the next place, the statuettes known as the "Mentone," or better the Grimaldi statuettes. They were found at different times by the collector, Julien, in the Barma Grande cave; and although their authenticity was for long disputed, it is now admitted by all competent observers. The statuettes from this locality, instead of being made of ivory, are carved from a soft mineral—steatite or soap-stone.[1] I here reproduce the most interesting of them, from photographs kindly lent me by M. S. Reinach (Fig. 193). Along with a very rude model of a head and a statuette of a man, there are five almost complete female statuettes, all remarkable for the great development of the breasts, the hips, and the genital parts.

Two no less interesting discoveries were made about ten years ago. In 1909, Szombathy[2] published the photograph of a curious figure from Aurignacian settlements in the Willendorf loess, 20 kilometres from Krems in Lower Austria. The statuette, 0.11 metre [4⅓ ins.] in height, was carved from a piece of limestone, and its surface retained some traces of red painting. It represents a nude woman of massive proportions, with enormous breasts, protruding belly, and full thighs (Fig. 194). The head is covered by a mop of hair, represented by concentric lines, and re-divided by markings at right angles to these primary lines. This coiffure almost completely conceals the face, no part of which is even indicated. The arms, which are extremely slender and are ornamented by bracelets, are folded over the chest. The thighs and legs are thick, short, and fat, and the genital region is distinctly portrayed. The general appearance is very realistic, the workmanship most skilful.

Two years afterwards, in 1911, Dr Lalanne,[3] who for

[1] Reinach, S., "Statuette de femme nue découverte dans une des grottes de Menton" (*L'Anthropologie*, ix., 1898). Piette, E., "Gravure du Mas d'Azil et statuettes de Menton" (*Bull. de la Soc. d'Anthrop. de Paris*, 1902).

[2] Szombathy, J., "Die Aurignacienschichten im Loess von Willendorf" (*Korrespondenzblatt der deutsch. Gesellsch. für Anthrop.*, xl., 1909).

[3] Lalanne, G., "Bas-reliefs à figurations humaines" (*L'Anthropologie*, xxii., 1911, and xxiii., 1912).

several years had been excavating the great deposit at Laussel in the Dordogne, discovered there a very curious bas-relief, representing two persons lying down and facing each other (a representation of birth or perhaps rather of mating). In the following year three other bas-reliefs were brought to light; a fourth was stolen from Dr Lalanne by his head workman, and

FIG. 194.—Willendorf Statuette, front view and profile. Half natural size. Photograph from a cast.

sold to the Berlin Museum through the agency of Professor Verworn of Bonn. I here reproduce the two finest of these sculptures.

The first bas-relief is very skilfully executed. It measures 0.46 metre [18⅛ ins.] in height and represents a nude woman, holding in her right hand a bison's horn (Fig. 195). The head is represented only by a vague outline, but, on the other hand, the remainder of the body is carefully treated. Here, again, we find enormous, elongated, and pendulous breasts, a prominent but well-modelled abdomen, with folds of fat and carefully drawn *mons veneris*. The hips are stout, with marked iliac and femoral prominences; the general appearance is fleshy and adipose. The thighs are full, the legs slender and short. The arms are slender and well modelled; the fingers on the hand are indicated. The whole surface of this bas-relief,

FIG. 195.—Bas-relief from Laussel, representing a Nude Female, seen full face. About one-quarter natural size. (After G. Lalanne.)

apart from the head, which seems to have been crushed, is delicately worked and even polished, and some traces of red paint are still to be found upon it.

The second, and more damaged figure, again represents

a nude woman. The head is vaguely worked, the hair is executed in the manner of the female statuettes from Brassempouy and Willendorf. The greatly enlarged breasts resemble those of the preceding bas-relief, while the prominent belly ends below in a triangular projection. The hips must have been very large, but they are very much abraded, as are also the upper parts of the thighs. A portion of the arms may still be distinguished.

The third bas-relief, 0.4 metre [$15\frac{3}{4}$ ins.] high, is a three-quarter length figure of a man, contrasting in its graceful form with the massive figure of the woman (Fig. 196). Unfortunately this work of art is also incomplete. The head is lacking, as well as the greater portions of the arms and the feet. The proportions are graceful. The trunk and loins are bent, the legs are placed as though the individual had been in the act of drawing a bow. Two parallel lines mark a belt around the body.

In the summer of 1922, M. and Mme. de Saint-Périer discovered, in a cave in the Pyrenees, at Lespugue, a fine ivory statuette of Aurignacian age. It exhibits the general morphological characters of the works of art just discussed, at the same time reproducing several of their peculiar features. (For details and illustrations, see Appendix, p. 473.)

Brought together from districts far apart, and undoubtedly of one and the same age, corresponding to the most ancient phase of the Reindeer Age (Aurignacian), this small collection of artistic productions is remarkable for the association of characters which give it a general appearance of considerable uniformity, almost, indeed, a family likeness testifying to the sincerity of the artists. These statuettes or bas-reliefs must, therefore, translate with some fidelity the general characters of the form of their models. We may, then, place a certain amount of reliance on the information they contribute.

Their Interpretation.

I say "a certain amount," because occasionally writers have gone too far in the interpretation of them, as, for example, when Piette attempted to calculate the cephalic index of the "hooded figure" from Brassempouy.

FIG. 196.—Bas-relief of a Man, from Laussel. About one-third natural size. (After G. Lalanne.)

The following facts may be definitely recognized.

In almost every case the head is scarcely blocked in; the hair is often represented, and always by the same method of criss-crossing strokes forming a chequered pattern, for example, at Willendorf as well as in the Dordogne, on the Côte d'Azur as well as in the Pyrenees. We may interpret this feature, also found in primitive Egyptian art, as a type of hair-dressing, an arrangement in narrow plaits, or in short tresses or small tufts, such as are found in modern negro or negroid races, particularly of the Bushman and Hottentot tribes whose hair is grouped in little bunches. Or, on the other hand, the pattern may be looked upon as representing a net, similar to the hair-nets made of shells found on several of the Grimaldi skeletons. Perhaps both types were in use.

The features of the face are always obliterated or roughly represented, even in the hooded figure from Brassempouy. Dr Lalanne thought he traced, on his bas-reliefs, elongated faces, with prominent cheeks and pointed chins. Such an interpretation may perhaps be admitted without difficulty, for in its vague terms it does not stand in any great contradiction with the osteological characters of contemporary skulls. On the other hand, however, the Grimaldi head has really a negroid appearance.

The female bodies are short and massive; the breasts very large, long, cylindrical, and pendulous; the belly is prominent, with rolls of adipose tissue sometimes overhanging the pubes. The hips are greatly developed, fleshy enough even to amount to true steatopygia; the thighs are likewise fleshy, sometimes with projecting ridges of fat; they are supported by slender legs. The upper limbs, on the contrary, preserve a certain degree of delicacy. The sexual parts are always strongly emphasized; the *mons veneris* is large and well defined. We can recognize in "La Poire," as Piette did, a development of the nymphæ (*labia minora*) similar to the "apron" of Bushman women.

On certain regions of the body, hair seems to have been well developed, where to-day it is much reduced.

Resemblance to the Bushmen.

Several authors, following Piette, have compared this association of characters with that now or formerly exhibited by Bushmen women, in whom steatopygia has been regarded as classic since the work of Cuvier on the "Hottentot Venus" in the French National Natural History Museum. Quaternary male figures likewise agree in their general morphology with the male Bushman, in possessing a slender body. Together with Dr Lalanne in France, Professor Sollas in England, Dr Péringuey at the Cape and others, I have several times had occasion to lay stress on this resemblance, which is of considerable interest.

We now know that the ethnography of South African tribes presents many striking similarities with the ethnography of our populations of the Reindeer Age. Not to speak of their stone implements which, as we shall see later, exhibit great similarities, Péringuey tells us that in certain burials on the South African coast "associated with the Aurignacian or Solutrian type of industry," rows of circular discs made from ostrich egg-shell, and round plates of bone or of shells, perforated for stringing, accompany the skeletons of women and of children; one of these necklaces, adds the learned Director of the Cape Museum, exactly resembles those found with the Mentone skeletons.[1]

Bushman mural art is extraordinarily like that of our caves. As in France and Spain, the pictures of animals are also of higher quality than the human figures, and many of the latter wear an animal mask and a long tail. The two centres are united by a long, connected series of works of art, from France to the Cape by way of Spain, North Africa, the Soudan, the Tchad States and the Transvaal. This almost uninterrupted series leads us to regard the African continent as a centre of important migrations, which at certain times may have played a great part in the stocking of southern Europe. Finally, we must not forget that the Grimaldi negroid skeletons show many points of resemblance with the Bushman skeletons.

[1] Péringuey, L., "The Bushman as a Palæolithic Man" (*Transactions of the Royal Society of South Africa*, vol. v., 1915).

It is difficult in the present state of our knowledge to decide whether the Bushmen were descended from our Aurignacians or the latter from the ancestors of the Bushmen, who are considered by all anthropologists to be the survivals of a very ancient race. But, in my opinion, the relationship of these two groups cannot be denied, however far apart they may be in time and space. The most reasonable course, it would seem, is to admit that they are descended from a common primitive stock, which must have spread towards the centre or the north of the African continent, and whose branches have developed in different directions, morphologically and geographically, while retaining a common fund of ethnographical survivals. In any case, no explanation based simply on phenomena of convergence seems to be sufficiently far-reaching to cover such a mass of consistent data.

Magdalenian Sculptures.

Let us now glance at the anthropomorphic images belonging to a more recent period, the Upper or Magdalenian period of the Reindeer Age. Their value from an anthropological point of view is very small, as may be seen from the examples here reproduced, selected from the best or, let us say, the least bad specimens.[1]

The two statuettes, one from Laugerie-Basse (Vibraye Collection in the French National Natural History Museum), the other from the Mas d'Azil (Piette Collection at Saint-Germain), resemble certain idols, fetishes, or amulets of savage tribes and could never be regarded as portraits. The first, known as *Venus impudique*, is in ivory (Fig. 197, No. 3), and differs markedly from Aurignacian statuettes in its more slender outlines. The second, carved from the root of an incisor tooth of a horse, is an even coarser work (Fig. 197, No. 1). Piette has laid far too much stress upon the study of the features.

Another production from Laugerie-Basse is an engraving in low relief on a sunk background, representing a pregnant woman beside the hind legs of a reindeer (No. 2). It is the "Woman with the Reindeer" of Abbé Landesque. The creator of this bas-relief was certainly very skilful, for the legs of the reindeer are perfectly rendered; but the drawing of the

[1] See Piette, *L'art pendant l'âge du Renne.*

woman's body is quite mediocre. As it is highly probable that the two subjects were engraved by the same artist, we have here yet another proof of the lack of skill of the artists of the Reindeer Age in reproducing the human body. The woman is adorned with a necklace and bracelets. The only anthropological fact of interest relates to the development of an arrangement of hair, clearly indicated on all the abdominal surface.

The line engravings are still poorer or more rudimentary; that discovered by E. Lartet at La Madeleine, representing a man between two horses, may be taken as an example (No. 7). Such also is the famous "Hunting the Aurochs," discovered by Massénat at Laugerie-Basse, where the Aurochs is much better drawn than the man with the absurdly pointed skull (No. 6). G. de Mortillet, however, considered he possessed an "air as gay as it was intelligent." Other drawings are more grotesque still; such as the odd creature carved on a round bone disc from the Mas d'Azil, in which Piette attempts to see the figure of a monkey, because of its muzzle-like face (No. 5). It is probable that here, as at Gourdan (No. 4), at Marsoulas (No. 8), at Altamira and elsewhere, we have a man wearing a mask in the form of an animal's head, either in order to indulge in dancing or in certain ceremonies, or more readily to stalk game, as certain native peoples still do or formerly did, such as the Australians, Bushmen, and others. There are also some simple caricatures recalling childish drawings.

To the laborious investigations of Breuil and his disciples at Madrid,[1] Spain has yielded many human representations

[1] Breuil, H., and Cabré, J., "Les peintures rupestres du bassin inférieur de l'Ebre" (*L'Anthropologie*, xx., 1909). *Id.*, and Serrano, P., "Les abris del Bosque à Alpéra" (*L'Anthropologie*, xxiii., 1912). Breuil, H., "Les roches peintes de Minateda, Albacete" (*L'Anthropologie*, xxx., 1920).

Legend, Fig. 197.

1, Incisor of horse, the root of which has been carved into a female figure, from the Mas d'Azil in Ariège; 2, engraving on reindeer antler, known as "The Woman with the Reindeer," from Laugerie-Basse in Dordogne; 3, the "Venus impudique" in ivory, from Laugerie-Basse; 4, engraving on reindeer antler from Gourdan in the Haute-Garonne; 5, engraving on disc of bone from Mas d'Azil; 6, portion of an engraving on reindeer antler known as "Hunting the Aurochs," from Laugerie-Basse; 7, engraving on reindeer antler from La Madeleine; 8, engraving on the wall of Marsoulas cave in the Haute-Garonne.

Nos. 1, 2, 4 and 7 are after Piette; No. 3 is a photograph of the original in the French Nat. Hist. Museum; No. 8 after Breuil.

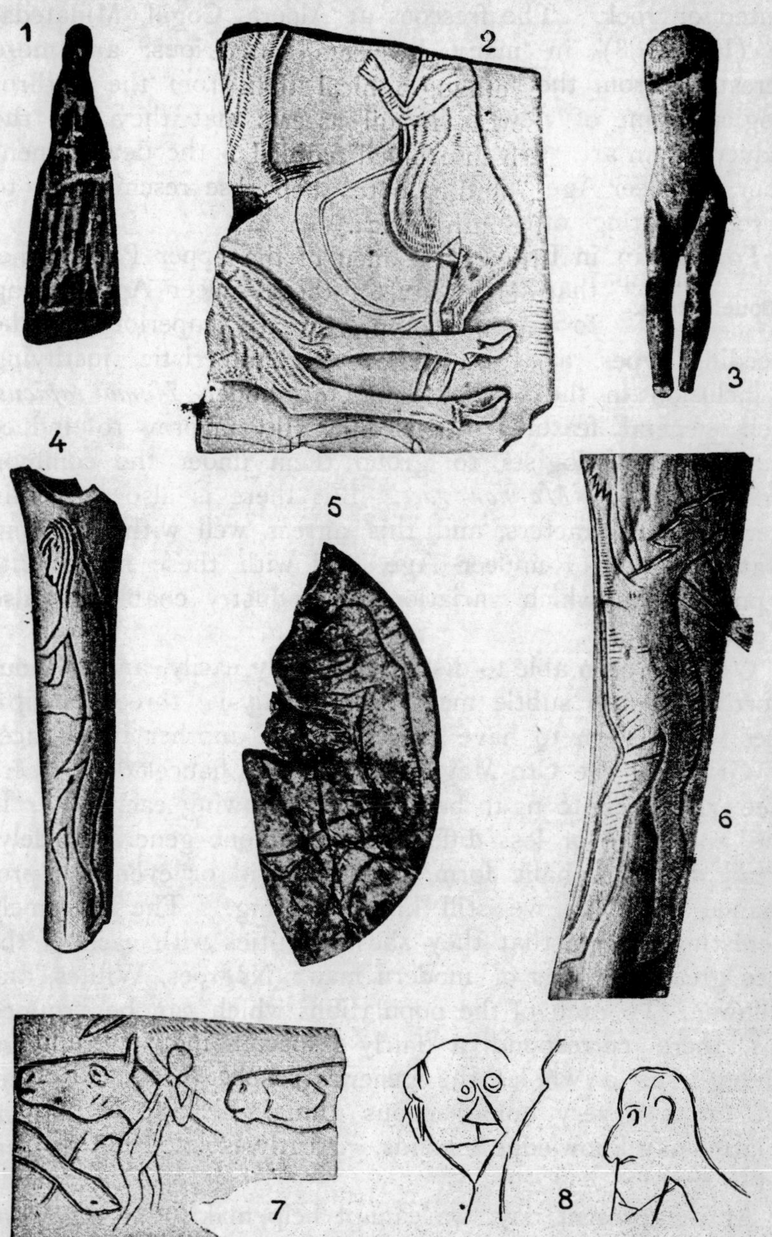

FIG. 197.—Magdalenian Human Representations.

painted on rock. The frescoes at Alpéra, Cogul, Minateda, etc. (Fig. 198), in many respects so curious, are more interesting from the ethnographical than from the anthropological point of view. Breuil asserts that they are the products of an art "which evolved parallel to the development of our Reindeer Age," and he notes their close resemblance to the rock paintings of South Africa.

Conclusions.

To sum up, in France the men of the Upper Pleistocene, that is to say, of the Reindeer Age, belong to a new type, greatly superior to the preceding types, and by its every characteristic, justifying its inclusion in the general mass of modern *Homo sapiens*. Their general features are sufficiently uniform to induce certain anthropologists to group them under the common name of the *Cro-Magnon race*. But there is also a certain diversity of characters, and this agrees well with the long duration of the Reindeer Age, and with these movements of populations which variations in industry enable us also to envisage.

We have been able to distinguish fairly easily, and without resorting to too subtle methods of analysis, three principal types which seem to have succeeded one another in France: the Grimaldi, the Cro-Magnon, and the Chancelade. These three types seem to us to be varieties, following each other in time, and more or less differentiated, of one general, widely-spread dolichocephalic form, of the origin or even the provenance of which we still know nothing. The extremely interesting fact is that they show affinities with each of the three great divisions of modern man: Negroes, Whites, and Yellows. To each of the populations which can be grouped thus, there correspond a fairly special industry and art, although, as a whole, the general aspect of the Reindeer Age forms a very homogeneous unit which appears, so far as latter-day knowledge extends, greatly isolated both in time and in space.

At the present day, we cannot help making comparisons between the men of the early Reindeer Age, that is, the Aurignacians or at least the first of the Aurignacians,

198.—Panel from the frieze at Minateda in Spain, representing a fight, and various ubjects belonging to other, preceding and succeeding, periods. One-sixth natural size. After H. Breuil.)

and certain modern groups, and in so doing we are led to admit that they are African in origin. The presence and the persistence of a plentiful Aurignacian industry in Africa and particularly in North Africa (where it has been designated the *Capsian* or *Getulian* industry); the uninterrupted chain, across the whole of the Dark Continent from north to south, of an art developed on cave-walls or rocks; the really extraordinary resemblances between the art of the South Africans and our prehistoric art; the evidence, obtained by Breuil's remarkable work and observations in Spain of transition stages between Aurignacian and Magdalenian art on the one hand and on the other the less naturalistic and more diagrammatic art of more southern regions, as well as the neolithic art of Susa;—all these facts plead in favour of our hypothesis, all are at one in breaking down the isolation of our Reindeer Age and leading us to admit the existence of broad relationships with countries and peoples whose prehistory has hardly yet begun to be unveiled.

The Grimaldi Negroids are certainly African—a fact which is not in formal contradiction with the opinion of M. Verneau who considered them as indigenous, for indigenousness must necessarily have a beginning somewhere.

The Cro-Magnons established themselves in the Mediterranean countries, and spread widely throughout western and southern Europe. If they no longer correspond exactly to any living ethnographical group, their various characters are to-day still to be recognized, more or less clearly, scattered here and there among the populations of many countries. They seem to represent an ancient stock which is not yet exhausted.

The Chancelades represent a group already evolved in a somewhat different direction, probably beneath more northern skies. Towards the end of Pleistocene times they seem gradually to have supplanted, more or less, the Cro-Magnon group, and later, at the dawn of modern times, to have withdrawn northwards at the same time as the Reindeer under the pressure of new invaders.

We cannot hide from ourselves the fact that these conclusions are still vague and uncertain. It will be necessary to submit them to the test of new excavations, and these must be undertaken with an absolutely clear understanding of the great anthropological problems to be solved, and not with the sole aim of making collections of archæological objects.

CHAPTER IX

FROM FOSSIL MEN TO LIVING MEN

THE men of the Reindeer Age, as we have just seen, alread belonged to modern mankind both in structure and in ment characteristics. So that even as early as the dawn of ou present geological, or *Holocene*, period (see p. 50), modern Ma was definitely established, and had been so for a long peric of time; and now the role of palæontology is ended. Henc forth the task falls to anthropologists and archæologists pursue the study of human groups throughout the countle migrations and changes which preceded the historical even related in ancient manuscripts.

Difficulty of the Task.

This linking-up of Palæolithic times with Neolithic time of Neolithic times with the Metal Ages *Protohistoric* times, and of these again wi Historic times, is full of the most difficult problems. At t present time these problems can only be approached, and eve that with hesitation, in central and western Europe, the o region regarding which sufficiently abundant data has bee accumulated. Further, in regard to the relatively recent perio of our prehistory, as well as in regard to more ancient geologic periods, we barely know more than the final results of event the origins or starting points of which, at least in the majori of cases, are to be found in more distant countries, still almo unexplored from our point of view. The light which will o day illumine the great questions raised by a study of pr historic times, will come only after we have gained sufficie knowledge of Asia and of Africa; for to the palæontologist the two continents represent the great laboratories of life in t Ancient World. They must also have been the great centr of development of successive humanities, from those mc primitive types which were still akin to the animal, to those which saw the dawn of the great civilizations.[1]

This book is essentially a palæontological work, and I could now draw it to a conclusion, ignoring everything which relates to the Neolithic and Metal Ages ; for these, though to the historian they represent prehistory, are to the palæontologist only the conclusion of a history, that of the evolution of the human kind. It seems to me, however, that many of my readers might be disappointed did they fail to find here a summary of the known facts bearing on the transition from the vague Humanity of geology to the Humanity of history. And I decided to write this chapter in an attempt to join "the two ends of the chain," without having fully weighed the many and diverse difficulties to be encountered. It is a relatively easy matter to make a catalogue or to arrange a scheme (more or less formal or artificial) of observed facts or of relics, to arrange them geographically or in objective categories. But to give a synthetic account of all such scattered facts with a view to reconstruct a series of events, and that in the spirit of a naturalist, is a risky business. In excusing my effort, I wish to say that I myself am the first to realize the daring and the imperfection of the attempt I am about to lay before the reader.

A Preliminary Study. Races and Peoples.

Before taking up again the chronological thread of human prehistory where we left it, I must, in order to simplify the continuation of my account, summarize the state of our anthropological knowledge regarding the modern peoples of Europe with whom this account will finally land us.

I have just written "*anthropological* knowledge," because, as a matter of fact, historians and geographers consider only peoples or nationalities ; but anthropologists, who are first and foremost naturalists, must deal only with *races*. This word

[1] In thus expressing myself, I am viewing things broadly and from a naturalist's point of view. The movements of the tides of humanity have been, at least in the blind alley of the European peninsula, movements of flux converging on a centre. These movements have sometimes been followed by a sort of backwash or drift from the centre, but such movements, more limited in extent, can only be directly observed in protohistoric or historic times, when the populations of Europe had reached a higher degree of culture.

I use, not in the literary sense, which is generally metaphorical in meaning, but in its true sense, the general biological sense and the physical sense, that of a variation of the species more or less fixed by heredity.

For a long time now in France certain bright spirits in the historians' camp, and in the naturalists' as well, have laid stress on the extremely vexatious confusion which has arisen from the use of the words race, people, nation, language, culture or civilization, but the distinctions and the appropriate uses of these different expressions have not yet penetrated into the minds even of the enlightened public. Even to-day the most distinguished and most academic authors, when dealing with human classifications, use quite indiscriminately and in a completely wrong sense the word *race*, when they would express themselves more correctly in speaking of their horses or their dogs.

We must really impress on our minds the fact that the *race*, by which we mean the continuity of a physical type transmitting blood relationships, represents an essentially natural group, possibly having and, as a rule, actually having nothing in common with the people, the nationality, the language, and the customs, which correspond to purely artificial groupings of no anthropological significance and connected only with those historical events of which they are the products. Thus there is no Breton *race*, but a Breton *people*: no French *race*, but a French *nation*; no Aryan *race*, but Aryan *languages*; no Latin *race*, but a Latin *civilization*. De Quatrefages wrote: "A people changes its language, its customs, its crafts, sometimes in a relatively short period; it cannot with the same rapidity lose its stature, its colour, and the form of its skull."[1] The result is that maps, essentially multi-coloured

[1] An example of the extraordinary confusion which may result from a purely literary terminology is shown by the use of the word *Celtic*, which signifies to some people a language, to others a special civilization, and which is often used, rightly or wrongly, as synonymous with Gallic; to the mind of certain authors, it represents the blonde type, of great stature and elongated skull, found in the north; according to others, it should be applied to the dark type, of small stature and round head, belonging to the central plateau or the Alps. Anthropologists agree that it would be best to leave this expression to the archæologists and historians. "Ethnographical names are the bugbear of anthropology," said Salomon Reinach, and with good reason.

and changing, of peoples, of nations, or of languages, cannot bear and never do bear any resemblance to a map of races.

It is really very difficult to make a racial map, because of the multiple movements of peoples which have taken place over the whole of Europe since the end of Palæolithic times, and because of the innumerable crossings of every kind which have resulted.[1]

"The anthropologist who undertakes a history of the human races," said de Quatrefages again,[2] "is faced with a task exactly similar to that of the zoologist who attempts to set forth the races of one of our domestic species." It must be added that in the case of Man the problem is even more difficult, because crossings take place by chance and no longer under the influence of rational selection. The commingling of races has been both more intense and more haphazard. So it is that in anthropology, more than in systematic zoology, and more in Europe than elsewhere, the race can scarcely represent anything but an abstraction, a sort of ideal type, around which are grouped variations resulting from combinations, or rather from associations of characters borrowed from different primitive elements. "Neither the type nor the race, in the present state of mankind, are actual realities," said Topinard.[3] That is not to say, however, that the search for this ideal physical type is altogether an idle pursuit. We can still perceive individuals or groups of individuals who have remained more akin to their original stocks, either through having escaped, by isolation, from outside influences, or through possessing in themselves a combination of dominant characters similar to that of the ideal type.[4]

[1] There can be no doubt that great movements of peoples took place during Palæolithic times, and that in great part they were correlated with the migrations of faunas, for, in these remote ages, Man was less independent of the influence of physical phenomena than he later became. But we are unable to trace these movements. In Neolithic times, which are much less remote, the great migrations stand out more clearly. Thanks to the development of his industry, Man shook himself free more readily from physical conditions; his massed movements hardly ever originate except from his own will or from that of his leaders.

[2] Quatrefages, A. de, *Histoire générale des races humaines*, p. 188.

[3] Topinard, P., *L'Homme dans la Nature*, p. 43.

[4] We must, however, distinguish permanence of type from permanence of characters, the latter being much more frequent than the former.

In practice, we may define as races those human groups which possess in common the same physical characters, taking as some of the most important, stature of body, form of skull and face, and colour of eyes and of hair.

The Races of Europe.

For half a century, anthropologists have devoted themselves to laborious investigations in an attempt to rediscover or reconstruct those primitive physical types which represent in Europe the common stock, a stock now overlaid by successive ethnographical alluvials, reshuffled again and again. Thousands of living individuals and of skeletons have been examined, studied, measured, with a view to deciding the blood affinities and natural descent of each. In France, Broca, de Quatrefages and Hamy, Topinard, Collignon, and Deniker have published important works on this subject. England, Germany, Italy, and Scandinavia have been the centres of similar enquiries.[1] The American anthropologist, Ripley, has written a book of the highest value, illustrated by many maps, on *The Races of Europe.* These maps, like those of Deniker, present a medley of colours or of figures, which clearly reflect at first glance the extraordinary complexity of the human mosaic in the European region.

The conclusions of our most eminent anthropologists differ from each other as regards the number of divisions or sub-divisions adopted, as well as in the terminology used and in certain other details of secondary importance; but they agree in distinguishing three main groups, which are likewise admitted by Retzius, the Swedish worker. These three great sub-divisions of the species or of the race of *White Man,*[2] the only divisions which we need examine for the purpose we have in view, are as follows:

HOMO SAPIENS ALBUS.	Dolichocephalic (long-headed)	Fair, of large stature	1. *nordicus.*
		Dark, of small stature	2. *mediterraneus.*
	Brachycephalic (round-headed)	Dark, of small stature	3. *alpinus.*

[1] For bibliography, see Deniker, J., *Races et peuples de la terre* (Paris, 1900); and particularly Ripley, W. Z., *The Races of Europe* (London, 1900).

[2] All Europeans are white, except the Laps. They all belong to *Homo sapiens albus* (see p. 72). It is this group, species, or race which we must sub-divide into races or sub-races.

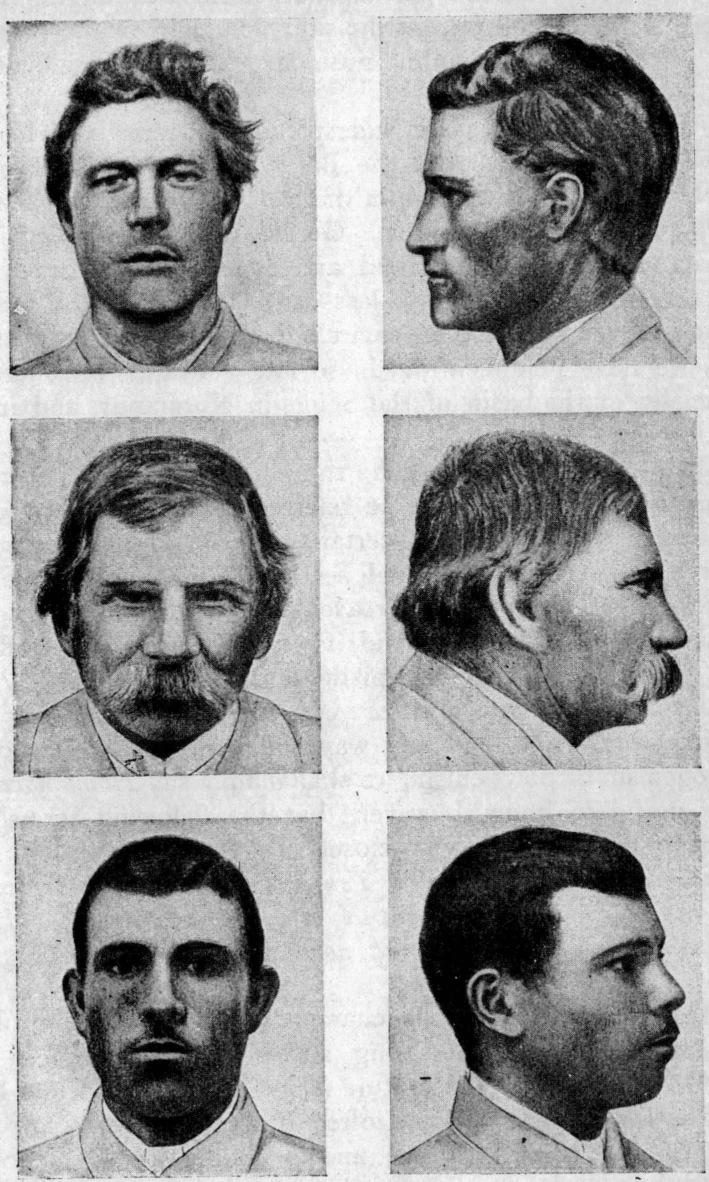

FIG. 199.—Representatives of the three Physical Types of Europe, full face and in profile. Above, Nordic race (Norwegian). In centre, Alpine race (Austrian). Below, Mediterranean race (Sicilian). (After Ripley).

The Nordic Race.

The first of these types is characterized by having a long dolichocephalic skull, long narrow face, straight fine aquiline nose, large stature, blue eyes, fair hair, and rosy skin.

At the present day it is widespread in the north of Europe, round the North Sea and the Baltic Sea, in Scotland, in the north and east of England, in the east of Ireland, in Flanders, Holland and Denmark, North Germany, the Baltic Provinces, Russia, the coasts of Finland, and especially in Norway and Sweden. It is also found, less pure or in sporadic groups, throughout a wide belt encircling the outer limits of the area we have just mentioned; so far as France is concerned, especially in the basin of the Seine, in Normandy, and in the south-west.

This is the type which the great Swedish naturalist, Linnæus, had in view when he referred to his *Homo europæus.* Following the example of certain anthropologists, we might reserve this term for the small section of mankind I have just described. But for various reasons, I consider it better to use Deniker's term *nordicus.* And if we wish accurately to define this first type, in a monogenistic sense, we must say: *Homo sapiens albus nordicus.* For convenience in speaking and writing, and without in any way prejudicing the question of monogenism or polygenism, we shall simply say *Homo nordicus.* It is useful to know, however, that the following terms also correspond more or less closely to this designation: the *Scandinavian* race, Ripley's *Teutonic* race, or the *Germanic* race of numerous authors, the *Kymric* race of Broca, the *Homo indo-europæus dolichomorphus nordicus* of Giuffrida-Ruggeri, etc.

The Mediterranean Race.

The second type is also characterized by a dolichocephalic skull and long and narrow face, but in this case the stature is small or medium, the body slender, the nose larger and often tiptilted, the eyes are very dark, the hair black or brown, and the skin swarthy.

At the present day it occupies the whole surrounding area as well as the islands of the Mediterranean Sea: the Iberian Peninsula, the south of France, Provence, Corsica and Sardinia,

Italy southward of Rome, Sicily, the shores of the Ægean Sea, and North Africa. Mixed or in a sporadic state, it is found on the Atlantic coast of France and in the west of the British Isles, particularly in Wales.

It is often described by the very expressive name of *Homo mediterraneus*, which corresponds to the *Mediterranean branch* of Sergi, the *H. meridionalis* of Wilser, the *Homo indo-europæus dolichomorphus mediterraneus* of Giuffrida-Ruggeri, the *Ibero-insular* race of Deniker, and to the Iberians of historians.[1]

The Alpine Race.

The third type is also of small or medium stature, but thickset and round-headed. The face is broad and round, the nose is rather large, the eyes are light or dark-brown, and the hair black or chestnut.

Driven like a wedge between the two other races, it occupies at the present day the greater part of Russia, Asia Minor, the Balkans, Bohemia, Switzerland, the western Alps, the central massif of France, Brittany, the south-west and Cantabrian coast, and North Italy.

This is the *Homo alpinus* of modern anthropologists, the *Homo indo-europæus brachymorphus alpinus* of Giuffrida-Ruggeri, the *Occidentale* or *Cévenole* race of Deniker, the *Celtic* or *Rhætic*, *Celto-Slav*, *Arverne*, *Laponoid*, and *Armenoid* races of other authors.

Needless to say, this modern geographical distribution of the three principal physical types which we have succeeded in disentangling from the European ethnographical medley, is only diagrammatic. Their limits are far from being clearly defined; there are imperceptible transitions from one region to the next; there are also intrusions of all sorts, islets all but swamped in a different mass. There are in particular innumerable combinations of characters, producing gradual transitions from one type to the other, and we have seen that this could not but be so. But, in view of the goal we are striving to

[1] Along with Deniker, we can distinguish as a sub-race or sub-type, an Atlanto-Mediterranean group, mesocephalic and of large stature, limited to certain points on the Atlantic and Mediterranean coasts of France, Spain, and northern Italy. I mention it here because it has been thought that in it may be seen a survival of the Cro-Magnon race.

reach, we must treat the subject broadly. From this standpoint, I consider that the disentangling and elucidating of these three principal types is a veritable conquest of European anthropology.

To sum up, as the accompanying map (Fig. 200) shows, the distribution of these races follows the general lines of the physical geography of our continent, while it is quite independent of its political geography. The three names

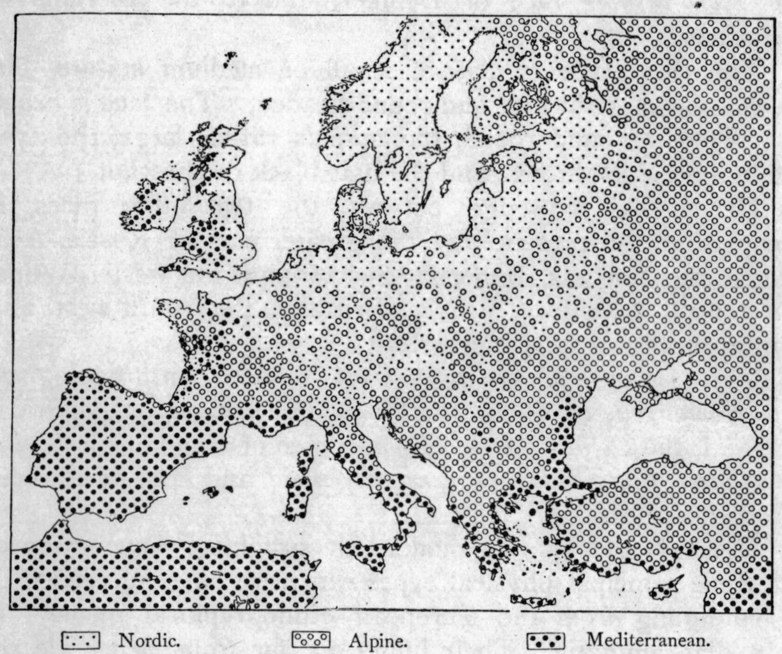

FIG. 200.—Very diagrammatic Map of the distribution of the three principal European human types. (After Ripley and Madison Grant.)

which we have adopted, Nordic, Mediterranean, and Alpine, describe this distribution well. The Nordic is the man of the northern countries with their harsh climate, misty horizons, and long, pale nights; the Mediterranean type, of the southern countries, is also the product of his environment, of an easy life in a warm climate and bright sunshine; the Alpine type is the result of an age-long adaptation to mountainous and barren regions. Each one of these three physical types also possesses psychical characters on which we shall not lay stress here, but which will emerge in the course of this account.

* * *

After this digression, the utility of which the reader will readily perceive, let us return to our main subject.

Holocene Period. Its Divisions.

The *Holocene* period, which we have now reached, well deserves the other name of *Recent*, also given to it by geologists. Its topography is that of to-day; it possesses the same wild animals and plants, to which are now added the domestic animals and cultivated plants. In the new civilization we actually see the dawn of the civilizations of history. It coincides with the arrival in Europe of the first Round-headed or Brachycephalic Men.

The sub-divisions of the geologist, the palæontologist and the anthropologist are now in perfect agreement, and from the outset of prehistoric studies, as we have seen, they were clearly understood and established. The Pleistocene of the geologist is the Palæolithic of the prehistorian; the Holocene of the geologist begins at exactly the same time as the Neolithic of the archæologist. The Holocene period is divided in the following manner:—

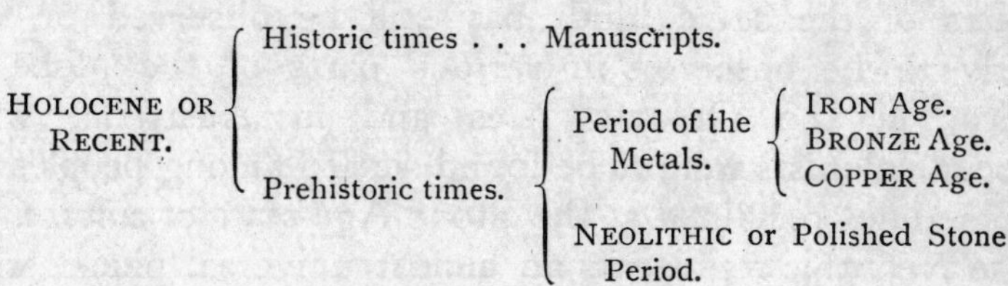

I should greatly exceed the limits of my subject were I to enlarge on the archæological characters and sub-divisions of the main stages of prehistoric times. For such information I refer my readers to special works on the subject, particularly Déchelette's excellent *Manuel*.[1] It will suffice if I draw attention to certain important points.

[1] [Or, for English readers, Sir John Lubbock's *Prehistoric Times;* Boyd Dawkins, *Early Man in Britain;* Rice Holmes, T., *Ancient Britain and the Invasions of Julius Cæsar*, Oxford 1907, contains many references to apposite literature.]

Comparative Chronology of Prehistoric and Historic Times.

At the outset it must be clearly understood that the expressions *historic times* and *prehistoric times* have a different chronological value in different countries. We, in France, consider that history really begins only with Julius Cæsar—that is to say, but a short time before the Christian era; in the northern countries, however, its beginnings are much later, while in the East, on the other hand, they date back for some 6000 years.

So it comes about that, during the prehistoric stages of the Polished Stone and Metal Ages, there cannot be, even in Europe, chronological agreements between the successive archæological fashions of different countries, such as the correlations we have been able to allow, not, however, without the strictest reservations, in Palæolithic times. Thus in the East, the Neolithic Age, at least in its beginnings, seems to be contemporary with the end of our Palæolithic. Copper, bronze, and iron had been known and made use of in Egypt thirty centuries before they were employed in Central Europe. Yet in the valley of the Nile, the Metal Ages overlap historic times, while with us their beginnings are lost in the darkness of prehistoric times.

Facts of the same kind may still be observed or were formerly to be observed in various parts of the globe: in America, in the Oceanic Isles and in Australia, where European colonists were to be found settled among peoples who had sometimes remained at the Stone Age state of culture.

The Neolithic represents an almost universal phase, which everywhere seems to have been of long duration. Montelius considers that it dated back 20,000 years in Egypt; Arthur Evans believes that in Crete its beginnings recede for 14,000 years; whereas, in our regions, we may place it at 7000 or 8000 years ago. Here it persisted for 4000 or 5000 years, for the Metal Ages, which began in the East at least 4000 years before our era, did not originate in the West till about 2500 B.C. The Iron Age, also very ancient in the East, lasted in Gaul from 900 B.C. till the Roman conquest. The accompanying table (Fig. 201), the dates in which have been

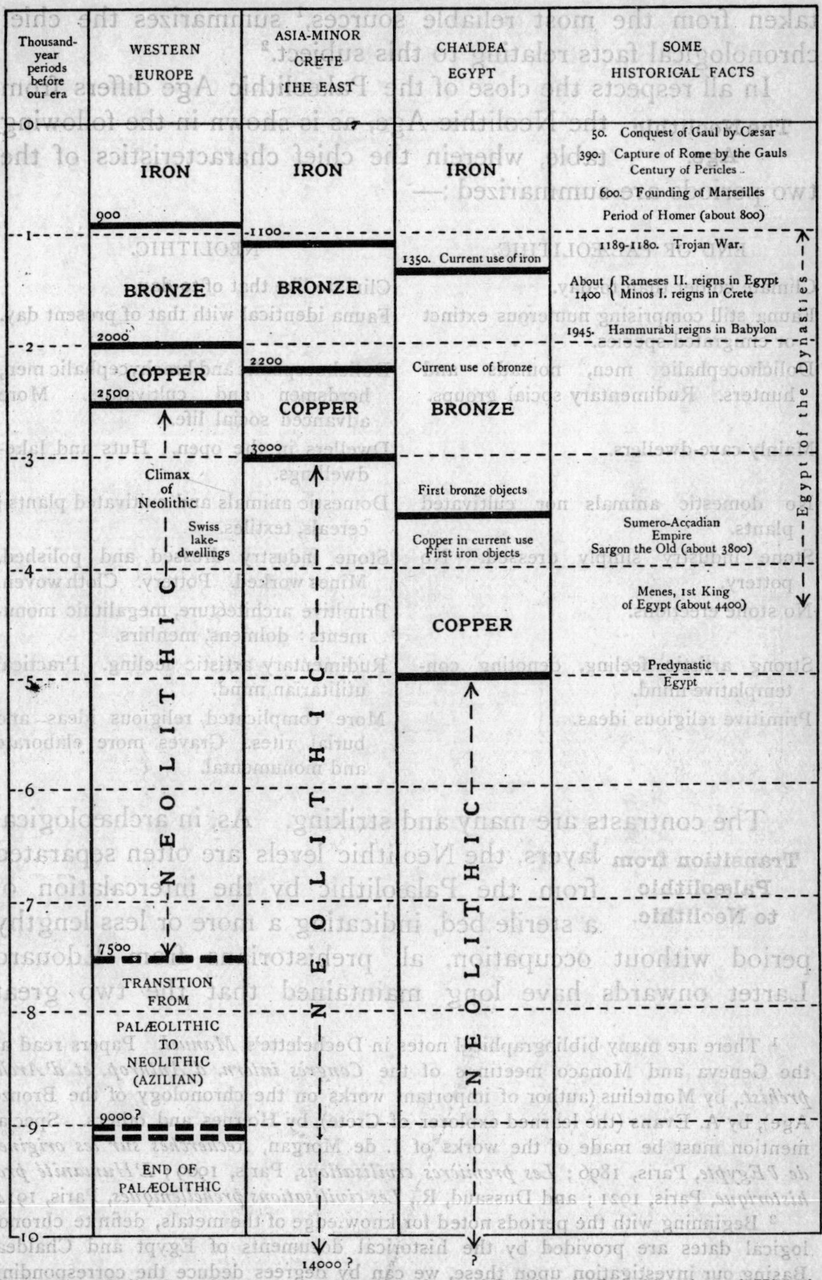

Thousand-year periods before our era	WESTERN EUROPE	ASIA-MINOR CRETE THE EAST	CHALDEA EGYPT	SOME HISTORICAL FACTS
0	IRON 900	IRON 1100	IRON	50. Conquest of Gaul by Cæsar 390. Capture of Rome by the Gauls Century of Pericles 600. Founding of Marseilles Period of Homer (about 800)
1	BRONZE 2000	BRONZE	1350. Current use of iron	1189-1180. Trojan War. About 1400 { Rameses II. reigns in Egypt Minos I. reigns in Crete 1945. Hammurabi reigns in Babylon ← Egypt of the Dynasties →
2	COPPER 2500	2200 COPPER 3000	Current use of bronze BRONZE	
3	NEOLITHIC Climax of Neolithic Swiss lake-dwellings	NEOLITHIC	First bronze objects Copper in current use First iron objects	Sumero-Accadian Empire Sargon the Old (about 3800)
4			COPPER	Menes, 1st King of Egypt (about 4400) Predynastic Egypt
5			NEOLITHIC	
6				
7	7500 TRANSITION FROM PALÆOLITHIC TO NEOLITHIC (AZILIAN)			
8				
9	9000? END OF PALÆOLITHIC		?	
10		14000 ?		

FIG. 201.—Indication of the comparative chronology of Prehistoric, Protohistoric, and Historic times in Western Europe, the East, Egypt, and Chaldea.

taken from the most reliable sources,[1] summarizes the chief chronological facts relating to this subject.[2]

The Neolithic Age.

In all respects the close of the Palæolithic Age differs from the Neolithic Age, as is shown in the following table, wherein the chief characteristics of the two periods are summarized :—

END OF PALÆOLITHIC.	NEOLITHIC.
Climate colder than to-day.	Climate like that of to-day.
Fauna still comprising numerous extinct or emigrated species.	Fauna identical with that of present day.
Dolichocephalic men, nomads and hunters. Rudimentary social groups.	Dolichocephalic and brachycephalic men, herdsmen and cultivators. More advanced social life.
Mainly cave-dwellers.	Dwellers in the open. Huts and lake-dwellings.
No domestic animals nor cultivated plants.	Domestic animals and cultivated plants ; cereals, textiles.
Stone industry simply dressed. No pottery.	Stone industry dressed and polished. Mines worked. Pottery. Cloth woven.
No stone erections.	Primitive architecture, megalithic monuments : dolmens, menhirs.
Strong artistic feeling, denoting contemplative mind.	Rudimentary artistic feeling. Practical utilitarian mind.
Primitive religious ideas.	More complicated religious ideas and burial rites. Graves more elaborate and monumental.

Transition from Palæolithic to Neolithic.

The contrasts are many and striking. As, in archæological layers, the Neolithic levels are often separated from the Palæolithic by the intercalation of a sterile bed, indicating a more or less lengthy period without occupation, all prehistorians from Edouard Lartet onwards have long maintained that the two great

[1] There are many bibliographical notes in Dechelette's *Manuel*. Papers read at the Geneva and Monaco meetings of the *Congrès intern. d'Anthrop. et d'Arch. préhist.*, by Montelius (author of important works on the chronology of the Bronze Age), by A. Evans (the learned explorer of Crete), by Hœrnes and others. Special mention must be made of the works of J. de Morgan, *Recherches sur les origines de l'Egypte*, Paris, 1896 ; *Les premières civilisations*, Paris, 1909 ; *L'Humanité préhistorique*, Paris, 1921 ; and Dussaud, R., *Les civilisations préhelléniques*, Paris, 1914.

[2] Beginning with the periods noted for knowledge of the metals, definite chronological dates are provided by the historical documents of Egypt and Chaldea. Basing our investigation upon these, we can by degrees deduce the corresponding dates in other countries, by means of archæological comparisons.

divisions of the Stone Age are separated by a blank, sometimes described as a gulf, corresponding to a complete transformation or revolution. More moderate opinion on the subject is content to regard it as a mere hiatus in our knowledge. "After the age of La Madeleine," said M. Cartailhac, "there is—in our knowledge of the facts—a break in continuity, a very lengthy and as yet very obscure period of transition. And when we once more emerge therefrom into the light, great changes have taken place; progress of prime importance has been effected, the sum of introductions seems large. Thus the Reindeer has entirely disappeared; domestic animals are plentiful, the populations are settled and are engaged in agriculture; implements and weapons in stone are often polished, pottery is known . . . monuments are erected, and art no longer portrays living nature."[1]

In point of fact every one is right. It is clear that if the Neolithic Age shows a completely new order of things, particularly the arrival of peoples with quite different industries and customs from those of the last Palæolithic peoples, then, in accordance with the general principle of continuity, the break could not have existed everywhere; sometime or other the transition period will appear less obscure. Indeed the hiatus has been bridged, in great part at least, by the excellent researches of Piette in the Mas d'Azil Cave in Ariège.

During the years 1887-1889, this eminent prehistorian explored a series of river deposits of archæological interest on the left bank of the Arize, a rapid stream which flows in a kind of exaggerated tunnel forming the Cave of Mas d'Azil. The accompanying figure illustrates a section of these deposits as I recorded it, along with Piette, in 1889. From top to bottom there is revealed (Fig. 202):—

5. Masses and blocks of rock detached from the roof of the cave. Here were found Gaulish objects, Bronze Age objects, and, towards the base of the layer, polished stone axes and Neolithic pottery. Depth varied from 2 to 3 metres.

4. Banded masses of ashes, with beds of charcoal and innumerable snail-shells. Polished stone implements, pebbles worn with usage at the points. Fragments of pottery. Variable depth.

---

[1] Cartailhac, E., *Les âges préhistoriques de l'Espagne et du Portugal* (Paris, 1886, . 47)

3. Bed of cinders, reddish incinerated earth, and charcoal, with modern wild fauna. Red Deer and Beaver abundant; not the slightest trace of Reindeer. Flat perforated harpoons of deer-horn (Fig. 203). Painted pebbles (Fig. 204); some pebbles polished at ends, etc. The upper part consists of an ancient surface (*as.*) marked by a line of stones; in the lower part are beds of ashes (*ch.*). Depth, 0.10 metres to 0.80 metres.

2. Great deposit of silt from the neighbouring river, made up of hundreds of laminæ, and cut up by several ashy lines of incinerated earth. In these layers of occupation Red Deer are abundant, Reindeer rare; harpoons of Red Deer and Reindeer antlers; needles and artistic engravings. Total depth about 3.50 metres.

1. Earth mixed with pebbles, almost barren, resting on the calcareous rock (R) of the cavern. Depth, about 1.40 metres.

The lower layers (1-2) represent the later Pleistocene deposits; they belong to the end of the Reindeer Age, to the Magdalenian period. The upper formation (5) corresponds to the modern period, from Neolithic times to our own day. The intermediate layers (3 and 4) cover the well-known hiatus between Palæolithic and classic Neolithic times: layer 4 represents the dawn of Neolithic times, while layer 3, containing coloured pebbles, is really the transition layer. It corresponds to a particular period well named by Piette[1] the *Azilian* period.

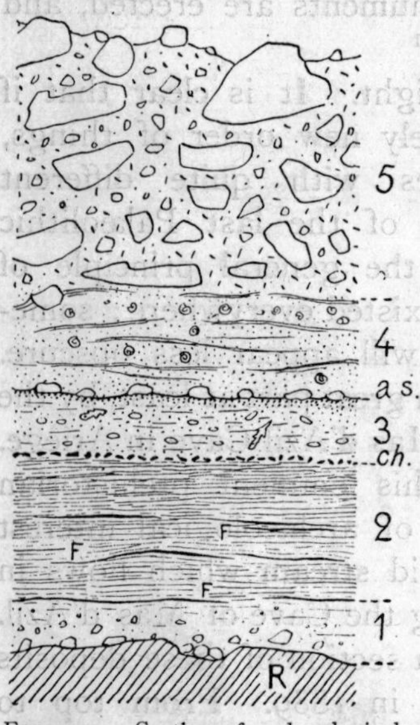

FIG. 202.—Section of archæological layers in the left bank of the Mas d'Azil Cave. (For description of the layers, see accompanying text.)

To my mind the majority of archæologists are mistaken in wishing to place this Azilian period in Palæolithic times, and in calling it *Epipalæolithic*. As I pointed out, so long ago as the year 1889,[2] the Azilian, though not yet Neolithic, is no longer Palæolithic; it is a period apart, with a special facies, and the facies is that of transition.

[1] Piette, E., "Un groupe d'assises représentant l'époque de la transition entre les temps quaternaires et les temps modernes" (*Comptes rendus de l'Acad. des Sc.*, 25th Feb. 1889); and numerous other papers, particularly in *L'Anthropologie*, vii., xiv.

[2] *Congrès intern. d'Anthrop. et d'Archéol. préhist.* (Paris, 1889, p. 209).

From the geological or stratigraphical point of view, this transition is obvious; from the palæontological point of view it is no less clear, since the fauna of the Azilian layer, from which the Reindeer is absent, is identical with the wild fauna of the present day and does not yet contain any trace of domestic animals. Transition—I do not say direct succession—is also evident from the archæological point of view, for along with flint implements resembling those of the Magdelenian period, we find the first products of stone-polishing. Though harpoons of deer-antler still persist, their form and substance are different. There are no longer present products of artistic expression, for the colouring of the pebbles has nothing in common with Palæolithic paintings. It would be difficult to imagine a set of conditions more expressive of transition than those I have just enumerated. All the evidence leads us to believe that we are here dealing with a contribution from the Mediterranean area.

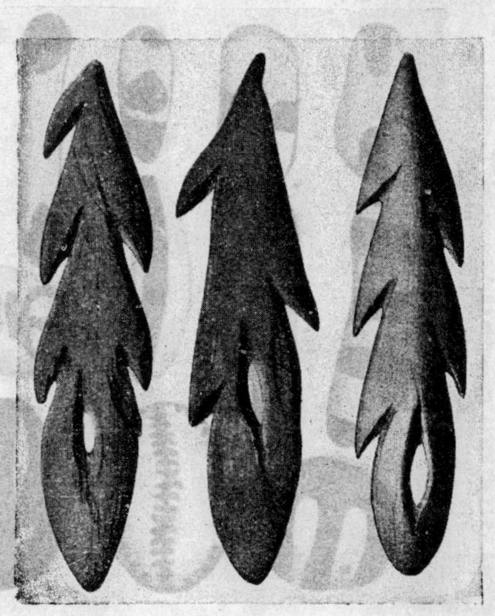

FIG. 203.—Flat Harpoons of Red Deer Antler from the Mas d'Azil Cave. Three-fourths natural size. (After Piette.)

Piette's conclusions were at first not well received by the prehistorians whose ideas they upset. It was not long, however, before these conclusions were confirmed. The *characteristic fossils* of the Azilian period, flat harpoons and coloured pebbles (Figs. 203 and 204), were soon after discovered in other deposits hailing from the Pyrenees, the Gard, the Dordogne, from Spain, Switzerland, Bavaria, and Denmark, even from Scotland and perhaps from Russia.

Traces of the transition period are to be found elsewhere under a slightly different aspect.

In France and other countries of Europe, in Asia and in Africa, certain settlements are characterized by an implement industry of small flints with geometrical outlines, known as pygmy flints (Fig. 205). Messrs de Mortillet considered that this industry, which they named *Tardenoisian* (from Fère-en-Tardenois, where it is well represented), belonged to the Lower Neolithic.[1] The most competent prehistorians now

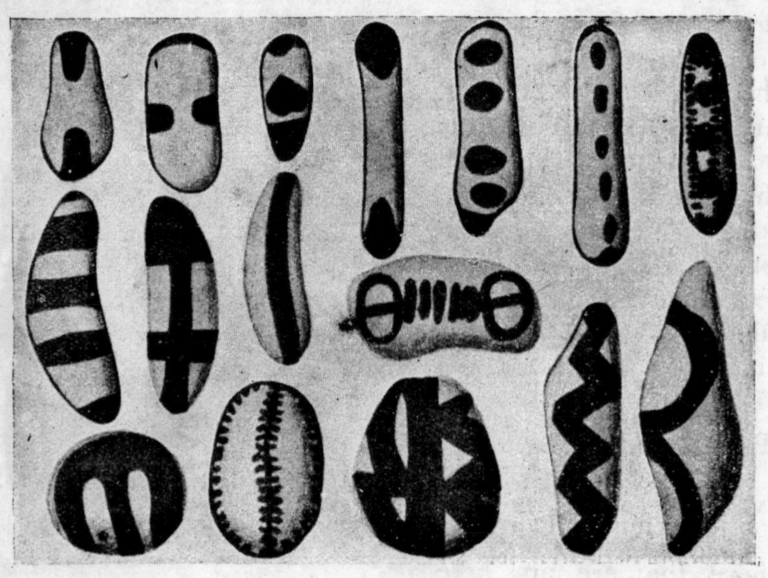

FIG. 204.—Coloured Pebbles from Mas d'Azil. (After Piette.)

agree in regarding the Tardenoisian as almost contemporaneous with the Azilian period, or as slightly more recent.

The close relationship which exists between these two archæological facies is shown in the curious settlement at Maglemose, in the island of Zeeland in Denmark, which has been studied by M. Sarauw. Here, along with Tardenoisian geometrical flints, were found deer-antler harpoons like those from Mas d'Azil, but no pottery or polished axes. As was the case in the Midi, this northern Azilian deposit may be placed, from the geological point of view, exactly between the

[1] Mortillet, A. de, "Les petits silex taillés à contours géométriques" (*Revue de l'Ecole d'Anthrop.*, 1896).

end of the Pleistocene period and the beginning of the Holocene.[1]

On the other hand, on the coasts of Denmark, France, and

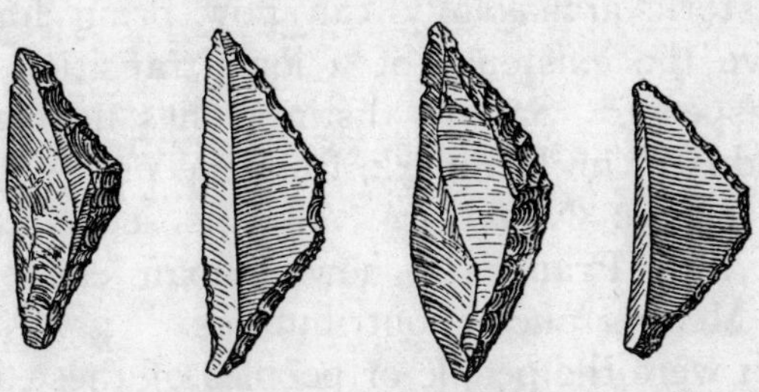

FIG. 205.—Tardenoisian Flints from Valle, Spain. Twice natural size. (After Breuil and Obermaier.)

Portugal, we have long been familiar with artificial mounds, composed of earth, cinders, and floors of occupation containing, scattered everywhere, innumerable sea-shells, bones of animals, implements of flint, particularly "chisels" (*tranchets*), and of bone, as well as pottery (Fig. 206). They are the sites of ancient settlements, to which Danish archæologists, who carefully studied those in their own country, have given the name *kjökkenmoddings* (or better, *kjökkenmöddinger*), which is to say kitchen middens. The industry of these curious deposits is already Neolithic, but it is characteristic of very early Neolithic times, for domestic animals, with the exception of the dog, are not yet present.[2]

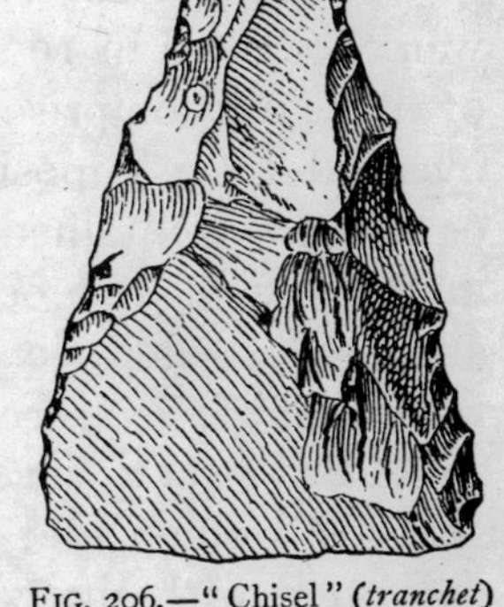

FIG. 206.—"Chisel" (*tranchet*) from Campigny. Natural size. (After Capitan.)

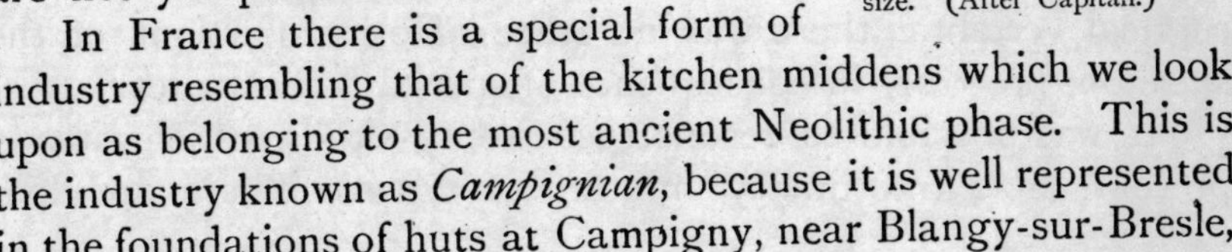

In France there is a special form of industry resembling that of the kitchen middens which we look upon as belonging to the most ancient Neolithic phase. This is the industry known as *Campignian*, because it is well represented in the foundations of huts at Campigny, near Blangy-sur-Bresle,

[1] Sarauw, G. F. L., "Maglemose" (*Præhistorische Zeitschrift*, iii., 1911). See also *L'Anthropologie*, xxiv., 1913, p. 64.

[2] Old works of Steenstrup, J., and Worsaae. More recent and fine work by Madsen, Muller, S., Neergaard, Pétersen, Rostrup, Steenstrup, R. J. V., Winge, H., *Affaldsdynger fra stenalderen i Danmark*, Copenhagen, 1900.

in the Seine-Inferieure. The characteristic implements are chisels (*tranchets*) and picks; there are no polished axes, but coarse pottery is known.[1]

So prehistoric archæology can now bring forward many facts to prove the existence of a long transition period with many local aspects.[2] Sarauw distinguishes no fewer than six phases of industry in Denmark, from the end of the Reindeer Age to the classic Neolithic. All this seems to imply the introduction into France of new human elements, and in particular of Mediterranean contributions.

Azilian Races. Arrival of the First Brachycephalic People.

Who then were the people or peoples of these intermediate ages? Unfortunately we have not sufficient osteological evidence to answer this question. At Mas d'Azil, Piette and Cartailhac collected in the Azilian layer a few bones coloured red. They are too fragmentary to enable us to determine the chief characters of the race they represent.

At Ofnet, near Nordlingen in Bavaria, an interesting cave was excavated in 1907 and 1908 by R. Schmidt. The deposits of this cave comprise a series of layers of the Reindeer Age, overlaid by a deposit of Azilian Age. Two shallow ditches or trenches contained a large number of human skulls, enveloped in a mass of red ochre, having their lower jaws intact, and arranged concentrically close to each other, as if in a nest, with their faces turned towards the west. The large trench contained twenty-seven skulls, the small one six (Fig. 207). The skulls of women and children, which were most numerous, were decorated with the canine teeth of deer and perforated shells, similar to those from the Mas d'Azil. Apart from some cervical vertebræ, there was no trace of other portions of the skeleton, probably in consequence of special funeral rites.[3]

It was possible to reconstruct twenty skulls, which already present an extraordinary mixture of types. There are doli-

[1] Salmon, P., d'Ault du Mesnil, G., and Capitan, L., "Le Campignien" (*Revue de l'Ecole d'Anthrop.*, 1898).

[2] J. de Morgan names this period the *Mesolithic.*

[3] Schmidt, R. R., "Die diluviale Vorzeit Deutschlands, Stuttgart, 1912; the third part containing anthropological notes by Schliz, A. See also Breuil, H., "Le gisement quaternaire d'Ofnet" (*L'Anthropologie*, xx., 1909).

chocephalic forms, brachycephalic forms, and intermediate forms (the cephalic index ranging from 70 to 89). The first have long faces, harmonious in their proportions; they thus differ from the dolichocephalics of the Cro-Magnon race. Schliz regards them as belonging to the group of *Homo mediterraneus.* The second represent the most ancient brachycephalics known with certainty, the first arrivals of *Homo alpinus.* The third are already products of crossing.

FIG. 207.—Burial in the smaller trench at Ofnet. (After R. R. Schmidt.)

The Maglemose settlement has yielded no human skeletons and we know very little of the men of the Danish kitchen middens. Osborn's[1] theory that the Maglemose men must have belonged to the tall fair race of the North, *Homo nordicus,* is so far quite without foundation, although very plausible.

On the other hand, the kitchen middens at Mugem in Portugal, containing Tardenoisian products, are exceedingly rich in human skeletons, which have been studied by M. de Paula.[2]

[1] Osborn, H. F., *Men of the Old Stone Age*, p. 488.

[2] In Emile Cartailhac's fine work, *Les âges préhistoriques de L'Espagne et du Portugal*, Paris, 1886.

Two quite distinct types have been recognized. The first, much the more numerous, is dolichocephalic, small in stature (about 1.60 metres [5 ft. 3 ins.] in height), and of poor cranial capacity; the face is long, of harmonious proportions, with slight sub-nasal prognathism. This is the "Mugem race," as de Quatrefages named it, but it would really seem that we have here simply very old representatives of *Homo mediterraneus* and in no sense descendants of the Cro-Magnon type (see Appendix, p. 474).

Only a few skulls are brachycephalic; one of them, with a broad face, possesses Mongoloid features.

Thus, from the anthropological as well as from all other points of view, the Azilian and Tardenoisian deposits present transition features in the persistence of physical types more or less related to the Men of the Reindeer Age, and in the appearance, at first rare then becoming more and more frequent, of a brachycephalic type, newly arrived in the countries of Europe. We already see unfolding a picture of the distribution of modern European races, especially through the diffusion of the Mediterranean type. Examination of Neolithic remains will help us to envisage the development of this new state of things, by showing us its increasing complexity, as well as an increasing diversity of types according to the country of their habitation.

Neolithic Races. Relationships of the Three Great Types.

Unfortunately these evidences, although very numerous, are not so informative as we could wish. We have in particular the skulls, but the indices of these do not always allow of a precise definition of the race to which they belong. Complete skeletons, which would have enabled us to determine the stature of the peoples, are much rarer. And we can hardly learn anything of the characters founded upon the skin or nature of the hair.

In a general review of the great accumulations of statistical data collected by anthropologists, but neglecting details and adhering only to the main lines, we find two general facts emerge:—

1. In the first place, we note the simultaneous presence in every country of varied physical types, long heads, short heads

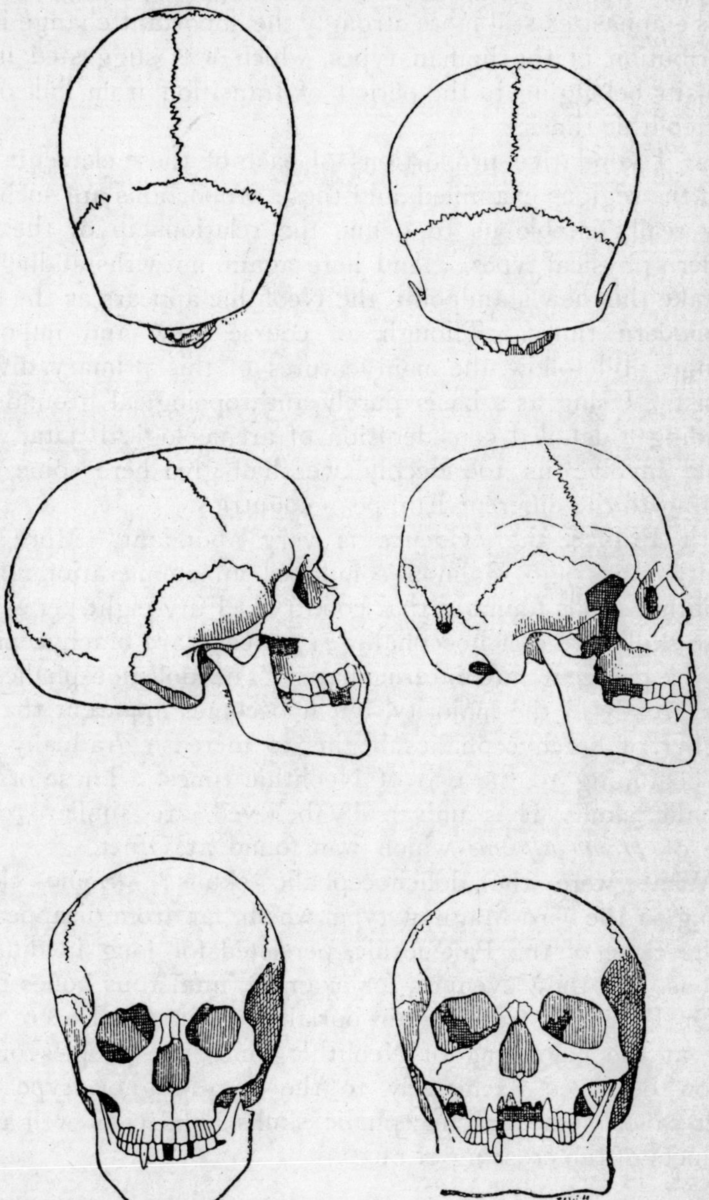

FIG. 208.—Dolichocephalic and Brachycephalic Skulls seen from above, in profile, and full face. From covered passage in the Mureaux (Seine-et-Oise). One-fourth natural size. (After Dr Verneau.)

(Fig. 208), and intermediate (or mesocephalic) forms of head. This emphasizes still more strongly the important change in the distribution of the human types, which was suggested in the deposits belonging to the period of transition from Palæolithic to Neolithic times.

2. The relative proportions of each of these elements vary with the regions examined, and these proportions are such that they really enable us to define the relationship of the chief modern physical types. And here again, notwithstanding that we take this new standpoint, the Neolithic appears as the dawn of modern times. Though, of course, new and important changes still follow, the main features of this primary division persist. Using as a basis purely anthropological grounds and avoiding a detailed consideration of archæological data, which would involve us too deeply, we shall give here some facts relating to the different European countries.

In France, the evidence is very abundant. More than twenty years ago, Salmon completed an enumeration of 688 Neolithic skulls found in that country.[1] Fifty-eight per cent. of those skulls are dolichocephalic, 21 per cent. are brachycephalic, and 21 per cent. are intermediate. The dolichocephalics are thus greatly in the majority, but it becomes apparent that the number of brachycephalics begins to increase gradually from the beginning to the end of Neolithic times. These brachycephalic skulls, it is universally believed, are similar to the type of *Homo alpinus* which was found at Ofnet.

What were the dolichocephalic skulls? Some clearly belong to the Cro-Magnon type, which, far from disappearing at the close of the Palæolithic, persisted for long in different regions. In the Cevennes, for example, numerous bones taken by Dr Prunières from certain burial caves in the Lozère show that, at the beginning of Neolithic times, the peoples of this region belonged exclusively to the Cro-Magnon type. In other caves, some brachycephalic skulls appear, as well as the products of the crossing of races.

[1] Salmon, P., "Dénombrement et types des crânes néolithiques de la Gaule" (*Revue de l'Ecole d'Anthrop.*, 1896). See also Hervé, G., "Les brachycéphales néolithiques," *ibid.*, 1894 and 1895.

There can be no doubt regarding the contemporaneity of these brachycephalics and the old Palæolithic dolichocephalics, since several of the skeletons exhumed by Dr Prunières still bear, in their bones, the Neolithic arrows which pierced them (Fig. 209), and which show that they must have been the victims of the Neolithic invaders whose skulls are found in the dolmens.[1] But the invasion of these mountainous regions must have been slow and incomplete, for, speaking generally, Neolithic brachycephalic skulls are relatively rare in the Cevennes; they become numerous only in the Bronze Age. A little farther south, at Montouliers in Hérault, Neolithic skulls, according to M. Mayet, are still Cro-Magnon in type, with a slight brachycephalic admixture.[2]

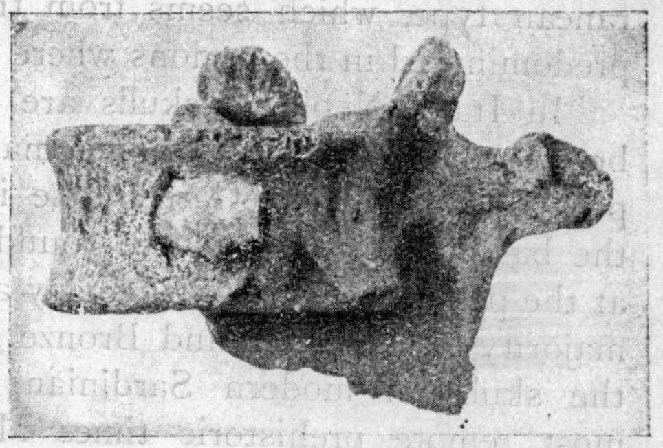

FIG. 209.—Human Vertebra pierced by a Neolithic Arrowhead, from a cave in the Lozère. Natural size. Prunières' collection in the Anthropological Gallery of the National Natural History Museum of France.

The artificial caves of the Marne, explored by M. de Baye, show the same association of types. Among the skulls examined by Broca and de Quatrefages many still resemble the Cro-Magnon type, others are brachycephalic and these show mixed characters resulting from crossing.[3]

In the Duruthy cave at Sorde, in the department of Landes (see p. 260), the Neolithic skulls from the upper portion of the bed are similar to the skull from the base of the deposit of Reindeer Age; while at Monaco, the Neolithic skulls examined

[1] Dr Prunières has scattered his numerous notes throughout different publications, particularly those of the *Association française pour l'avancement des Sciences*. The anthropological material has been studied chiefly by Broca (*Congrès intern. de Bruxelles*, 1872, p. 182). See also Quatrefages, A. de, *Hommes fossiles et Hommes sauvages* pp. 99 and 105.

[2] Mayet, L., "Les Néolithiques de Montouliers" (*L'Anthropologie*, xxiii., 1912).

[3] De Baye, *L'archéologie préhistorique*, Paris, 1880. Contains a contribution by Broca on the skulls from Baye's caves. See also Quatrefages, A. de, *Hommes fossiles et Hommes sauvages*, p. 107.

by M. Verneau[1] are mainly brachycephalic, accompanied by rare individuals still exhibiting affinities with the Cro-Magnon type, so well represented in this region in Palæolithic times.

Along with dolichocephalic types having short faces (disharmonic), which may be regarded as more or less modified descendants of the Cro-Magnon race, other Neolithic dolichocephalic types, with long faces, form in France a new contribution, which, as we have seen, commenced in the Azilian period. These ought rather to be classified with the Mediterranean type, which seems from this period onwards to have predominated in the regions where it still holds sway.

In Italy, Neolithic skulls are very mixed, sometimes the brachycephalic, sometimes the small-sized dolichocephalic type predominating. The latter case is particularly noticeable in the basin of the Po, where round heads are most numerous at the present day. But in Sicily and in Sardinia, a very large majority of Neolithic and Bronze Age skulls exactly resemble the skulls of modern Sardinians and Silicians.[2] From the most remote prehistoric times Malta and Crete were also peopled by the Mediterranean type and Sergi declares that the Ancient Egyptians, who were identical with the Libyans, are only a branch of the Mediterranean family.[3]

We have seen this same type appear in the Iberian peninsula at the period of the kitchen middens of Mugem. M. de Paula, already cited in this connection (p. 333), tells us, moreover, that in the caves and Neolithic burials of Portugal the dolichocephalic type of Mugem still predominates, with, however, very wide individual variations; although in certain beds of the same age, the Cro-Magnon type is found. The brachycephalic type of Mugem likewise appears in certain Neolithic stations, where it shows all the characters of the Neolithic brachycephalics of France.

On the other hand, M. Verneau has found representatives of the Cro-Magnon race in the Neolithic deposits of Oviedo

[1] Verneau, R., and Villeneuve, L. de, "La Grotte des Bas-Moulins" (*L'Anthropologie*, xii. 1901).

[2] Ardu-Onnis, E., "Restes humains préhistoriques de la grotte de San Bartolomeo, près Cagliari" (*L'Anthropologie*, xv., 1904).

[3] Sergi, *Congrès de Moscou*, ii., 305.

and Segovia, as well as of Andalusia. M. Jacques, on examination of the material yielded by the excavations of M. Siret in the south-east of Spain,[1] distinguished, in a series of Neolithic and Bronze Age skulls, a first group resembling the Cro-Magnon type, a second group composed of brachycephalics, a third group resembling the Mugem men (Mediterranean type), and finally a fourth group, resembling the Basque peoples of Spain. As a whole, the dolichocephalic Mediterranean type appears to predominate in the Iberian peninsula.

In Belgium and Holland, there is the same mixture of round heads, long heads, and mixed forms produced by crossing (such as the Furfooz skulls). The dolichocephalics are classified by various authors as belonging sometimes to the Cro-Magnon type, sometimes to the Mediterranean type.[2]

In the British Isles a very clear distinction has long been recognized. The Neolithic burials in long barrows enclose only dolichocephalic skulls: "Long barrows, long skulls," we say in England. As soon as bronze makes its appearance, the tumuli become round in form and the skulls found are brachycephalic: "Round barrows, round skulls." Here, in the extreme west of Europe, for reasons readily understood, the first brachycephalic migrations were later in taking place. According to Sir William Turner,[3] the Neolithic dolichocephalics, small in stature, with long faces and narrow noses, probably belong to the Mediterranean type. They still play an important part in the ethnographical constitution of Great Britain and Ireland. The brachycephalics, who may temporarily have invaded them, were of variable stature, with broad faces; they belong to the Alpine stock. Nowadays, their traces have been almost wholly obliterated.[4]

[1] Jacques, V., "Les races préhistoriques de l'Espagne." *Congrès de Paris*, 1889, p. 451.

[2] Houzé, "Les crânes néolithiques des cavernes d'Hastières" (*Bull. de la Soc. d'Anthrop. de Bruxelles*, viii., 1890). Fraipont, J., "La Belgique préhistorique et protohistorique" (*Bull. de l'Acad. Roy. de Belgique*, 1901).

[3] Turner, W., "A Contribution to the Craniology of the People of Scotland," pt. ii. (*Trans. of the Roy. Soc. of Edinburgh*, li., 1915).

[4] Keith, A., "The Bronze Age Invaders of Britain" (*Journal of the Anthrop. Institute*, xlv., 1915).

In Switzerland the facts are somewhat different.[1] In the archaic Neolithic deposits, brachycephalics predominate almost exclusively. In the mid-Neolithic, the dolichocephalics or the mesocephalics gradually come to balance the brachycephalics. In the period of transition from the Polished Stone to the Bronze Age, the dolichocephalics, according to Schenk, predominate. The important series of burials at Chamblandes, near Lausanne, recalls the Cro-Magnon type, but the skeletons are of small size and in certain characters resemble the Grimaldi negroids. These burials, moreover, present a new element. Side by side with the skeletons of the small dolichocephalics, two skulls exhibit the characters of the northern race; and M. Pittard tells us that, towards the end of the Neolithic period, there arrived in Switzerland men of high stature, with elongated skulls and narrow noses, whom he is able to classify as belonging to the type of *Homo nordicus*. It must also be added that there were found at Schweizersbild, near Schaffhausen, the bone-remains of some individuals of very small stature, almost pygmies, who Kollman would have us believe played a considerable part in the development of the human races,[2] but who are regarded by the majority of anthropologists as being only dwarfs whose skeletons lay side by side with those of other individuals of normal stature.

Nevertheless, the fact remains that even in the Neolithic period or rather in the end of the Neolithic period, the anthropological complexity which now exists in Switzerland becomes clearly noticeable. Here, for the first time, we recognize the presence of representatives of the big dolichocephalic people of the North. These become more and more numerous as we pass towards Central and Northern Europe.

In Germany, Neolithic human remains are generally dolichocephalic. In the south-west, according to Schliz, the dominant type is at first the long-faced, Nordic type, but later

[1] Pittard, E., "Sur l'ethnologie des populations suisses" (*L'Anthropologie*, ix., 1898); "Deux nouveaux crânes humains de cités lacustres" (*Ibid.*, xvii., 1906). Schenk, A., "Les sépultures et les populations préhistoriques de Chamblandes" (*Revue de l'Ecole d'Anthrop. de Paris*, 1904); *La Suisse préhistorique*, Lausanne, 1912.

[2] Kollman, J., "Die Pygmaen und ihre systematische Stellung innerhalb des Menschensgeschlecht" (*Vorhand. der Naturforsch. Gesellschaft in Basel*, xvi., 1902).

we see the arrival of brachycephalics and dolichocephalics of Mediterranean type.[1] In Bohemia and Silesia, the men of the Polished Stone Age usually belong to the Nordic race.[2] So it is also in Hungary. According to Giuffrida-Ruggeri, in the Illyrian and Danube regions, dolichocephalic skulls are very numerous in the most remote periods ; they gradually decrease in number and almost entirely disappear in many regions on the descent of the Alpine peoples from the mountains to the plains.[3]

If we penetrate into Russia, we see that in the south-west, in Ukraine, Volhynia, and also in Poland, dolichocephalics of great stature are more or less dominant in the Neolithic "kourganes" or tumuli ; and farther north and east, we find only dolichocephalics. Bogdanov has shown that the most ancient race of Central Russia (inhabitants of settlements on the shores of Lake Ladoga described by Inostranzeff) had long heads and faces like the modern inhabitants of Sweden.[4]

Finally, in Scandinavia and Denmark, the phenomenon is still more marked. According to Montelius[5] some skulls from Neolithic burials are brachycephalic and resemble in form the skulls of Laplanders, but the majority resemble those of modern Swedes. The skeletons indicate a high stature and robust frame. Scandinavia at this period was already populated by the direct forebears of the modern populations who best represent the ideal type of *Homo nordicus.*

Thus, even at the end of the Neolithic period, it is possible to recognize, in its broad outlines, the geographical distribution of the three principal physical types which anthropologists have been able to distinguish by studying the numerous varieties or sub-races, the amalgamation of which forms the ethnographical constitution of present-day Europe.

[1] Schliz, A., "Die vorgeschichtlichen Schädeltypen der deutschen Länder" (*Archiv für Anthropologie*, vii., 1908).

[2] Reche, O., "Zur Anthropologie der jüngeren Steinzeit in Schlesien und Böhmen" (*Archiv für Anthrop.*, vii., 1908).

[3] Giuffrida-Ruggeri, V., "Contributo all' antropologia fisica delle regioni dinariche e danubienne . . ." (*Archivio per l'Antrop. et la Etnol.*, xxxviii., 1908).

[4] Bogdanov, A., "Quelle est la race la plus ancienne de la Russie" (*Congrès intern. d'Anthrop.*, Moscow, 1912).

[5] Montelius, O., "Les temps préhistoriques en Suède." French translation by S. Reinach, Paris, 1895.

The Metal Ages.

Contemporaneous with the Metal. Ages, whether they be protohistoric or historic, there occurred other movements of peoples and other interminglings, which have increased the complexity apparent to us to-day.

We know very little of the Copper Age, or *Eneolithic Age* as the Italians term it, though many burials and monuments from the close of the Neolithic period must belong to it.

In the Bronze Age the practice of cremation was gradually substituted in many regions for that of inhumation, and this practice is highly prejudicial to anthropological study. Speaking generally, the Bronze Age in Western Europe, particularly in France, may be characterized (from the point of view of our studies) by the arrival of new and extensive inundations of brachycephalic peoples. The bronze industry seems then to have been imported into our country by men of the Alpine type, though this does not necessarily mean that they were the inventors of it.[1] These new floods of brachycephalic peoples penetrated at this period into the British Isles, where they are commemorated by round barrows, containing round-headed individuals.

On the other hand, the tall dolichocephalics of the Nordic type survive in Russia and in the Scandinavian peninsula, which may already be regarded as their ancient fatherland, and in which tombs of Bronze Age still sometimes contain their fair hair. Furthermore, we see them penetrating in numbers into the Rhine valley, into Switzerland and Southern Germany, where already they are found in those Iron Age tombs which are arranged in rows, the so-called *reihengraeber*.

Towards the end of the Bronze Age, the brachycephalics

[1] The origin of metallurgy is one of the most controversial problems of prehistoric archæology. There is, *a priori*, no reason to suppose that the discovery of metals and their alloys may not have been made independently at different periods and in different countries. But this hypothesis is not readily supported by our present knowledge. All the evidence seems to be in favour of an Asiatic origin (at Chypre and in the mountains of Eastern Asia, according to J. de Morgan and Déchelette). The dates of the appearance of the metals in the different countries of the East and in Europe (see p. 325) show that they spread, generally speaking, from the east westwards, and from the south northwards. Step by step with this diffusion, one after another the mining centres of each country, as it was invaded, contributed in their turn towards the new manufactures. (See Morgan, J. de, "Note sur les origine de la métallurgie" (*L'Anthropologie*, xxxii., 1922, p. 487).

reappear in Switzerland and the north of Italy; there they are very pure, and from the outset they predominate.

On the other hand, the Eastern Mediterranean region, Crete, where the Ægean civilization flourished, Malta, Sicily, Southern Italy and the Iberian peninsula remain the strongholds of the Mediterranean element, which continues largely to predominate there.

The Iron Age[1] corresponds to the maximum expansion of the Nordic race: "a new metal in the hands of a new race," said Hamy,[2] speaking of France, where the tall fair dolichocephalics imported with them the civilization of the first Iron Age, termed *Hallstattian*, from the name of the famous necropolis at Hallstatt in Austria. The Nordic peoples now penetrated everywhere along the great rivers, and thus hemmed back and isolated the massive brachycephalic mountain races. These warlike invaders landed on the coasts of the British Isles, and spread over Belgium and the north of France; attracted by the sunny, wine-producing regions, they reached Spain by the valleys of the Loire and Garonne, and Northern Italy through Switzerland. They prospered in the valley of the Danube, which they overflowed in various directions as far as Macedonia, Greece, Asia Minor, and perhaps even as far as Turkestan and India.

On the other hand the brachycephalic peoples, thus bounded and constrained, annexed territory elsewhere, more especially in Russia, where they drove back the Nordic peoples, even reaching the Norwegian coast. And the Mediterranean races quite maintained their territories, in spite of some pressure from the Nordic advance guards.

Up to this point, all these peoples remain nameless to us.

The Ancient Peoples in History.

But now, by means of the oldest texts, confused and vague though these too often are, we can attempt to identify the first historical groups from an anthropological point of view. It must be

[1] The origin of iron is no less obscure than that of bronze. G. de Mortillet regards it as African. The most ancient objects of iron we know of were yielded by a predynastic Egyptian tomb; so that they date from at least 4000 years before our era. But we have seen that, before 1500, there could be no question of the existence of an iron age in Egypt.

[2] Hamy, E. T., "Les premiers Gaulois" (*L'Anthropologie*, xviii., 1907, p. 137).

clearly stated, however, that the ancient peoples whose names have come down to us, such as the peoples of ancient Gaul, were in many cases already highly mixed as regards physical type. Camille Jullian, in his excellent *Histoire de la Gaule*, says eloquently: "Away behind the tribes of the seventh century [B.C.] there lies a great mass of human lives, of sex relationships, of formations and disintegrations of States, and of invasions by land and sea, a confusion of languages, types, and customs which baffles all analysis."

The task, therefore, is full of difficulties and pitfalls. And yet we may make some comparisons which both anthropologists and historians are able to accept, at least provisionally, as a result of their respective studies.[1]

We may consider as belonging to the type *Homo nordicus* the whole series of peoples who, during a long series of centuries, successively and periodically invaded Great Britain, France, and Central and Southern Europe. And, first of all, I think, come the great majority of Celts[2] or Gauls, who formed the powerful "Celtic Empire" of prehistorians and dominated Europe. It was they who seized Rome in 390 B.C., who invaded Thrace, whence, though checked for a short period by Alexander, they soon invaded Macedonia and Greece, laying waste everything in their path, seizing the treasures of the Delphian temple under the leadership of their chief Brennus (in 279 B.C.), and who finally, "coveting Asia," crossed the straits and installed themselves in Phrygia (Galatia).

Later, the invaders came definitely from the north, and were called Belgians, Cimbrians and Teutons, Germans, Goths, Franks and Normans. These peoples represent each a fresh thrust of that race in which may still be

[1] On the one hand, see Deniker, *Races et peuples de la terre*. Ripley, *The Races of Europe* (copious bibliography). And, on the other hand, Arbois de Jubainville, D', *Les premiers habitants de l'Europe*, 1894. Jullian, C., *Histoire de la Gaule*, i. Dottin, G., *Les anciens peuples de l'Europe*, 1916. Mention must also be made of the curious work by Madison Grant, *The Passing of the Great Race*, New York, 1916.

[2] We must not forget that by the majority of French anthropologists the Celts are regarded as brachycephalic. I have already drawn attention to this confusion (see note, p. 316). But there is no doubt that Celt and Gaul are really synonymous from the historian's point of view.

grouped the Umbrians, the Achæans, the Dorians, the Cimmerians, and the Scythians, the latter having penetrated far into Asia.

To the type *Homo mediterraneus* must be attributed the old peoples of Northern Africa, the Egyptians and Libyans, as well as the Phœnicians, the Pelasgians, Ægeans, Etruscans, the oldest Ligurians, the Phœcians of Marseilles and the Iberians.

To *Homo alpinus* belong the ancient peoples of Western Asia, the Accadians and Sumerians (?), and the Hittites, to whom perhaps may be added the Sarmatians; the Slavs, whose invasions took place from the fourth to the ninth centuries of our era, and who, along with the Mongols, have finally expelled the Nordic peoples from the greater part of the Russian region.

Once more it must be recalled that these are and can only be approximate determinations, for the majority of the names just mentioned apply to ethnographical groups which are already intermixed, and which we can only describe from the anthropological standpoint by the predominant element. We are all aware that, after the first Celtic invasions, there soon arose peoples known as the Celto-Iberians, Celto-Ligurians, Celto-Scythians, Celto-Thracians, Gallo-Greeks, and so on. Sometimes even the old ethnographical names signify a more complex mixture. It is clear, for instance, that the "Celtic Empire" of prehistorians was not composed solely of representatives of *Homo nordicus*, and that, later on, the Gauls or Celts described by Cæsar, included types other than those of the original Gauls; the armies of Vercingetorix were composed of very heterogeneous masses of humanity, and included in their ranks representatives, more or less pure or more or less mixed, of the three principal physical types now blended together in the French nation.

The Mongoloid or Hunnish invasions on the one hand, and the Arab or Saracen invasions on the other, occurred later, still further complicating this extraordinary human mixture, although they have left only relatively slight traces in France.

Origin of the Three Great Races.

How much can we learn or hazard regarding the probable origin of the three great races? We have seen that, at the end of the Palæolithic period, the countries of Western and Central Europe were peopled by dolichocephalics, which are sometimes grouped under a general term as the "Cro-Magnon Race," but which, in addition to possessing common morphological features, even already showed notable differences, such as we see in the Grimaldi Negroids and the types of Cro-Magnon, Chancelade and Combe-Capelle. In this stock close search must be made among characters still somewhat generalized, in order to discover the ancestors of the two groups of dolichocephalics which subsequently become differentiated.

Homo nordicus is a product of the north, but he can hardly have originated beyond Europe, or, at farthest, Western Asia. As a matter of fact, his centre of dispersal really appears to be Scandinavia; but this, none the less, is a pure illusion, for during the Palæolithic period, Sweden, being covered with ice, was inaccessible. The cradle of the Nordic race must be sought in countries which have always remained free of ice, and geological and palæo-geographical considerations lead us to confine our attention to Central, Southern, and Eastern Russia, which, with perhaps also Western Siberia, alone fulfil this condition. Beyond this area we come up against the masses of the yellow races. Some anthropologists have considered that the Cro-Magnons were the ancestors of the Nordic peoples.[1] But these two groups have nothing in common apart from their great stature; in other respects they differ greatly, and must not be confounded.—Giuffrida-Ruggeri believes that the great Neolithic dolichocephalics of Central Europe and Russia, who are the true ancestors of the modern Nordic peoples, represent "the Mediterranean type transported into the north," and modified by environment. This is possible, but hardly probable. In the Reindeer Age men of the Nordic or pre-Nordic type must already have existed somewhere, and that can only have been in the regions indicated above. From

[1] Wilser, L., "L'origine des Celtes" (*L'Anthropologie*, xiv., 1903), and other publications by the same author.

the Russian plains, where, during the Neolithic period, they seem to have held exclusive sway, they must, step by step, as the ice retreated, have reached the shores of the Baltic, Denmark, and the Scandinavian peninsula where they established themselves so firmly that this country now gives the mistaken impression of being their original home. It seems probable that the Aryan languages were transported by these men, and linguists, agreeing on this point with anthropologists, have replaced the old and abandoned Oriental theories by the theory of their European source. M. Jullian likewise regards what he calls "the religious centres of the Aryan languages"[1] as being in the Baltic regions. It is interesting to note the agreement.

Homo mediterraneus of European countries is a product of the south; he belongs to the main mass of the dark dolichocephalic peoples who occupy North Africa, a large part of Europe in Asia, and the shores of the Mediterranean, and who sometimes exhibit certain Ethiopian affinities where their regions border on those of the black races.[2] Related to the Aurignacian Cro-Magnon men, whose industry seems to be of African origin, they were not long in becoming differentiated according to their habitats; and from the period of transition from Palæolithic to Neolithic, we can testify to the presence in our part of the world of a type very much akin to the modern type. It is very probable that Europe owes the importation of Neolithic civilization, of megalithic structures, and perhaps also the discovery of primitive industries in metal, to *Homo mediterraneus* of North Africa and Asia Minor. According to Giuffrida-Ruggeri, *Homo mediterraneus* must have arisen through the crossing of an equatorial or proto-Ethiopian type with a more northern type like that of Cro-Magnon.[3] This is pure supposition, for everything leads to a belief in the high antiquity of the Mediterranean type.

Homo alpinus can only be of Asiatic origin. He belongs to the great brachycephalic stock of Central Asia, which

[1] In conversation with the author.

[2] Sergi, G., *The Mediterranean Race*, London, 1901.

[3] Giuffrida-Ruggeri, "Quatro crani preistorici dell' Italia meridionale e l'origine dei Mediterranei" (*Archivio per l'Antrop. e la Etnol.*, lxv., 1916).

comprises both white and yellow races (Mongols). It is probable that the starting-point of the early brachycephalic peoples on their march to Western Europe was the Ural-Altai region. They then possessed certain mongoloid characters, which they seem gradually to have lost in their progress westwards. Their migration must have commenced after the end of the glacial period, at the same time as that of the fauna of the steppes of their native country. At first this migration took place slowly, rather by a gradual infiltration than by a regular invasion. Later, towards the end of the Neolithic period, it seems to have taken place in greater mass. Dark, round-headed men suddenly become very numerous in France in the Bronze Age,[1] so much so that, in the west, they played the part of initiators and propagandists of the Mediterranean culture received from the south. Their present distribution, the result, as we have seen, of repeated migratory movements, clearly indicates their origin. They form, as it were, a vast procession, enormous at the point of departure, that is to say in Asia, but, on its way from East to West, gradually decreasing and tapering off altogether as it approaches French Brittany. Here is pitched the apex of the triangle, hemmed in between the region of the fair dolichocephalics from the North (*Homo nordicus*) and that of the dark dolichocephalics from the South (*Homo mediterraneus*).[2]

Conclusions.

Such, reduced to its main features, is the picture which, with the slight resources at its disposal, modern anthropology has been enabled to sketch of the intermingling and transformations undergone by the groups of peoples who contended for the possession of the territories of Europe, from the end of Palæolithic times onwards.

[1] Hervé, G., "L'ethnologie des populations françaises" (*Revue de l'Ecole d'Anthrop.*, iv., 1896).

[2] I must add that the views I have here expressed are not shared by all anthropologists. Certain workers consider the brachycephalic Europeans as indigenous. According to Bogdanow, Ranke, Lissauer and others, the form of the head may vary fairly rapidly under the influence of environment and geographical conditions. According to Giuffrida-Ruggeri, *Homo alpinus* did not come from Asia; he developed on the spot, in mountainous regions.

This bird's-eye view, in spite of its sketchiness, is probably inaccurate in many points. But, if it has no other merit, at least it is exclusively based on the idea of race, in the true sense of that word, and not in the sense too often attributed to it by historians and even by certain anthropologists; it is drawn from the naturalist's point of view, and so, being more in conformity with biological laws, it may perhaps also throw more light on the subject.

We are thus led to view from another angle, from a standpoint presenting fresh perspectives, the succession of events which prehistory and history seek to reconstruct regarding humanity. So we can better appreciate, among the many factors which shape the evolution of peoples, those which arise directly from general biological principles, and in especial the double influence of heredity and environment, those profound and persistent forces which often remain concealed under an accumulation of influences more purely human in origin, perhaps of equal importance, yet undoubtedly more superficial and ephemeral.

CHAPTER X

FOSSIL MAN BEYOND EUROPE

Now that we have discussed Fossil Man in Europe, it would seem that a still greater task remained to be accomplished, for the surface of our own continent is but a very small portion of the total land surface of the globe. But the truth is that beyond our own country and its neighbours, the palæontology of mankind is miserably deficient.

It is indeed the case that, all over the world, there have been collected archæological evidences of a remote past prior to historical times, often dating even from geological times, but, beyond Europe, no country has yielded a collection of facts comparable to that which has just been described. The majority of the archæological evidences found outwith Europe consist of isolated and sparse finds; and these come from deposits hardly studied at all from the stratigraphical point of view, so that their relative chronology is far from being established, except in certain rare localities. In such circumstances, we must be doubly careful in interpreting ethnographical facts, and must keep in mind that resemblance does not always signify contemporaneity or descent. In certain parts of a continent the Stone Age seems to date as far back into the past as in Europe; in other places, even in the same continent, this Age persists to the present day.

Palæontological evidence is even less abundant than archæological evidence—in fact, hardly any exists respecting Fossil Man. Fairly frequent discoveries of skulls or skeletons have been made, especially in the two Americas, but they possess neither the high antiquity nor the importance attributed to them. Indeed the majority have not been able to withstand critical examination; so that, leaving Europe out of count, human palæontology can present only an extremely poor catalogue.

I shall set out this catalogue on broad lines, examining, in turn, Asia, Australia, Africa and America.

Asia and Malaysia.

In all the habitable countries of the great continent of Asia we find mementoes of the Stone Age, and everywhere legends and superstitions are attached to them; just as in Europe, stone weapons are looked upon as products of the heavens, as residues of thunder, or are made use of in magic rites to cure diseases.

Western Asia.

The first scientific discoveries were made in Asia Minor, where the sea-board is simply a continuation of our Mediterranean sea-board, and where, consequently, it is not surprising to discover facts similar to those of the most ancient European prehistory.

In 1864, Louis Lartet[1] rediscovered in Libya a prehistoric settlement found thirty years before by Botta, the contents of which, animal bones and dressed flints, resembled exactly those of the Périgord settlements, which his father, Edouard Lartet, was just then beginning to make known. Since that time, Palestine and the whole of Syria have been visited by many archæologists. Excavations have been carried out in caves or in deposits in the open, and collections have been made. To-day we can number by hundreds the prehistoric localities already mapped out or described in this part of Western Asia by Richard, Cazalis de Fondouce, Morestin, Arcelin, Chantre, de Morgan, Zumoffen, Blanckenhorn, Arne, Desribes, Neophytus, and others.

From the archæological point of view, all the forms of weapons and implements of our Stone Age have been found there: Chellean and Acheulean (Fig. 210), Moustierian, Aurignacian and even Magdalenian or Azilian types. The Neolithic

[1] Lartet, Louis, "Note sur la découverte de silex taillés en Syrie" (*Bull. de la Soc. Géolog. de France*, 2nd ser., vol. xxii., 1865).

deposits are no less rich in polished axes and arrowheads similar to our own.[1]

The undisturbed Palæolithic settlements, as in Europe, are accompanied by an ancient fauna, including extinct or emigrated species. The bone breccias, containing dressed flints, are sometimes composed of hard rocks, like those of our French caves, and in the days of classical antiquity these breccias were

FIG. 210.—Dressed Flints from Syria. (After F. J. Arne.)

used as building stones. The Romans constructed a road across the breccias from Ras el Kelb in Phœnicia. In the plain of Raphaim, to the south of Jerusalem, Chellean or Acheulean dressed flints were found in the ancient gravels. Their entire geological bearing points to the fact that the deposits date from Pleistocene times.

[1] Zumoffen, G., "L'âge de la pierre en Phénicie" (*L'Anthropologie*, viii., 1897); *La Phénicie avant les Phéniciens*, Beyrout, 1900. Blanckenhorn, M., "Ueber die Steinzeit und die Feuersteinartefakte in Syrien-Palestina" (*Zeitschrift für Ethnologie*, xxxvii., 1905. Neophytus, Frère, "La Préhistoire en Syrie-Palestine" (*L'Anthropologie*, xxviii., 1917).

At the commencement of the Neolithic period the physical conditions were those of the present day. Neolithic deposits, which are very abundant, are not always on the surface, but to find them at any depth, one must seek at the base of the *tells*, artificial hills formed by the rubbish of towns and villages of classic antiquity. These facts, and others besides, successfully refute the obsolete ideas of classical archæologists, who declared that our Palæolithic could not be more ancient than the old Chaldean or Egyptian civilizations.

Discoveries, at first localized in Syria, were not long in spreading and increasing in number. As early as 1878, Cartailhac[1] was able to compile an inventory of the Stone Age discoveries in Asia, and since that date the inventory has greatly increased. Here I can describe only the most important facts.

Asia Minor and Persia, with their mountains and high plateaux, in great part covered during the Pleistocene period with ice and snow, are very poor in Palæolithic remains. Here J. de Morgan has explored, without result, great quantities of ancient alluvial deposits carried down from the mountains. But he has been more fortunate in the low plains, where some settlements contain both Palæolithic and Neolithic remains. The latter are to be found everywhere, at Sinai, in Arabia, in Mesopotamia, and as far as Susa, where they have been discovered in the fine excavations made by J. de Morgan and Mecquenem.[2]

Central Asia. Siberia.

In Turkestan, Pumpelly in his researches[3] discovered traces of five successive civilizations, ranging from Neolithic times to the present day. The most ancient dated from 8000 years before our era. In Eastern Turkestan, Dr Vaillant has described rock drawings and polished axes.

Towards the north, we have to record the various discoveries made by Russian explorers in Siberia. The most important is that at Aphontova, near Krasnoïarsk, to the north

[1] Cartailhac, E., "L'âge de la pierre en Asie" (*Congrès des Orientalistes*, 3rd meeting, Lyons, 1878).

[2] Morgan, J. de, *Les premières civilisations*, Paris, 1909.

[3] Pumpelly, R., *Explorations in Turkestan* . . ., Washington, Carnegie Institute, 1908.

of the Altai massif, a district rich in bone caves. The deposit, described by Savenkov, de Baye, and Volkov,[1] is a Palæolithic settlement *in situ* on a high terrace of the Yenisei, 15 to 18 metres above the river. The gravels of this terrace, used for ballast, are overlaid by a deposit of silt or loess, which contains a rich Pleistocene fauna, including the Woolly Rhinoceros, Mammoth and Reindeer. Worked stones are found in abundance at the bottom of the silt-deposits, where they lie upon the gravels; and these include implements fashioned of quartzite pebbles, dressed on one or both surfaces

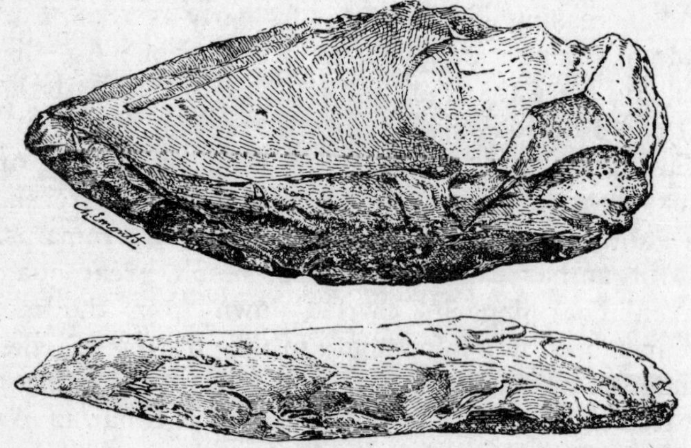

FIG. 211.—Scraper of Quartzite, from Aphontova-Gora. Natural size. (After de Baye and Volkov.)

in the Moustierian or Chellean manner (Fig. 211). Along with these stone objects were others in bone, in reindeer antler and mammoth ivory. This layer, essentially Palæolithic in character, and similar, from a geological point of view, to our European alluvial beds, is quite independent of the Neolithic of that region, of which characteristic evidences may be found in the layers of vegetable soil covering the loess. Krasnoïarsk, situated in about lat. 56° N., seems to be the most northerly Pleistocene Palæolithic settlement known. Here the great ice-sheets did not spread so far south as in Europe.

[1] Savenkov, "Sur les restes de l'époque paléolithique dans les environs de Krasnoïarsk" (*Congrès intern. d'Anthrop. et d'Archéol. de Moscou*, vol. i., 1892). Baye, Baron de, and Volkov, "Le gisement paléolithique d'Aphontova-Gora" (*L'Anthropologie*, x., 1899).

Savenkov recorded the occurrence, on rocks in Western Siberia, of curious engravings and paintings, which are sometimes of fine workmanship recalling the products of our Palæolithic art.

Among other localities in Siberia, interesting from our present point of view, may be mentioned the neighbourhood of Tomsk, Irkutsk, and the shores of Lake Baikal. In the province of Irkutsk, Vitkovsky has excavated tombs with Neolithic relics of an archaic character, where the human skeletons were covered with red ochre, like those of our Reindeer Age. Farther north, in Eastern Siberia, near Olekminsk, the deposits bordering the Lena contain arrowheads and implements in mammoth ivory, simple objects of a Stone Age which still exists or formerly existed among those circumpolar populations whose territories we have just reached.

Returning towards the centre of the continent, we must record the researches of M. and Mme. Torii in Eastern Mongolia, a region covered with incredible quantities of remains and ruins. Here the stone objects, axes, knives, scrapers and arrows, are Neolithic. According to these authors, this region and also Manchuria, Korea, Japan and even China, seem to have no Palæolithic remains. In Asia, the Polished Stone Age seems to be less clearly separated from the older Age of Dressed Stone than it is in Europe.[1]

But all these countries are rich in prehistoric relics ranging from the Stone Age to the Iron Age. The seaboard of Manchuria (Port Arthur), of the Amur and of Japan have many shell-mounds and kitchen middens similar to those of Denmark. Sometimes raised high above the present sea-level, they contain Neolithic implements of a special character.

These countries, as well as Korea, are dotted with many megalithic monuments.

The Chinese prehistoric period is still almost unknown; but we are aware that it includes a Neolithic phase.

[1] Torii, R., and Torii, Kimico, "Études archéologiques et ethnologiques" (*Journal of College of Science*, xxxvi., Tokyo, 1914).

Southern Asia is much richer. The old stone ages are represented by many beds, some of which date from very remote geological times.

Southern Asia.

An English geologist, Nœtling, has recorded the presence

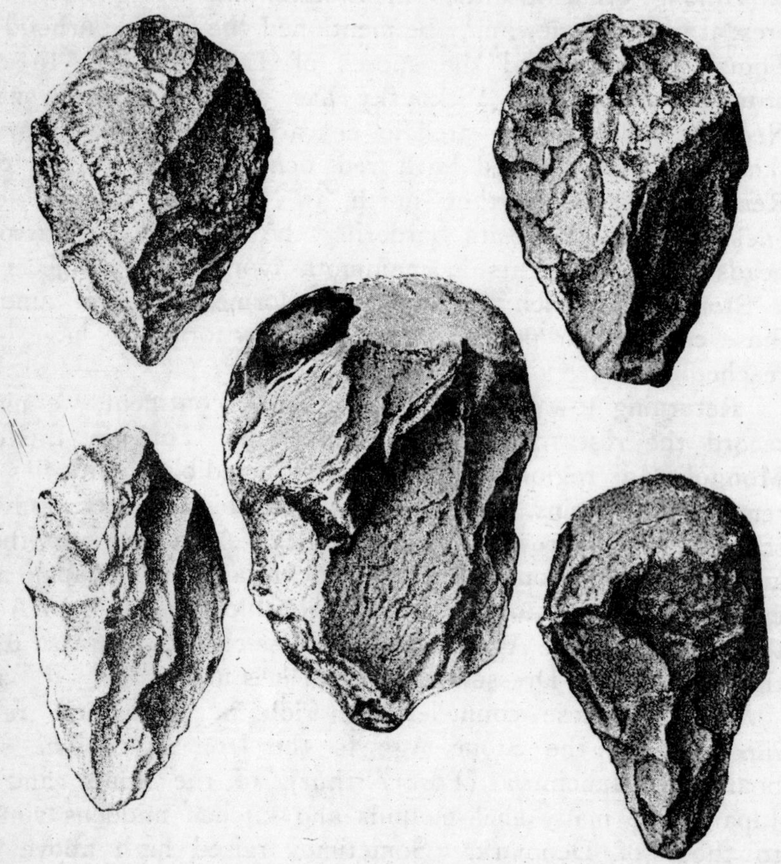

FIG. 212.—Dressed Quartzites from India. (After Ball.)

of dressed flints in a Pliocene conglomerate from Central Burma. But these are probably only natural eoliths (see p. 128). Nevertheless, over almost the whole of India, stones dressed in the Palæolithic fashion are to be met with on the surface of the soil. In 1866, Foote[1] discovered dressed quartzites of almond

[1] Foote, R. Bruce, Various publications, principally in the *Congrès intern. d'Arch. et d'Anthrop.*, 1868, p. 224.

shape in the laterite[1] deposits of the suburbs of Madras, the formation of which dates at least from Pleistocene times. Since that time, many discoveries of this kind have been made. Seton-Karr puts the age of the quartzites from the laterite deposits at 250,000 years.

Other dressed stones have been found *in situ* in the ancient alluvials of many streams, such as the Nerbudda, the Godavery, the tributaries of the Kistna and others, as well as in a cave in the Karnul district. The fauna accompanying the relics is for the most part composed of extinct species, several of which date from Pliocene times. Here the problem of the antiquity of Man confronts us under the same aspect as in Europe.[2]

In Banda, a mountainous district in the north-west of India, there are caves and rock shelters where stone implements have been found, particularly small flints of Tardenoisian form (see p. 331), as well as drawings in red ochre, similar to some of the productions of our Reindeer Age. Recently, prehistoric paintings, representing hunting scenes and extraordinarily like those at Cogul in Spain, have been discovered in a cave near Raigarh, in the Central Provinces of British India. (For further details and figure, see Appendix, p. 476).

Neolithic relics are to be found throughout almost the whole of Southern Asia. In the north-east of Baluchistan there are artificial mounds containing stone implements and bronze weapons. Polished stone axes, arrow-heads, etc., have been collected in abundance in India, the Malay Archipelago, in Cambodia, Tonkin, and other parts. As for the many megalithic monuments in the Deccan, and those which are to be found even in the mountains of Assam, I shall do no more than record their presence in passing.

[1] Laterite is a red clayey earth, formed by the decomposition of underlying rocks, and is very widespread in tropical regions.

[2] Ball, V., "On the Forms and Geographical Distribution of Ancient Stone Implements in India" (*Proceed. of the Royal Irish Academy*, 1878). Logan, *Old Chipped Stones of India*, 1906. Oldham, R. D., *A Manual of the Geology of India*, 2nd edit., Calcutta, 1893. Foote, *The Foote Collection of Indian Prehistoric and Protohistoric Antiquities: Notes on their Ages and Distribution* (Madras Museum, 1916).

The island of Ceylon has been studied from an anthropological standpoint by F. and P. Sarasin, who found in the caves where the wretched Veddas still live, or used to live, quartz implements resembling those of our own Magdalenian Palæolithic industry. According to Wayland the Palæolithic here comprises Chellean, Moustierian, and Aurignacian forms, and belongs to a period previous to the arrival of the Veddas.[1]

The same explorers have carried out investigations in Celebes and Sumatra. Here again the caves have yielded many stone relics similar to those of Ceylon, and comparable to those of our Reindeer Age. These various islands would seem to have been inhabited, at a remote period, by a primitive race, small in stature, whose descendants, more or less mixed, form the relict groups of the Veddas of Ceylon, the Kubus of Sumatra, the Toalas of the Celebes, and the Senois of the Continent.[2]

To sum up, the little we know of Asiatic prehistory enables us to state that it dates, at least at certain points, as far back as in Europe. The few facts I have just enumerated already show a diversity which indicates, in a general way, a succession of periods or phases comparable to that in Europe: Palæolithic, Neolithic, Bronze Age, and Iron Age. There is, perhaps, no exact time correlation or synchronism, but there is an affinity of similarity (*homotaxia*). The day will come when science will be able to establish, in each great tract of Asia, a more detailed succession with parallels and time correlations between one region and another. An Indian scholar, Mr P. Mitra, has most interestingly made such an attempt in connection with the prehistory of India from the most remote times.[3] All the facts point to the belief that it is in Eastern and Southern Asia, so rich in Tertiary and Quaternary fossils, that the most valuable if not the most conclusive discoveries will be made.

[1] *Spolia Zeylandica*, vol. ii., pt. 41, 1919.

[2] Sarazin, Paul and Fritz, *Ergebnisse naturwissenschaftlicher Forschungen auf Ceylon*, vol. iv., "Die Steinzeit auf Ceylon," Weisbaden, 1908. *Versuch einer Anthropologie der Inseln Celebes*, 1905. *Neue-lithochrome Funde im Innern von Sumatra*, Bâle, 1914.

[3] Mitra, Panchanan, "Prehistoric Cultures and Races of India" (*The Calcutta University Journal of the Department of Letters*, 1920). "Prehistoric Arts and Crafts" (*Ibid.*, vol. iii., 1920.)

Osteological Evidences. The richness of the fossil deposits of the Himalayan foothills, known as the Siwalik Hills, in mammals of all kinds and particularly in remains of anthropoid apes, is well known. It is surprising that there has not yet been found any bone remains which can be attributed to Man or to his immediate forerunner. Pilgrim, indeed, believed that *Sivapithecus* was an ancestral form, but we have seen that this opinion was without foundation (see p. 86).

It is to be hoped, and it is quite possible, that the future

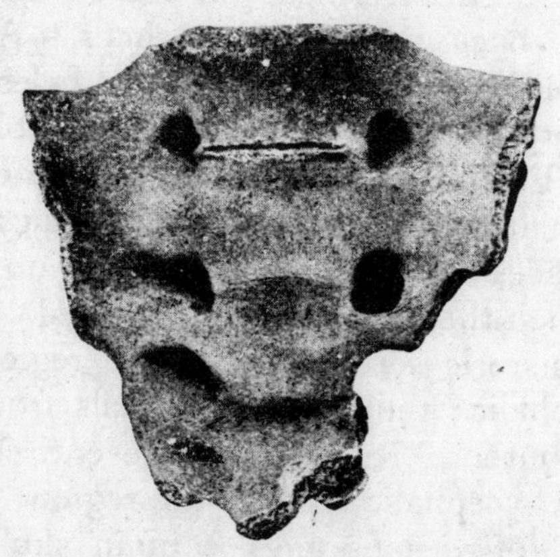

FIG. 213.—Human Sacrum from Loess at Ho-nan. Half natural size. (After Matsumoto.)

may have fortunate surprises in store for us, for, at the present moment, with the exception of *Pithecanthropus* from Java, to which I have devoted a special chapter, there fall to be recorded only two discoveries of real palæontological value from the whole of Asia.[1]

Zumoffen extracted, from the Antelias Cave in Phœnicia, some human bones belonging to several individuals, and these are possibly of Aurignacian age.

In 1915, the Japanese palæontologist, Matsumoto,[2] described a human sacrum, found in a loess deposit at Ho-nan in China

[1] In one of M. Mitra's memoirs, referred to above, mention is made, without details, of human remains found by M. Réa at Adichannanliu.

[2] Matsumoto, H., "On some Fossil Mammals from Ho-nan, China" (*Science Reports of the Tôhoku Imperial University*, 1915).

along with bones of animals, particularly those of an elephant related to the Mammoth. The most remarkable features of this sacrum would seem to be its slight curvature and a gradual decrease in the size of the sacral vertebræ from the first to the last (Fig. 213). In these respects it differs from the sacrum of modern Man and resembles that of Neanderthal Man. The main interest of this discovery lies in the anticipation it arouses of future discoveries. It proves that there *are* human fossil remains in Pleistocene China.

A certain number of the skulls collected in ancient tombs and prehistoric deposits have been described, but these relics do not seem to date further back than modern times. At most they belong to the Neolithic period. They nevertheless suggest the interesting theory, which archæology was able to anticipate, that Asia, like Europe, has been the theatre of great shiftings of the population. The majority of these skulls actually belong to races differing from those now occupying the countries in which they have been discovered.

Thus two skulls from prehistoric settlements in the region beyond Baikal are dolichocephalic; they resemble skulls from the khurgans or tumuli of Southern Russia, and are entirely different from the modern brachycephalic types of the region.

The khurgans of Siberia, of various ages, contain skulls differing greatly in form. Along with specimens resembling those of the Mongols and Ostiaks, there are dolichocephalic forms similar to those from the Russian khurgans.[1] The American anthropologist, Hrdlička,[2] is convinced that in the khurgans of Siberia, Mongolia, and Tibet, skeletons have been discovered representing the race which peopled America.

Professor Verneau[3] described three skulls from deposits of Neolithic appearance in a cave in Tonkin. These skulls, in their lack of balanced development, or "disharmony," resemble those of our Cro-Magnon race. They did not belong to Mongols, and pertain rather to the white than to the yellow races.

[1] Zaborowski, *Bull. de la Soc. d'Anthrop. de Paris*, 1898.

[2] Hrdlička, A., *Congrès intern. d'Anthrop. et d'Archéol.*, Genève, 1912.

[3] Verneau, Dr, "Les crânes humains du gisement préhistorique du Pho-binh-Gia, Tonkin" (*L'Anthropologie*, xx., 1909).

Lastly, according to the brothers Sarasin, Ceylon, Sumatra, and the Celebes seem to have been inhabited, at a remote period, by a primitive race, of small stature, probably a simple variety of the Negrito type, upon the antiquity and wide distribution of which, in ancient times, de Quatrefages has laid such stress in his writings.

Dr E. Dubois has recently published illustrations and a description of two human skulls found at Wadjak in Java. These he regards as of Pleistocene Age. The skulls present, along with a rather more robust appearance, all the characters of Australian and Tasmanian skulls. For the first time such skulls have been recorded from beyond the confines of Australia. *Homo wadjakensis* is a Proto-Australian, whose origin must thus appear to have been in Eastern Asia. (For further details and illustrations of this find, and for more recent discoveries in Japan and the Phillipines, see Appendix p. 478.)

Australia.

General Notes.

Australia is the smallest, but most peculiar of the continents. Everything in it strikes the naturalist as strange and archaic in character. Its vegetation, comprising tree-ferns, cycads, araucarias, palms, mimosas, eucalyptus and prickly scrub, recalls that of the Secondary era. In the seas surrounding it are found the corals, the Trigonias, and the Nautili of our Jurassic and Cretaceous seas. In its rivers still lives *Ceratodus*, that curious amphibious fish which was first discovered in the Triassic deposits of Europe. The dry land is peopled by a very peculiar mammalian fauna, which includes the Monotremes, belonging to a primitive and still reptilian type, and is almost wholly composed of forms remaining at the Marsupial stage. This fauna, therefore, forms a legacy from Secondary times, which, nevertheless, has been considerably enlarged and diversified. The indigenous human populations likewise belong to one of the most primitive of modern races.

The majority of these general features of Australia may be explained by its geological and palæogeographical history. The larger part of its land surface remained above water during the Primary and Secondary eras. At first Australia was united to a vast Antarctic continent, including South Africa, Madagascar, India, and what geologists call the *continent of Gondwana.* The latter soon began to break up; it seems that, during the Cretaceous period, there was still some temporary communication with Asia through the solid lands of the Malay Archipelago, but after the end of the Cretaceous period, or at the commencement of the Tertiary era, Australia became isolated in much its present form. Thus there have remained, imprisoned on an immense island, the flora and fauna which were then dominant throughout the world. As this isolation seems to have lasted, more or less completely, up to our own time, the organized life of Australia has been compelled to continue its own evolution in its own area, borrowing very little from the rest of the world, and gradually assuming its present aspect. This independent evolution, carried on in a very special direction, has produced the extraordinary diversity of marsupial mammals, some of which, not so long ago, attained to gigantic size.

Modern Australians.

It is very difficult to explain the settlement of Australia by Man, who can only have been one of the latest comers, unless we believe, with Schœtensack, that Australia is the place of origin of our species, a supposition which seems hardly admissible.[1]

When the first European navigators disembarked in Australia and Tasmania, they found there men of strange and miserable appearance, cannibals, and they immediately compared them with monkeys: "tailless Chimpanzees," said the English. The Tasmanians no longer exist; they were annihilated by the "black war" waged upon them by the settlers, and also by alcohol, syphilis, and pulmonary phthisis. The Australians are still numerous. They have been carefully studied from every point of view, and appear to us now to be

[1] Schœtensack, "Die Bedeutung Australiens für die Heranbildung der Menschen aus einer niederen Form" (*Zeitschrift für Ethnologie*, xxxiii., 1901).

far removed from the wild brutes spoken of by the first explorers.[1]

In spite of many points of difference, corresponding to the

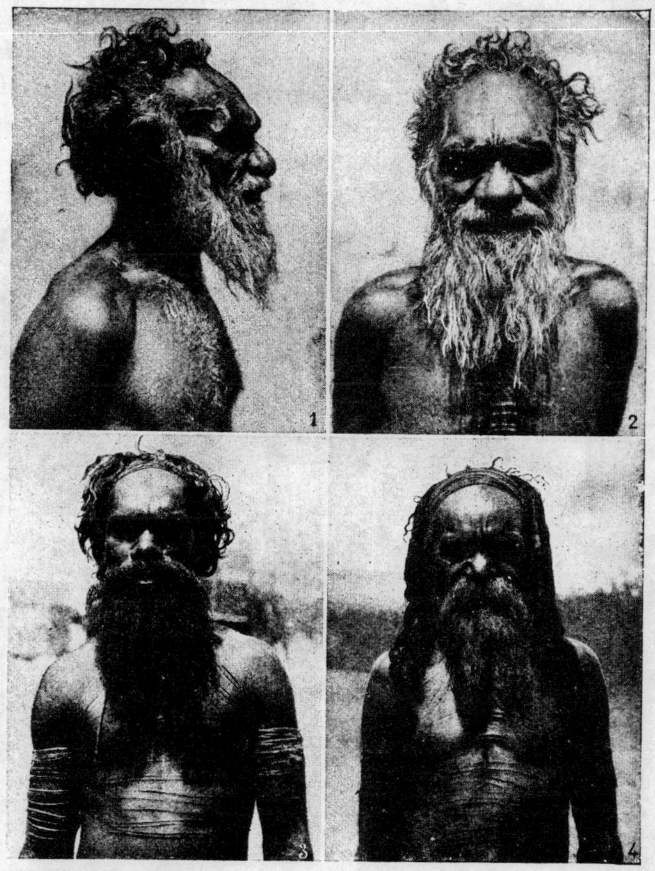

FIG. 214.—Different Types of the Tribe of Aruntas. (After Spencer and Gillen.)

various regions of Australia, the common features of Australians are sufficient to stamp a type forming a very distinct sample of primitive humanity (Fig. 214). It would seem that the

[1] See particularly Spencer and Gillen, *The Native Tribes of Central Australia* London, 1899; *The Northern Tribes of Central Australia*, London, 1904; *Across Australia*, London, 1912, 2 vols. Spencer, *Native Tribes of the Northern Territory of Australia*, London, 1914. See also Brough Smith, *The Aborigines of Victoria*, London, 1878. Thomas, *Natives of Australia*, London, 1906. Howitt, *Native Tribes of South-East Australia*, London, 1904, etc.

Tasmanians were of purer race, more closely resembling the first arrivals. The Australians of to-day are probably Tasmanians intermingled with Papuan and Malay elements, and it is because of this intermixture that we have the very considerable diversity to be observed among the modern population.

We can readily understand with what eager interest anthropologists have studied these men, for in certain of their

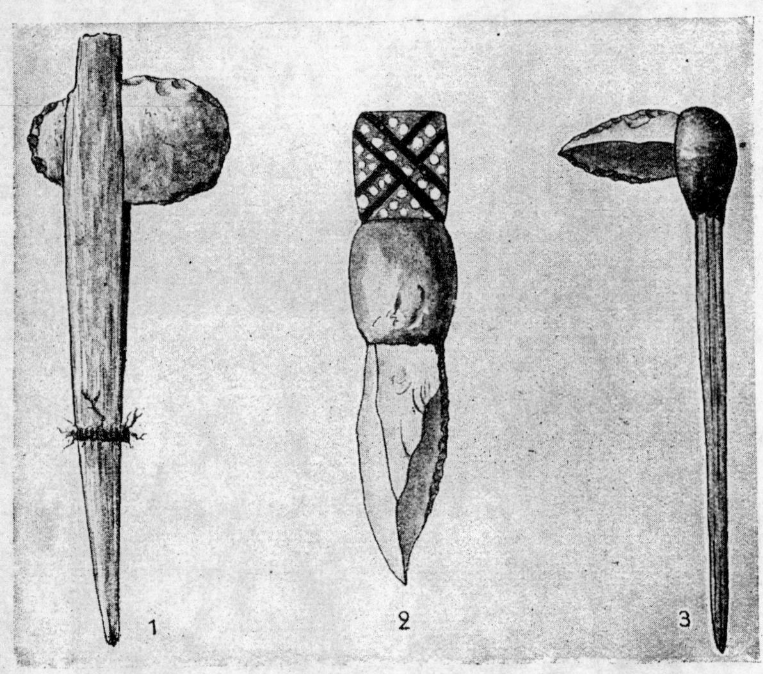

Fig. 215.—Weapons from Central Australia (Warramunga tribe). 1. Stone axe fitted into the branch of a tree, split and tied by means of a thong of horsehair. 2. Stone knife with shaft of decorated wood, held together with resin. 3. Stone pick with shaft of wood, split and held together with resin. (After Spencer and Gillen.)

physical characters they recall our fossil men, and they live in conditions similar to, if not the same as, those in which our Palæolithic peoples must have lived. Their stone weapons often resemble our dressed flints from Saint-Acheul and Le Moustier (Fig. 215); their throwing weapons resemble those of our hunters of the Reindeer Age (propellers); they use religious or magical objects, "churingas," which may be compared to the coloured pebbles from the Mas d'Azil. They ornament their rock shelters or caves with paintings, not unlike those of our prehistoric caves; and similar drawings of hands, sometimes with the same mutilation of the fingers, are to be found (Fig. 216). These tangible evidences of a primitive culture are related to the social customs, religious or magical, which may still be studied on the spot, and which, on the basis of analogy, have sometimes been attributed to our cave-dwellers.

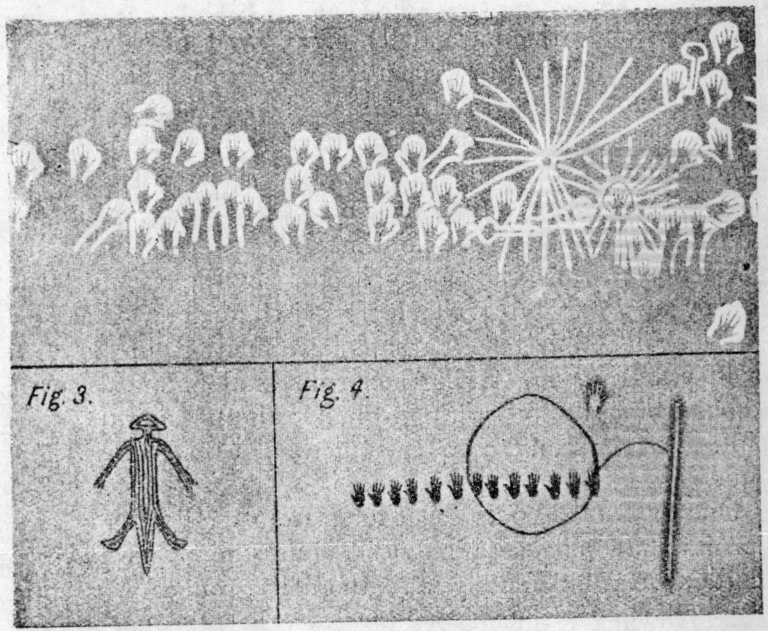

FIG. 216.—Australian Rock-paintings. Above, hands traced in white round the outline, and various signs. Below (Fig. 3) a Man, and (Fig. 4) red hands, joined by a line and a circle to a boomerang. (After Matthews.)

Thus prehistorians have endeavoured to draw extremely interesting comparisons between our Palæolithic ancestors and the modern Australians. Professor Sollas has traced these analogies in a most vivid picture.[1]

[1] Sollas, W. J., *Ancient Hunters and their Modern Representatives*, 2nd ed., London, 1915. Comparisons of this kind are always extremely interesting, but by themselves they are not sufficient to prove the close relationship of peoples who exhibit them. In fact, ethnographical resemblances often result simply from the fact that similar necessities everywhere require similar means to meet them. The Australians

The study of human palæontology in Australia ought therefore to be particularly informing. Unfortunately the data are still very meagre.

Let us examine the matter first from the archæological standpoint.

Archæological Evidences. Products of a stone industry are not lacking in Australia; on the contrary, they are very abundant. All the museums, especially that at Melbourne, are rich in stone implements, some polished, others simply flaked. Among them are to be found most of the types upon which our European historians based their classifications, from "eoliths"[1] to large and beautiful polished axes, passing through the Chellean to the Moustierian forms, Aurignacian and Magdalenian blades and knives, and even pygmy flints.

But in Australia these various forms are contemporary; they were all in use at the time of the arrival of the first Europeans, and as they are collected on the surface, there is scarcely any method available for distinguishing recent from more ancient objects.[2] Perhaps examination of the patina would justify us in making some distinction, but I have not seen anything in the writings of investigators relating to this point. And in any case this method would only allow of vague evaluations. (See Appendix, p. 480.)

One or two discoveries of axes from the depths of alluvial deposits have been announced, but the facts have been recorded very indefinitely, and there is ground for scepticism when, following Keane,[3] we consider that the exploration of auriferous gravels, carried out over hundreds of square miles, has not yet yielded any conclusive evidence.

On the eastern and southern coasts of Australia, there are many shell- and ash-mounds similar to our kitchen middens.

certainly resemble our Palæolithic peoples in their implements and customs, but we now know, contrary to the opinion long held, that they differ completely in physical characters (see p. 238).

[1] Nœtling, F., "Notes on the Tasmanian amorpholithes" (*Pr. Roy. Soc. Tasmania*, 1906-7, and *L'Anthropologie*, xix., 1908, p. 645).

[2] Etheridge, R., *Has Man a Geological History in Australia?* 1890. Giglioli, E. H., *Le Eta della Pietra nell' Australasia*, 1894. Klaatsch, "Die Steinartefakte des Australier und Tasmanier" (*Zeitschrift für Ethnologie*, xl., 1908).

[3] Keane, *Ethnology*, Cambridge, 1896, p. 95.

In them are found rudely worked stones and bones, but neither pottery nor arrowheads. And yet, since the extent and the bulk of these artificial mounds are sometimes considerable, we may suppose that their formation took some considerable time and that they date from a fairly remote period.

The study of the art of the rock-paintings leads to the same conclusions. Engravings and paintings are very common in the districts where caves and rock-shelters are found. Many of these drawings, which represent all kinds of animals and men, sometimes grouped in definite scenes, are the work of the contemporary tribes. But there are some which are obviously more ancient, and Matthews[1] believes it possible to distinguish two kinds of artistic productions, which must be attributed to two different races. According to Basedow,[2] the rock paintings of the Flinders Mountains must be very ancient, for they are as patinated as the rest of the rock upon which they are engraved. They have besides a special character, and the present native tribes of the region believe that they are the work of their ancestors. Having compared this patina with that of certain Egyptian monuments of 5000 years ago, Basedow finds it to be at least equally developed. Further, certain drawings represent foot-marks which can only be attributed to extinct animals: to *Genyornis*, a giant bird, to *Diprotodon*, a gigantic marsupial complete skeletons of which have been taken from the deposits of Lake Callabonna (Fig. 217). Here, then, is a very ancient geological record.

Some of the discoveries are more definitely palæontological in character.

At one time it was thought that there were undoubted traces of a fossil man, dating perhaps from Tertiary times. In 1898, Archibald, Director of the Varnambool Museum in Victoria, announced the discovery of footprints on slabs of sandstone of marine origin, obtained from a depth of 20 to 60 metres. These concretionary sands represent an ancient beach, on which Kangaroos, Dingoes, and Emus had been wont to

[1] Matthews, "Rock Paintings and Carvings of the Australian Aborigines" (*Journal of the Anthropological Institute*, xxiv., 1895; xxvii., 1898).

[2] Basedow, H., "Aboriginal Rock Carvings of great Antiquity in South Australia" (*Journal of Anthropological Institute*, xliv., 1914).

wander. Among the tracks of these animals there appeared others which, it seemed, could only have been those of a Man, who, walking and seating himself upon the yielding sand, had left upon it the imprints of his feet and buttocks. These imprints were studied later by the German palæontologist, Branco,[1] who was struck by their narrowness. Some years afterwards, Noetling,[2] while travelling in Tasmania, observed on the snow long tracks, arranged in pairs, and so closely resembling the supposed human tracks on the Varnambool sandstones that they might have been mistaken for them. But these tracks, which were remarkably narrow, had been made by Kangaroos. This observation appears to be conclusive.

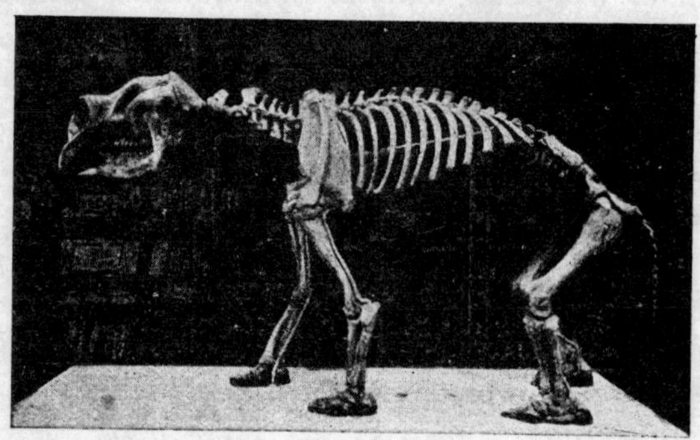

FIG. 217.—Skeleton of *Diprotodon australis*. One-fiftieth natural size. Palæontological Gallery of the National Natural History Museum of France.

For long it has been well known that in Australia there are fairly numerous deposits containing fossil animals, dating from Pleistocene and even from Pliocene times: bone caves such as the Wellington Caves, or alluvial deposits such as those of Lake Callabonna. These deposits have yielded a host of most curious creatures, particularly the giant marsupials, *Nototherium*, *Diprotodon* (Fig. 217), and *Thylacoleo*, but not the smallest dressed stone has ever accompanied the remains of these vanished creatures.

[1] Branco, W., "Die fraglichen fossilen menschlichen Fussspuren . . ." (*Zeitschrift für Ethnologie*, xxxvii., 1905).

[2] Noetling, F., "Bemerkungen über die angebliche Menschenspur . . ." (*Centralblatt für Mineralogie*, 1907).

Nevertheless, certain facts hold out grounds for hope. De Vis recognized on one side of *Nototherium* certain marks which may be attributed to human agency. The Dingo or native dog, which must have arrived in Australia at the same time as its master, has left fragments of its skeleton in these bone deposits. Finally, a human tooth has been obtained from cave deposits at Wellington.[1]

The Talgai Skull.

Matters stood at this unsatisfactory stage when, in 1914, at the time of the outbreak of the Great War, the British Association for the Advancement of Science met at Sydney, and was thrilled by Messrs David and Wilson's account of the discovery of a human fossil skull, found near Talgai, on the Darling Downs in Queensland. This fossil has since been the subject of various communications; it was recently described by A. S. Smith[2] in a memoir of which I here give a summary.

The discovery of the skull dates as far back as 1884. It was found by an old workman at a depth of about 2½ metres. The deposit which contained it was laid down by a brook, the Dalrymple Creek, and is formed of two layers. The upper layer consists of black vegetable soil, and rests upon a layer of reddish brown clay containing calcareous nodules. It seems that the skull was taken from the top of this second layer. No other fossils were found there, but bones of *Diprotodon*, *Nototherium*, *Megalania* and others, have been obtained from similar formations laid down by other streams in the neighbourhood of Talgai. The remains of these extinct animals are in the same state of fossilization as the human skull, which was entirely encrusted with ferruginous calcareous matter, within as well as without. When freed from the soil, and after having been treated and prepared for examination, it still seemed much fossilized, and was badly cracked and very fragile. The facial portion is better preserved than the cerebral, which has the appearance of a mosaic of bone fragments (Fig. 218). The density of a fragment of the parietal bone

[1] Etheridge, R., *Memoirs of the Australian Museum*, 1916.

[2] Smith, A. S., "The Fossil Human Skull found at Talgai, Queensland" (*Philosophical Transactions of the Royal Society of London*, Series B, vol. 208, 1918).

is 2.79, and it does not consist of more than 3.60 per cent. of organic matter.

The skull belonged to a male, aged from 14 to 16 years. The different profiles superimposed upon the profiles

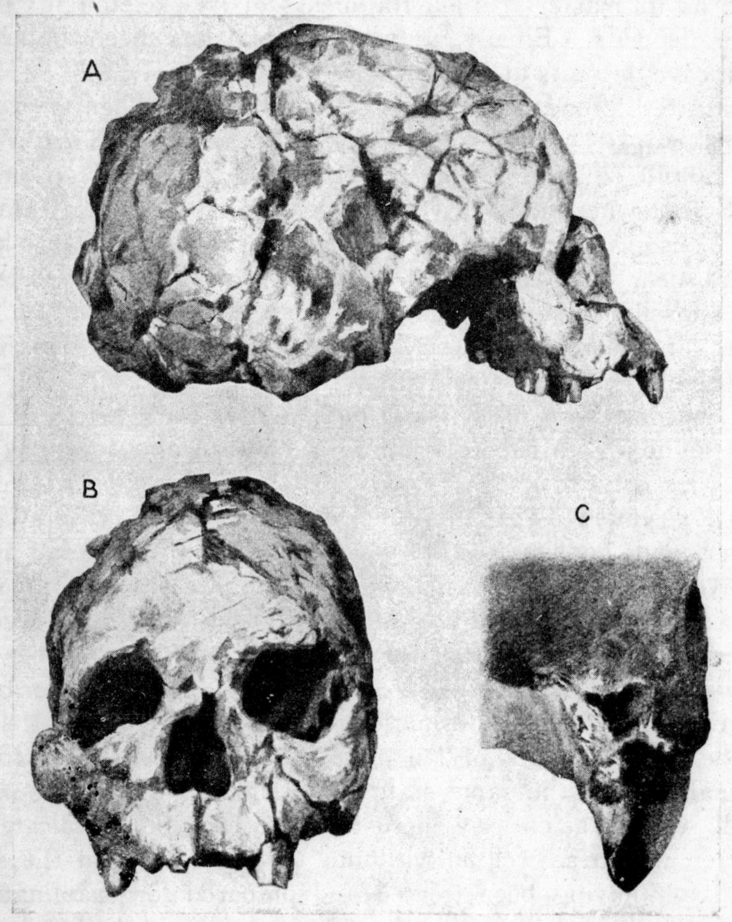

FIG. 218.—The Talgai Skull. A, seen in profile ; B, full face. One-third natural size. C, upper right canine tooth (enlarged). (After A. S. Smith.)

of various modern Australian skulls, corresponded with extraordinary accuracy. The Talgai skull is, therefore, essentially an Australian skull. Smith declares that, had it been found without the facial portion, no one would have regarded it as of any importance from the anatomical point of view.

The face also is Australian in type, but it presents more primitive features than modern skulls. The facial prognathism is striking. The forehead is receding, but has no pronounced superciliary ridges (probably on account of the youth of the subject). The orbit is large and quadrangular; the nasal bones arise from a deep hollow; the floor of the nasal fossæ gradually merges in the supra-alveolar surface of the maxillaries, which are very large.

The form of the palate is very primitive, and the molar series are almost parallel. The canines are enormous; the pointed summit of the crown exceeds by several millimetres the general level of the biting surface. Two worn facets have been produced by the contact of the canine with the lower premolar (Fig. 218). The arrangement of these facets differs from that normally seen in Man, and is more like that found in the skulls of young Orangs and young Gorillas, although there is no real diastema in the upper jaw to accommodate the lower canine. The premolars and true molars are very large. The majority of these characters may be encountered separately in some skulls of Tasmanians or modern Australians, but it would seem that they have never been observed simultaneously and developed to such a degree, in any individual case.

The Talgai Man is thus Australian and not Tasmanian in type; he represents a proto-Australian, who had long possessed a human brain, but whose face retained much more brute-like relics of his origin. Although the geological age of the skull is difficult to determine exactly from its condition of fossilization, it may be regarded as Pleistocene. It appears to be the first proof of the high antiquity of Man in Australia, apart perhaps from the human molar found in the Wellington cave in New South Wales. In association with the Wajak discovery it assumes fresh interest, since the origins of the ancient Tasmanian and Australian races become less obscure, but much yet remains to be discovered in Australia and Malaysia before the important anthropological problems they present can be solved.

Africa.

General Survey. Africa must have played as important a part as Eurasia in the origins and primitive history of the human kind.

Evidences of a long extended past increase in number with the progress of exploration. In all those parts where the European has penetrated, he has found traces of a Stone Age; in many places, he has collected weapons and implements similar to those of our most ancient Palæolithic. Furthermore, these objects are usually scattered on the surface of the soil. Yet in the regions which have long been colonized, or which have been more closely studied, deep beds have been found in stratigraphical or palæontological conditions allowing of chronological comparisons. Examination of these beds justifies this important conclusion, which may be applied to the whole continent, that the prehistory of Africa seems to be as old as that of Eurasia.

I cannot, without digressing too far from the main subject of this book, describe even in summary the archæological discoveries already made. Such a task would require a volume in itself, a labour of very considerable interest and usefulness; but a brief review of the principal facts known up to the present day must suffice.

In every respect North Africa differs greatly from Central and Southern Africa, from which it is separated by the great desert region of the Sahara. It resembles Southern Europe in its flora, in its fauna, and in its human population belonging to the white race. North Africa is a Mediterranean territory rather than an African territory. "Cross the desert solitudes," said Hamy, "and everything changes, plants as well as animals, and the Negro makes his appearance. This is the true Africa, the great Dark Continent."[1]

Geology, palæontology, and prehistory all tell us that there was a time when this division into *Holarctic* and *Ethiopian* regions did not exist, or was less pronounced. It was only by degrees that the various regions of Africa assumed the

[1] See Gsell, Stéphane, *Histoire ancienne de l'Afrique du Nord*, vol. i., Paris, 1913.

different aspects they now present. Before it dried up the Sahara was no desert, that is to say a barrier, but a bridge; the subtropical Ethiopian fauna extended over North Africa.[1] Man lived in every part: dressed stones, identical with those of the oldest European Palæolithic, are found exactly alike at a great many points, forming an almost continuous chain from the extreme north to the extreme south of the continent.

Observations were first made in North Africa. Egypt has been studied by archæologists from the days when the idea of prehistory scarcely existed.

Egypt.

In 1867 Worsaae recorded the first dressed flints, which Arcelin, Hamy, and Lenormant compared to European flints, confirming the existence of an Egyptian Palæolithic phase, far older than the oldest monuments of classical Egypt. Delanoue, John Evans, and Haynes supported this conclusion, which was for a long time contested by Egyptologists. The latter, unnecessarily impressed by the thousands of years of Egyptian history, a period which barely counts from the geologist's point of view, retained their mistaken views regarding questions of origin. To refute the idea of the Palæolithic phase, they laid stress on the fact, quite accurate in itself, that the use of flint implements had lasted for a long time after the first dynasties. But there are flints and flints. To our distinguished fellow-countryman, J. de Morgan, belongs the honour of establishing the distinction, and of proving definitely the existence, the importance, and the geological antiquity of the Stone Age in Egypt.[2]

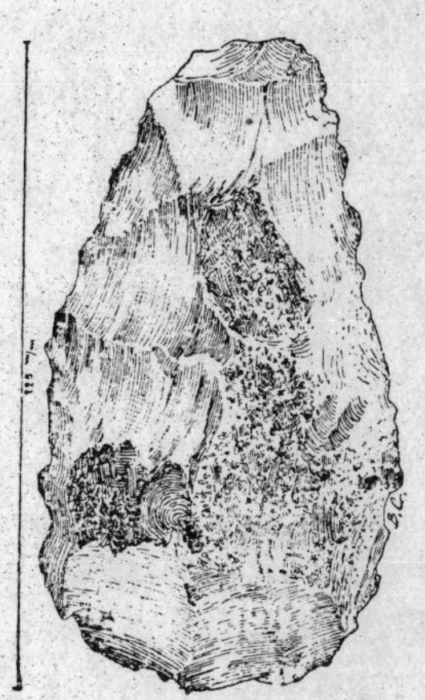

FIG. 219.—Dressed Flint from the Egyptian Desert. One-fourth natural size. Seton-Karr Collection. Musée de Saint-Germain. (After Reinach.)

It is true that the innumerable flints of Chellean, Acheulean,

[1] Boule, M., "Les Mammifères quaternaires de l'Algérie d'après les travaux de Pomel" (*L'Anthropologie*, x., 1899).

[2] Morgan, J. de, *Recherches sur les origines de l'Egypte: L'âge de la pierre et des métaux*, Paris, 1896. See also Cartailhac, "L'âge de la pierre en Afrique" (*L'Anthropologie*, iii., 1892).

and Moustierian form from Africa have not so far been found in association with a fossil fauna, but their physical character, their patina, their topographical distribution or the geological bearings of their beds are sufficient to enable us to date them back to Pleistocene times. (See Appendix, p. 480.) At that period the Nile Valley had not yet been completely hollowed out; whereas the most ancient monuments of classical Egypt and even of Neolithic Egypt, those of Negadah and Abydos, 6000 years old, correspond to a period relatively very recent, when the country had already exactly assumed its present-day aspect.

Dressed stones, of quite as archaic an appearance and type, have been collected in the Libyan Desert and in Upper Egypt.

Barbary.

Lesser Africa, or Barbary, has long yielded many clear evidences. In 1875, Bleicher[1] discovered dressed stones after the type of Saint-Acheul in a Pleistocene alluvial deposit at Ouzidan, near Tlemcen. Some years afterwards, Pomel, Tommasini, and Pallary described the Palikao or Ternifine deposit in the region of Mascara.[2] In this area stones dressed by the hand of man, Chellean and Moustierian in form, are to be found, along with numerous bone-remains of vanished or emigrated mammals, especially those of the extinct species, *Elephas atlanticus*. The deposit at Aboukir, near Mostaganem, and the rich settlement on Lake Karâr, also in the province of Oran, exhibit similar palæontological features.[3]

In 1887, Dr Collignon[4] observed, in ancient alluvials in the neighbourhood of Gafsa, in Tunisia, several superimposed layers containing relics of various stone industries, the oldest of which is identical with our Chellean industry.

Then and since, a great number of places, of caves, rock-shelters, or settlements in the open, have been investigated throughout the whole of North Africa. The literature on prehistory has become considerable. Among the works which have been published must be mentioned those of Flinders Petrie,

[1] In *Matériaux*, x., 1875, p. 196.

[2] *Bull. de la Soc. géol. de France*, 3rd series, vol. vii, 1878, p. 44; *Bull. de la Soc. d'Anthrop. de Paris*, 1883; *Matériaux*, 1888, p. 221.

[3] Boule, M., "Station paléolithique du lac Karâr" (*L'Anthropologie*, xi., 1900).

[4] Collignon, Dr, "Les âges de la pierre en Tunisie" (*Matériaux*, xxi., 1887).

Henri and Jacques de Morgan, Forbes, Bissing, Beadnell, Schweinfurth, Blanckenhorn, etc., in Egypt; those of R. Smith, in Libya; of Couillault, Schweinfurth, J. de Morgan, Capitan and Boudy, Dr Gobert, Raygasse, Latapie, etc., in Tunisia; of Pallary, Doumergue, Debruge, Flamand, etc., in Algeria; and of Pallary, Pinchon, Campardou, etc., in Morocco.[1] Particular mention must be made of the work of a humble schoolmaster in Oran, Pallary, who after having explored the whole of

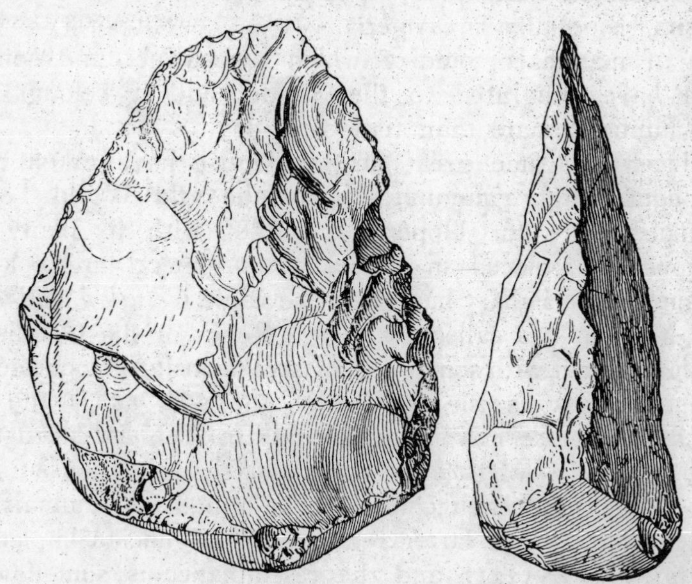

FIG. 220.—Dressed Flint from Diabet, Morocco. Three-fourths natural size. (After Pallary.)

Barbary, was able to collate the facts he had observed and put them forward in a comprehensive review.[2]

The oldest phase of Palæolithic culture in North Africa resembles in every feature that of the ancient Palæolithic of Europe and Asia. This would seem to indicate, if not a common race, at least the existence of a relationship between Africa on the one hand and Eurasia on the other, therein agreeing with geological and palæontological data. It is often

[1] For bibliography, see past numbers of *L'Anthropologie*.

[2] Pallary, P., *Instructions pour les recherches préhistoriques dans le Nord-Ouest de Afrique*, Algiers, 1909.

difficult to make a distinction in time between articles in the Moustierian style and those fashioned in the Chellean and Acheulean manner. Even in the ancient alluvials of Gafsa, where the distinction seemed quite clear to Dr Collignon's eyes, Couillault's and de Morgan's more recent researches have shown that these different forms are as "intimately" intermingled in the Pleistocene conglomerate as in the workshops on the surface. (See Appendix, p. 481.) This early archæological stuff is found everywhere in the deep layers of cave deposits, especially in Algeria. It is contemporary with a fauna of mammals some of which are extinct and some of which have emigrated to the South; and it belongs to a more humid climate than that of to-day.

Since that time great changes must have taken place. Inter-continental communication in the Old World became more difficult. Our Upper Palæolithic, with its many subdivisions, is replaced in Africa by a general archæological facies, to which Pallary has given the name *Getulian*, and which J. de Morgan has called *Capsian*. Relics of the Getulian or Capsian are met with sometimes in cave deposits or rock shelters, and sometimes at the surface of the soil, as is the case especially in those curious settlements in Tunisia called snail-shell mounds (*escargotières*), artificial deposits consisting of snail-shells mixed with cinders, bone-remains of animals, and dressed flints. The latter recall the Aurignacian implements of France (Fig. 221), and the resemblance is sometimes so close as to imply a certainty of close relationship between Europe and Southern Africa during the Upper Pleistocene. The great land bridges, particularly that uniting Tunisia to Sicily, had not yet been completely broken at the beginning of the Getulian period. It is probable that our Aurignacian period is African in origin. Aurignacian and Getulian are two geographical aspects of the same Mediterranean culture.

[1] Pallary, *loc. cit.* J. de Morgan, Capitan and Boudy, "Étude sur les stations préhistoriques du Sud tunisien" (*Revue de l'École d'Anthrop. de Paris*, 1910). Gobert, Dr, "Notes et recherches sur le Capsien" (*Bull. de la Soc. préhist. de France*, 1910). *Id.*, "L'abri de Redeyef" (*L'Anthropologie*, xxiii., 1912). *Introduction à la paléontologie tunisienne*, Tunis, 1914. Raygasse, M., "Études de palethnologie maghrébine" (*L'Anthropologie*, xxvii., 1916).

The extent of the Getulian industry, which at first seemed to be confined to Tunisia, is becoming daily more widely known, and is found to spread from the Libyan desert to Morocco. It is to be found, either in caves or on the surface, in many other regions of the continent and even in South Africa.

It is succeeded in Barbary by the *Ibero-Moorish* industry, so named by Pallary because it is found also in Spain. In addition to the Getulian implements there occur microlithic

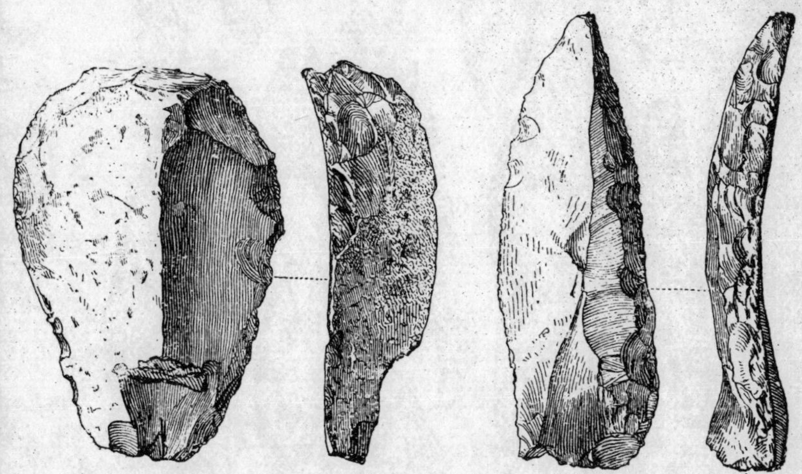

FIG. 221.—Dressed Flints from Tunisian snail-shell mounds, comparable to our Aurignacian types. Two-thirds natural size. (After Pallary.)

flints, grinders, and mullers indicating the use of colour materials. In the course of the Getulian and Ibero-Moorish periods, the fauna became impoverished by the disappearance of species; but it still contained many species which have now migrated to the south. The climate was damp, the country less barren than now.

The Ibero-Moorish phase passes imperceptibly into the *Mauretanian* phase, with many small geometrically shaped flints, similar to those of our Tardenoisian period. But already the presence of polished stones and of pottery indicates the commencement of the Neolithic phase. The latter was widespread, and was remarkable for the abundance and variety of its stalked or hafted arrowheads. Sometimes, according to Cartailhac, it presents interesting points of resemblance with

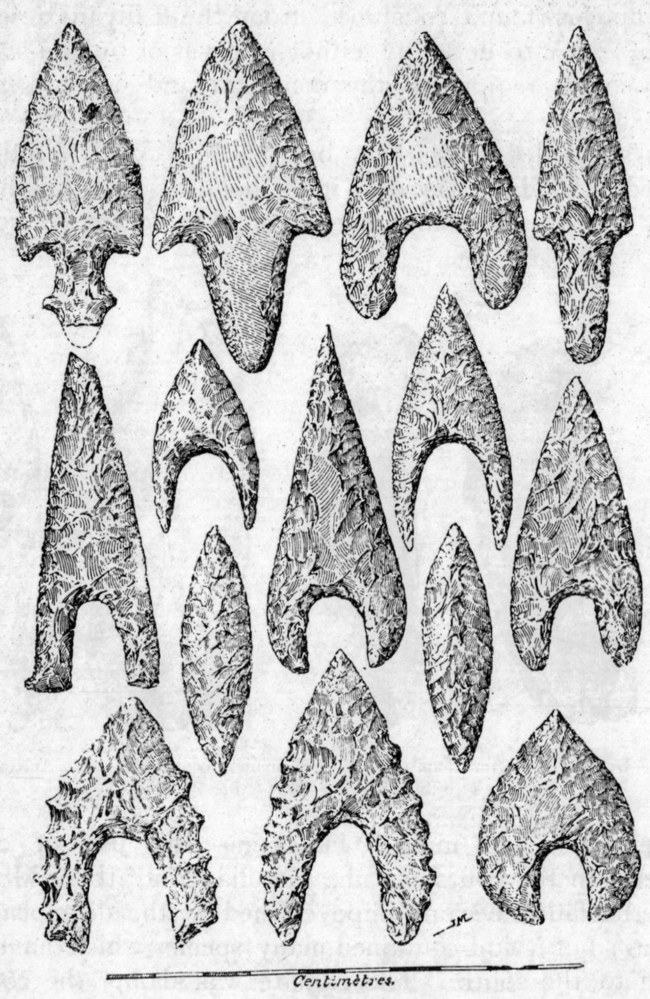

FIG. 222.—Flint Arrowheads from Fayoum in Egypt. Two-thirds natural size. (After J. de Morgan.)

the Egyptian Neolithic, which partially merges with the early Metal Ages.

The prehistory of North Africa may thus be divided into a succession of ages comparable to the European succession.

The Palæolithic exhibits several successive phases of long duration. Future researches will enable us to establish chronological comparisons and relationships; but up to the present, North Africa, and Barbary in particular, seems actually to form part of the Mediterranean world, so far as concerns its ancient human populations, as well as its fauna, flora, and physical characters. In Egypt the succession is less assured. There a great blank still exists between the very old Palæolithic of the plateaux and the Neolithic of the valleys, which is remarkable for the exceptional beauty of its dressed flints (fashioned in the Solutrean manner), and which imperceptibly grades into the civilization of the first dynasties (Fig. 222).

The Sahara. It is not without surprise that we find in the Sahara such extraordinary abundance of Stone Age antiquities. Since Abbé Richard, more than half a century ago, recorded the presence of dressed flints in Southern Algeria, such finds have been repeatedly made. Féraud, Thomas, Largeau, Weisgerber, Rabourdin, Foureau, Flamand, Laquière, E. F. Gautier, Mme. Crova, Arnon, de Zeltner, de Saint-Martin, Tarel, Roulet, Noël Cortier, Gruvel, Chudean, Aulner and other explorers have collected, in almost every part of this vast desert, innumerable stones shaped by flaking of various fashions, as well as polished axes, mortars, and very beautiful and delicate arrowheads (Fig. 223); worked ostrich eggs, cut into rings; pottery, etc. The deposits are chiefly found in the vicinity of springs, modern or ancient, and near river-beds now dried up, a fact which indicates that, in the Sahara, great changes of climate and of the conditions determining habitability have taken place at a relatively recent date. (For bibliographical references, see Appendix, p. 481.)

The material with a Palæolithic facies evidently extends over a long period of time, although the most diverse forms are often found at the same place and in the same bed. Amygdaloid implements of Acheulean type, first recorded by Rabourdin in the Algerian Sahara, have been found in many other places. The Sahara collections also contain implements exactly like our Moustierian, Aurignacian and Tardenoisian flints, and the Capsian types of North Africa. It is possible that implements of archaic form are sometimes contemporary with Neolithic implements, which seem to have persisted up to quite a recent period, for, according to Gautier, the Neolithic in the Sahara merges with the Iron Age. Nevertheless, certain conditions of deposit, supplemented by a study of the raw materials and of the patina, lead to the belief that the majority of the "Acheulean axes" must date from a past of true geological significance. Lastly, notwithstanding a few points of difference, the Stone Age of the Sahara is extra-

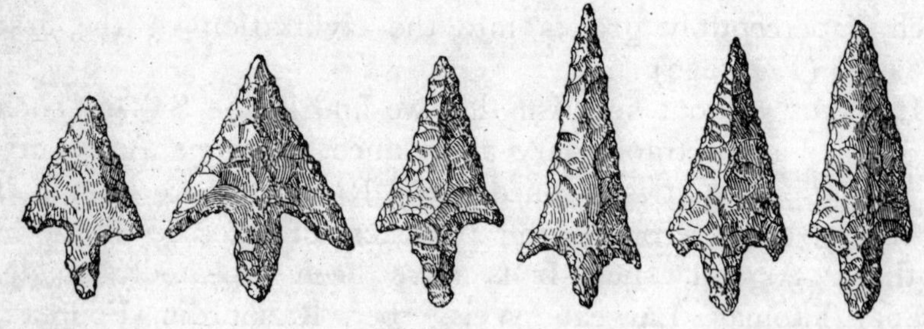

FIG. 223.—Neolithic Arrowheads from the Sahara. Natural size. (After E. F. Gautier.)

ordinarily like that of Lower Africa, and often presents points of resemblance with that of Egypt.

West Africa and the Soudan have also yielded many relics.[1]

Central Africa. Hamy has recorded some objects, mainly of Neolithic workmanship, from French Guinea (Konakry Cave), Portuguese Guinea, the Ivory Coast, the Gold Coast and the Gaboon.

Captain Duchemin has described tumuli and megaliths in the valley of the Gambia. In the Senegal valley in French Soudan, de Zeltner collected worked stones, some of Palæolithic workmanship, in superficial but usually clearly marked beds. Between Kayes and Timbuctoo, Dr Decorse has observed and recorded the site of many Stone Age settlements. The Central Nigerian plateau, explored by Lieut. Desplagnes, is rich in Neolithic monuments and relics. According to Welcome, the Southern Soudan seems to be no less rich.

[1] Bibliography may be found in F. de Zeltner's "Notes sur le Préhistorique soudanais" (*L'Anthropologie*, xviii., 1907).

Dupont, Cornet, Stainier, Taramelli, Jacques, Delisle and others have described many series of Stone Age implements found in the Congo basin.[1] Here, as in the other great African territories, the raw material, the workmanship and the forms of the weapons or tools show the greatest diversity. Palæolithic types, especially the amygdaloid, are not absent, but up to the present, it has been found impossible to distinguish them chronologically from the great Neolithic collections containing polished axes and fine arrowheads. According to Taramelli,[2] however, the character of the patina and the presence of dressed stones in the heart of alluvial beds seem clearly to prove the geological antiquity of the ruder material. Certainly, in the opinion of all the authors, even if the Stone Age came to an end in the Congo only five or six centuries ago, its beginnings are lost in the obscurity of time.

In East Africa, we have to record the discoveries of Seton-Karr and Paulitchke in Somaliland. There dressed stones are abundant, and John Evans[3] has pointed out their resemblance to those of Saint-Acheul, Chelles, and Le Moustier (Fig. 224). Yet here again, even though these specimens sometimes lie at a considerable depth in the soil, no fossils accompany them by which their age may be determined.

On the plains of Ethiopia two missionaries, Revoil and du Bourg de Bozas, found products of a Neolithic industry. Weyland collected, in the Portuguese territory of Mozambique, stones shaped by large flakes.

Southern Africa.

From the archæological point of view there still remain many blanks in the Dark Continent, particularly in the interior. South Africa, on the contrary, having been more thoroughly explored, is much richer.

Here the first discoveries of dressed stones were made more than half a century ago, and since that time they have vastly increased. Among the many publications on the subject, those

[1] Stainier, X., "L'âge de la pierre au Congo" (*Annales du Musée du Congo*, Brussels, 1889).

[2] Taramelli, A., *L'Anthropologie*, xii., 1901 p. 411.

[3] Evans, Sir John, *Proc. Royal Society*, lx., p. 19.

of Péringuey, Director of the South African Museum, and of the English geologist, Johnson,[1] must be placed in the first rank, as being most comprehensive.

Relics of the Stone Age have been collected in almost

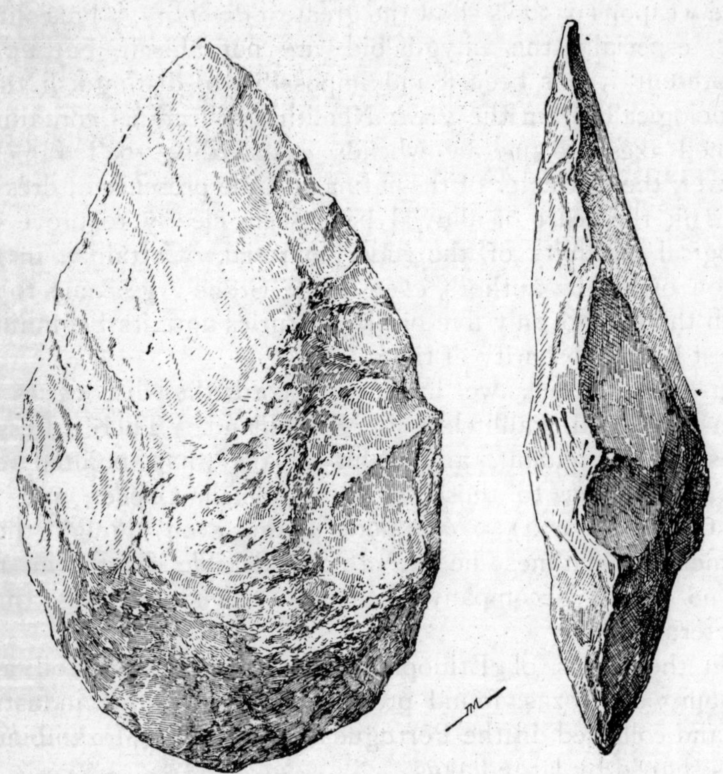

FIG. 224.—Quartzite Implement from Somaliland. Two-thirds natural size. (After J. de Morgan.)

every part of Rhodesia, of the Zambesi valley, the Transvaal, the Orange River State and Cape Colony. Some are of Palæolithic, others of Neolithic workmanship, among the former being many Chellean hand-hammers (*coups de poing*) (Fig. 225), almond-shaped Acheulean implements, and flints

[1] Péringuey, L., "The Stone Ages of South Africa" (*Annals of the South African Museum*, 1911). Johnson, J. P., *The Stone Implements of South Africa*, London, 1907, 2nd ed., 1908. *The Prehistoric Period in South Africa*, London, 1910, 2nd ed., 1912. See in *L'Anthropologie*, xxiii., 1912, p. 513, a list, drawn up by Péringuey, of works on the Stone Age in South Africa.

or quartzites dressed in the Moustierian manner on one surface only. There are also implements more delicately worked which may be compared to Aurignacian or Getulian types, and even, according to Johnson, to Solutrean types.

The beds are usually superficial, but some of them seem to indicate a high antiquity, using the term in the geological sense. Rupert Jones described stones, lanceolate in form, taken from gravels forming terraces above the present bed of the river Embalaan, in Swaziland. Leith has seen similar stones taken from the ancient gravels of the Vaal.

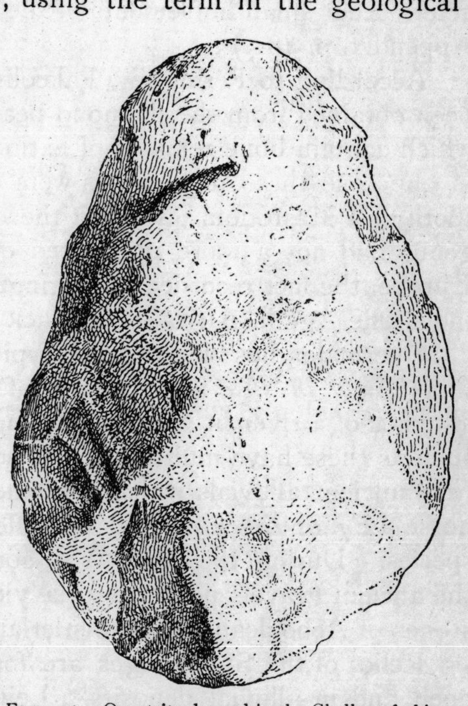

FIG. 225.—Quartzite dressed in the Chellean fashion, from Orange River. (After Hamy.)

Feilden has recorded the presence of many stone implements both above and below the famous Victoria Falls, in connection with the gravels of the great terraces of the Zambesi valley. The fact that these alluvials were laid down when the river flowed some 130 to 150 metres above its present bed seems beyond dispute. Lamplugh and Balfour have confirmed these observations. All the evidence points to the belief that the worked stones, generally of somewhat rude workmanship, often Chellean in character and highly patinated, date from a period prior to the protracted hollowing out of the gorge. But another English geologist, Codrington, does not believe in so great an antiquity.

Johnson found implements at the bottom of a lake deposit exploited as a brick-earth, near Robinson, in the Orange River Colony. He showed that the Palæolithic deposits of the Orange River Colony are often connected with the old alluvial terraces, and he saw, in the deep layers of a conglomerate in one of these terraces, implements which he was unfortunately unable to extract owing to the hardness of the matrix. At one point on the banks of a tributary of the Vaal, a low terrace contains Acheulean types in its basal gravels, while in its upper layers occur small implements of the "Solutrean" group. (See Appendix, p. 481.)

According to Péringuey, Palæolithic implements have often been obtained from the diamond-bearing deposits of the Vaal, which contain bone-remains of extinct or migrated species, and in one case these implements were accompanied by the molar tooth of a Mastodon. Were it the case that this proboscidian genus had not a more protracted existence in Africa than in Europe, the fact recorded by Péringuey would place the date of the Stone Age in Africa as far back as the Pliocene period.

Obermaier has described a typical "hand-hammer," which was found *in situ* at a depth of 5 metres in the ancient alluvials of a river in Natal. During the last few years, cases such as these have multiplied. Broom obtained dressed stones from ancient alluvials which contained also bone remains of a horse (*Equus capensis*) and of different ruminants of extinct species. During the Boer War, some of the trenches dug in the ancient gravels near Pretoria yielded at their base worked stones of Acheulean and Moustierian types.

Relics of the Stone Ages are found elsewhere than in the open and in alluvial deposits. Leith described the caves and rock shelters of the Stormberg mountains in Cape Colony, where he found a stone implement: other caves, at Cape Saint-Blaize, near the sea, contain accumulations of pebbles, shells, bone-remains, and ashes, along with many dressed stones, often carefully worked. Elsewhere on the shore, fragments of breccia with bones and shells contain also dressed flints. Shell mounds have been found at different places on the south coast; they are rich in dressed quartzites, in remains of ostrich eggs, bone needles, and shards of pottery.

Mennel and Chubb examined the deposits in a cave in Rhodesia, and these contained both dressed stones and fossilized bones of animals.

According to Johnson, the Stone Age in the Orange River Colony and in the whole of South Africa would seem to date from a period when the climate was more humid than it is to-day. Apart from some "eoliths" from Leijfontein, there seems to have been first of all an Acheulean type of industry, the products of which, worn and deeply patinated, are often found in gravels. A more recent phase, described as "Solutrean," seems to be found in connection with superficial deposits containing finely worked flints, which include long blades, scrapers, and pygmy flints. Along with these fine stone implements, other settlements contain ostrich eggs, shaped, incised, and sometimes cut in rings, as well as perforated stones, etc.

Péringuey also distinguishes several periods. To the first of these belong the Chellean, Acheulean, and Moustierian types, which in this case are absolutely contemporaneous, and can only have been produced by strong and vigorous men. It cannot be denied that some at least of these specimens are of an antiquity comparable to similar specimens from European deposits. The author believes that the Chellean type originated in Africa, whence it spread to Europe and Asia.

The second group comprises implements frequently of a more primitive character, and sometimes of more finished workmanship, after the Aurignacian, Solutrean, Magdalenian, and Tardenoisian forms, these expressions, it must be understood, being used in this connection without any chronological significance. Other objects also occur, such as mortars, millstones, pottery, bone implements, and decorative objects often made of the shell of ostrich eggs. In its general type this industry has lasted until recent years; it may be described as the "Neolithic industry of South Africa."

To the third group belong certain objects of workmanship similar to that of European Neolithic culture, especially small stalked arrowheads, laboriously retouched on both surfaces.

In South Africa the problem of the age of the different cultures is difficult to solve because of the continuity of the climatic and faunistic conditions of the country. The two industries, Palæolithic and Neolithic, are partly contemporaneous. There is no hiatus comparable to that which we find in Europe. But, even in a single group sharing the same structural peculiarities, the later implements by no means possess the patina and ancient appearance of the early ones.

There can be no doubt that the resemblances between our Palæolithic and that of South Africa are very great, and that they indicate an ancient and very close relationship between the two continents. Dr Péringuey is convinced that such

FIG. 226.—Wall Engraving in Algeria, representing a Buffalo. (After Pomel.)

relations must have existed between European peoples of the Reindeer Age, whether Aurignacian or Solutrean, and those of South Africa. On which side does the debt lie between the two continents? It really seems that we owe our Aurignacians to Africa (see p. 307).

Be that as it may, the Bushmen (or Boschimans) appear to be the descendants of men of the Upper Palæolithic, to whom they retained their resemblance even up to the time of their final extermination, which dates but from yesterday.

This conclusion is strengthened by a group of facts which till now I have passed over in this rapid review of the Stone

Age in Africa, but regarding which I shall now say a few words. I refer to representations on rocks, engraved or painted, which have long been known to exist at certain points far distant from each other, in the South African continent, and which are now known to form an almost continuous chain from north to south. They occur in Barbary (recorded by many authors), in Upper Egypt, Nubia, and Arabia, all over the Sahara (recorded by Flamand and Gautier), to the north of Lake Tchad, in Mauretania, Nigeria (Desplagnes and Chudeau), in the Soudan (de Zeltner), Somaliland (Carette-Bouvet and Neuville), in the Victoria Nyanza region (Koch),

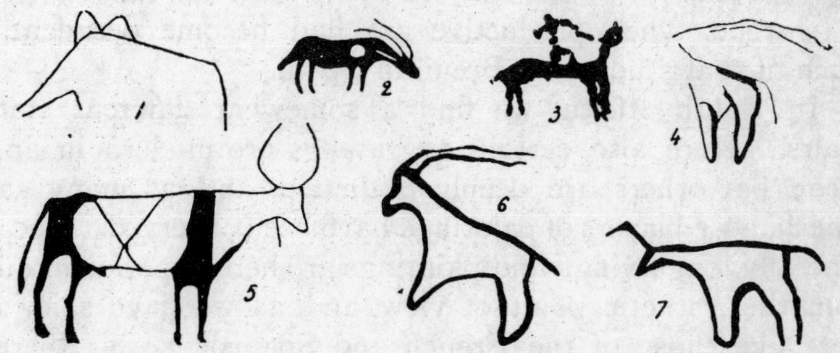

FIG. 227.—Rock Sculptures from different localities in the Sahara. One-twentieth natural size. (After E. F. Gautier.)

and lastly all over South Africa (Christol, Holub, Stow, Moszeik, White, Péringuey, Johnson, Theal and others).

The style of these more or less artistic works, their state of preservation, their patina, more or less deep, the kind of animals represented, and the relics found at the base of the engraved rocks, enable us to distinguish several periods. In North Africa there are engravings, large in outline and deeply cut, representing extinct species such as the Great Buffalo (*Bubalus antiquus*) (Fig. 226), or species which have emigrated such as the Rhinoceros, the Elephant, the Giraffe and others. These designs are sometimes well drawn, and have been described as prehistoric and classified as belonging to the Neolithic period; but there is no proof that they may not be even more ancient, and Flamand attributes the drawings of Buffaloes to Pleistocene times. Other drawings somewhat different in character, less artistic, small in size, stippled and often accompanied by inscriptions, are much more recent, and are described as Libyo-Barbaric. Many date from historical times. The distinction between the two sets is sometimes difficult, and they may have formed a continuous series. Flamand has pointed out the contemporaneity of a drawing of a herd of Buffaloes and of men armed with polished axes. The Neolithic period seems thus to be older in Africa than in Europe, and this, indeed, is not a matter for surprise.

The best drawings in North Africa are far from possessing the qualities and beauty of our own drawings of the Reindeer Age. Comparison can only be made with the inferior periods in Europe, when productive art had become decadent, like much of that studied by Breuil in Spain.

In South Africa, we find a somewhat different state of affairs. Here also, certain engravings are modern in appearance; but others are deeply patinated, and as many as five superimposed layers of paintings have been observed. Speaking generally, engravings and paintings are here superior in quality from the æsthetic point of view, and, as we have said, much more like those of the French and Spanish caves; there are the same preferences in the choice of subjects, animals being most often made use of, the same realism, fidelity of attitude, skill in reproduction and technique, and even the same weakness in drawing the human figure. Sometimes these artistic productions of South Africa are arranged in pictures (Fig. 228) or assume a hieroglyphic, perhaps symbolic nature, like those recently discovered in certain Spanish localities. These works of art are usually attributed to the Bushmen. This opinion, however, is not held by Péringuey,[1] who attributes them to a Hottentot tribe called the "Strand Loopers." But no one has seen a modern native in the act of drawing them.

Human Bone-Remains.

Bone-remains of fossil men in Africa are as scarce as the products of their industries are abundant. The bone-remains of the primitive Africans are almost quite unknown.

[1] Péringuey, L., "On Rock Engravings of Animals and the Human Figure . . ." two papers in the *Trans. South African Philos. Society*, 1906 and 1909.

In North Africa a fairly large number of skulls and skeletons have been exhumed from Stone Age deposits, but the majority of these specimens belong to the Neolithic period. It has never been proved that the skulls taken from the snail-shell mounds (*escargotières*) are really Getulian, that is to say, Palæolithic; for these artificial heaps usually served as Berber cemeteries.

Bertholon[1] believes it possible to distinguish a neanderthaloid type of skull in North Africa, particularly in Tunisia, whence it spreads into the region of the Sahara. Several

FIG. 228.—Black and white painting in a Cave in Baroaland, representing Bushmen engaged in hunting, being attacked by Kaffirs. Length of picture 1½ metres. (After Hamy.)

skulls, from the deep layers of the snail-shell mound at Mechta-Châteaudun in Constantine, seem to be representative of this African neanderthaloid race.[2] But we need only examine the photographs of the best preserved of these specimens to learn that it is but a case of mere superficial resemblance, and that the Mechta skulls exhibit none of the essential characters of *Homo neanderthalensis.*

Delisle, on the other hand, classified as Cro-Magnon a skull found by Debruge in the Palæolithic cave of Ali-Bacha, near Bougie. This record quite agrees with what we know of the wide distribution of the Cro-Magnon type in the

[1] Bertholon, Dr L., "La race de Néanderthal dans l'Afrique du Nord" (*Revue tunisienne*, 1895).

[2] Bertholon, Dr, in Mercier and Debruge, "La station préhistorique de Mechta-Châteaudun" (*Soc. archéol. de Constantine*, xlvi., 1913).

Mediterranean basin, and of its persistence in Barbary in a more or less pure form.

M. Pallary has kindly informed me that in the Faculty of Science at Algiers, there is a whole series of skulls and bone-remains obtained by him from the rock shelters at La Mouillah, near Lalla Marnia on the frontiers of Morocco. These settlements are Ibero-Moorish, that is to say Upper Palæolithic, and according to Dr Weber, of the Faculty of Medicine at Algiers, the skulls are negroid in character.

The museum at Oran contains several skeletons with skulls, obtained by Pallary and Tommasini from the Troglodytes' Cave near Oran, dating from ancient Neolithic times. According to Dr Bloch, these also seem to be negroid.

Bertholon[1] again found these "undoubtedly" negroid characters in skulls from snail-shell mounds at Tebessa, and especially in skulls which Dr Gobert obtained from Neolithic beds in the Redeyef rock-shelter. Here the men would seem to have been of small stature.

Bertholon and Chantre[2] came to the conclusion that in Barbary the Negroids formed an ancient ethnographical stock, while Neolithic skulls belong sometimes to the Mediterranean, sometimes to the Negroid or Nubian type.

In ancient prehistoric burial-places in Egypt, Fouquet recorded the existence of several types, which are again found at a later date among the Egyptians and Ethiopians. Elliot Smith[3] tells us that these prehistoric Egyptians or Pre-Egyptians were small, dolichocephalic, and less negroid in type than some of their successors; in short, he classifies them as Sergi did, among the Mediterranean peoples, who, from Egypt to France and even as far as the British Isles, were related by close ties. On the other hand, I do not know on what facts Sir Harry Johnston bases his assertion

[1] Bertholon, Dr, *Note sur quatre crânes humains trouvés par M. Debruge à Tebessa*; no place of publication nor date, probably 1912. Bertholon, in Dr Gobert's paper, "L'abri de Redeyef" (*L'Anthropologie*, xxiii., 1912, p. 167).

[2] Bertholon, Dr, and Chantre, E., "Recherches anthropologiques dans la Berberie orientale," Lyons, 1912.

[3] Elliot Smith, G., *The Ancient Egyptians*, London, 1911.

that the valley of the Nile was populated, 20,000 to 30,000 years ago, by men of the African pygmy type.

To sum up, the main fact seems to be that Man really formed part of the mammalian fauna, and that he accompanied it during its distribution. It would appear that in the Pleistocene period, men of the white race, more or less akin to the Cro-Magnon in type, were already in possession of North Africa, which may perhaps have been their cradle, but that men of Negroid type, of true African origin, often pushed northwards, like their contemporary mammals, and may have succeeded in crossing the Mediterranean and landing on the shores at Mentone, taking with them their Capsian or Aurignacian industry. Of the precursors of these men, of the dressers of the Chellean stones, we know absolutely nothing.

Many anthropologists, basing their arguments largely on linguistic grounds, have tried to unravel the peoples of Central and Southern Africa, and to discover the different ethnographical tides which here have succeeded each other. But the results are all somewhat uncertain, and do not bear on the problem of true Fossil Man, of whom we have only two evidences, and these of unequal value.

The Oldoway Skull.

Some months before War was declared, a number of German, English and French journalists announced the discovery of a very old human skeleton in the north-eastern region of German East Africa. Dr Hans Reck,[1] the discoverer, is a geologist-explorer from the Berlin Geological Institute. In the course of explorations carried out in the Oldoway Gorge, on the borders of the Seregenti plain, he found, together with many bone-remains of animals, an almost complete human skeleton lying in the midst of well-stratified volcanic tuffs at a depth of from 3 to 4 metres. The skeleton and the mammal bone-remains lay in exactly the same circumstances; like them it was enclosed in a tuff of such hardness and consistency that it was necessary to use hammer and chisel to free it. There could be no question

[1] Reck, Dr H., "Erste vorläufige Mitteilung über den Fund eines fossilen Menschenskeletts aus Zentralafrica" (*Sitzungsberichte der Gesellsch. naturforschender Freunde zu Berlin*, 1914).

of a burial; the human bones are as old as the deposit which contained them.

The determination of the age of the skeleton is therefore simply a question for the geologist and the palæontologist. It does not appear that the mammals at the different levels of the fossil-bearing tuff differ from present-day African species. But, taking the deposit as a whole, differences in the fauna may be noted, and these correspond to differences in climate. The stratigraphical horizon which yielded the human skeleton, and where many bone-remains of the elephant, rhinoceros, hippopotamus, crocodiles, and fishes abounded, points to the existence of a wooded region, with a damp climate which might be regarded as contemporaneous with the last great Glacial Age in Europe. Dr Reck is convinced that this stratum dates from a geological period prior to that of the present geological era.

As yet we have but few details regarding the characteristics of the skeleton. The skull would seem to be large, dolichocephalic, clearly of Negroid type. It was furnished with thirty-six teeth, several of which showed traces of filing similar to that still much practised by negroes at the present day. These are somewhat strange facts. We must await further and more detailed information in order to be able to judge of the antiquity of this discovery, and to determine, in consequence, its scientific value.

The Boskop Skull.

In 1914 also, a friendly letter from Dr Péringuey, Director of the Cape Museum, informed me of a discovery, made at Boskop in the Transvaal, of fragments of a human fossil skeleton, consisting of portions of the skull, of the lower jaw and of shafts of the long bones. This discovery was brought to the knowledge of the public through an article published in the English journal *Nature*, on 5th August 1915. The author, F. W. Fitzsimons, compared the Boskop skull-cap to that from Neanderthal. Examination of a cast, which Dr Péringuey was kind enough to send me, prevents me from adopting this opinion, which in any case was soon rejected by the naturalists of the South African Museum. In October 1915, one of these naturalists,

Mr Haughton, communicated to the Royal Society of South Africa a paper on the subject, the text of which appeared in 1917.[1]

Unfortunately it is impossible to determine the age of the

FIG. 229.—Skull-cap from Boskop, seen in profile and from above. One-third natural size. (After Haughton.)

Boskop human fragments. The bones, deeply impregnated with mineral matter, were found during the digging of a drainage trench in a field. The soil of the field merges gradually into a laterite[2] sub-soil containing no other fossils.

[1] Haughton, S. H., "Preliminary Note on the Ancient Human Skull-Remains from the Transvaal" (*Transactions of the Royal Society of South Africa*, vi., 1917), followed by a note by Elliot Smith on the endocranial cast of the Boskop skull-cap.

[2] For explanation of this word, see p. 357.

The exact bed of the skull-cap can no longer be definitely ascertained, and explorations subsequently carried out by the Cape Museum have yielded no bone-remains *in situ.* Mr Haughton considers that the skull was found at a depth of about 4 feet 6 inches, but he does not know if this interment is due to natural causes or to human intervention.

The brain-pan is remarkable for its flatness and great size (length, 205 millimetres; breadth, 154 millimetres). Its cephalic index is about 75. The total capacity of the whole skull is estimated at 1830 cubic centimetres, which is really enormous.[1]

According to Haughton, the Boskop skull-cap (Fig. 229) resembles in character the Cro-Magnon type: the form as a whole is pentagonal, with slight superciliary ridges, prominent glabella, straight forehead, the same kind of antero-posterior curve, and strong occipital prominence. From the Cro-Magnon type it differs mainly in possessing a depression, a kind of hollow, in the interparietal region, which I have noted on several negro skulls in the Anthropological Gallery of the National Natural History Museum in Paris, especially in the Namaqua skull, well known to all specialists.

The frontal bone is very narrow. The larger portion of the right temporal bone has been preserved; the mastoid apophysis is small but well separated; the petrus region is strongly developed; the glenoid cavity is large and shallow; and there is a post-glenoid apophysis. The most striking feature of this bone is the marked development of the supra-mastoid crest. Here we have a simian structure, resembling that of Neanderthal Man, although, in its general form, the skull-cap may be compared to the Negroid, Bantou, and Cro-Magnon types.

The Boskop find includes also the two horizontal rami of a lower jaw, unfortunately in a poor state of preservation. Their principal characters are their great strength, the presence of a slight chin and of small digastric fossæ. Only one tooth is in

[1] According to the palæontologist, Broom, even these figures are too small: the skull should be 220 mm. in length, 160 mm. in breadth, 148 mm in height. Therefore Broom attributes it to a new species, which he calls *Homo capensis* (*Anthropological Papers, American Museum*, 1918).

position, the second left true molar. Its crown is badly preserved, so that it cannot be decided whether or not it possessed the posterior cusp characteristic of many primitive types. This jaw is already of the reduced dimensions which occur in modern Man; it is more highly evolved than the skull, and may be compared with that of the Bantous or of the Bushmen.

Examination of the portions of shafts of long bones, which are deeply impregnated with a kind of laterite cement and which are much mutilated, yields no interesting results.

Professor Elliot Smith has examined an intra-cranial cast. Its flattened shape and certain other characters suggest resemblances between the Boskop man and the Neanderthal type. But the appearance and great development of the frontal protuberances point to closer relationship with fossil men of the Upper Palæolithic in Europe, of whom the Boskop type might seem to be an immediate ancestor.

Some pieces of sandstone, with sharp edges, found in the laterite in the vicinity of the human bones have been examined by Dr Péringuey. In spite of a few signs which seem to support the idea, no real trace of intentional working can be recognized.

Here again we are confronted by an incomplete find, which, because of its incompleteness, it would be difficult, nay, rash, to interpret. If, as we may believe, the Boskop human remains are really Pleistocene, they reveal the existence in Africa of a generalized type, not lacking affinity with certain men of our own Palæolithic, but already quite definitely Negroid in character. At the same time they seem to furnish a new proof of the high antiquity of the human race, even in the countries it inhabits at the present day.

Since the first French edition of this work appeared, some human remains have been found in the Broken Hill Cave in Rhodesia. Of these a skull, lacking its lower jaw, possesses morphological characters which are entirely different from those of any modern African type whatsoever, and which on the contrary bear an extraordinary resemblance to the skull characters of *Homo neanderthalensis* of our ancient European Palæolithic. It resembles the latter in its dimensions, and also in exhibiting orbital arches as prominent and a forehead even more bulging. This skull has been examined by Smith Woodward, who has named it *Homo rhodesiensis*. Unfortunately there are no data by which the age of these remains may be ascertained, and they themselves show no traces of fossilization. (For further details and illustrations, see Appendix, p. 481.)

The Two Americas.

The history of the New World begins only in the sixteenth century, but, on the other hand, it has an immense prehistory stretching far in time as well as in space. Apart from a few traditions, we find indications of it in monuments of all kinds distributed over vast areas. To the scientist, however, America still remains as great a mystery as it was at the time of the Conquistadores. In spite of considerable efforts, American anthropology, ethnography and archæology—which taken together now constitute a special group, Americanism—have not so far been able to elucidate any but minor points and matters of detail. None of the great problems they discuss has been definitely solved.

Have we to deal with one or with several great American races? Were the populations of the New World before its discovery indigenous or immigrant? And if the latter, whence came they? Must we suppose that in America there existed one or several special centres of origin and evolution for a section of Mankind, or must we believe that the country was peopled from the Old World? If that be the case, at what period did this settlement take place?

I cannot here discuss these important problems, to the solution of which geology, palæontology, archæology, the study of language and ethnography must all lend their aid. The literature on these various branches of the subject already forms a bulky library. I shall touch only on such of it as deals with fossil Man, but I think it may be useful, as a preliminary, simply to enumerate a few theories which, without being actually established as fact, are shared by the majority of students of American history.

From the anthropological point of view, we may say that it is all but unanimously agreed that the prehistoric populations of America, that is those now known as the *Amerindians*, belonged to the main stem of the Yellow Races. And it would seem that all these peoples came from the Old World. But their distribution, which took place at such a remote period of time, over the whole extent of the two Americas, the physical, linguistic, and social differences they exhibit or which they previously exhibited, lead one to think that the stocking of the New Continent from the Old must necessarily date from a very remote period.

Archæology teaches us indeed that the indigenous peoples, the Indians or Amerindians, who lived in such numbers in America at the beginning of the sixteenth century, must have descended from a long series of ancestors more or less distant, to whom all sorts of monuments must be attributed—the shell mounds and kitchen middens which are strewn upon the sea-shore everywhere, and sometimes occur even inland; the earthworks, tumuli, and ramparts known as "mounds," scattered over all the great plains of the United States; the strange habitations of the "cliff-dwellers," hollowed out or perched on the great steep cliffs of Colorado, Arizona, and New Mexico; the *pueblos*, stone or mud towns or villages of the same regions and also of Central America; the wealthy cities of Central America and of Peru rich in monuments.[1] The skulls and human skeletons exhumed from these various sources present the chief characters of the "Indians" of the same regions.

We now know that the general civilization of these Amerindians, which, if a comparison must be drawn, may be likened to that of our Neolithic (with many polished stone implements, finely dressed arrowheads, very varied pottery and

[1] For general works on the subject, see: Nadaillac, Marquis de, *L'Amérique préhistorique*, 1 vol. in 8vo, Paris, 1883. Cyrus Thomas, *Introduction to the Study of North-American Archæology*, Cincinnati, 1898. Moorehead, W. K., *The Stone Age in North America*, 1910. Beuchat, *Manuel d'archéologie américaine*, Paris, 1913. Joyce, *South American Archæology*, 1912; *Mexican Archæology*, 1914; *Central American Archæology*, 1916; Holmes, W. H., Handbook of aboriginal American Antiquities, Pt. 1 (*Smiths. Inst. Bureau Amer. Ethn.*, Bull. 60, 1919).

an absence or scarcity of such metals as copper), lasted a very long time, for everywhere it has left behind countless traces; and some shell-heaps and mounds are of such large size, or are found in such topographical circumstances, that they must date from a very ancient period, perchance from the end of Pleistocene times.

Away beyond this period reigns the darkness of geological times. Here, as in the Old World, the question becomes a geological and palæontological problem, connected only with the natural sciences; thus it comes within the scope of this work.

I cannot describe individually the great number of discoveries or finds brought forward to prove the existence of fossil Man in America. Many, indeed, are unworthy of note; I shall, therefore, describe only some of the most important, those which are worth discussing.

It is curious to note that in 1840, even before it had been definitely settled in Europe, the question of the co-existence of Man with the large extinct animals, such as the Mastodon, had already been broached in the two Americas.

North America.

General Description.

During Pleistocene times, North America curiously resembled Europe. Glaciers, originating from three main centres, the first in Labrador, the second to the west of Hudson Bay, and the third on the ridge of the Cordilleran Highland from Alaska in the north to Montana in the south, coalesced to form an immense continental ice-sheet, covering the whole of Canada and the north of the United States almost as far as lat. 37° N. The thickness of the ice varied in different regions, from 1200 to 3000 metres.

The Rocky Mountains, the Cascade Mountains, and the Sierra Nevada had also their own glaciers, the advanced moraines of which reached sometimes as far as the plains, just as in the case of the glaciers of the Alps or of the Pyrenees.

In many places, the moraine formations alternated with deposits of different origin containing fossils. As in Europe, so too in America several phases in the advance and withdrawal of the glaciers are to be observed; according to different authors these phases varied in number from three to six.

Thus Pleistocene times in America present a physical aspect akin to that in Europe. Does the same hold good in relation to the biological, and especially the human side? Let us note first of all that the fauna of large animals is very different in the two continents, except in the northern regions inhabited by circumpolar animals such as the Mammoth. Elsewhere, we meet with a Mastodon (Fig. 230), with several

FIG. 230.—Skeleton of *Mastodon americanus*. Frankfurt Museum.

genera of large Edentata, *Megatherium*, *Megalonyx*, and *Mylodon* from South America, and other mammals differing generically and specifically from European forms. These differences make it difficult to correlate the divisions of Quaternary times in European countries with those which American geologists and palæontologists have attempted to establish in their own country.

As regards Man himself, discoveries put forward in support of his geological antiquity are very numerous. But many have no scientific value, and even with reference to those which seem to have been found under the best conditions, opinion is far from being unanimous. At the present time in the United States there are several firm believers in the existence of American fossil Man, but he has also many invincible opponents. Let us examine the principal material evidences.

FIG. 231.—Flint Arrowhead found under a shoulder-blade of *Bison occidentalis*, in a Pleistocene clay deposit in Kansas. About one and a half times natural size. (After Wilson.)

In the first place an attempt was made to establish the contemporaneity of Man with the extinct animals. The association of axes and stone arrowheads with the bones of more or less complete skeletons of Mastodons or Elephants has been repeatedly recorded.[1] The first observation of this kind dates from 1839. It was followed by several others, the value of which it is somewhat difficult to judge to-day. One of the latter, made by Clarke in 1903, tended rather to prove that the Mastodon survived in America until the dawn of modern times.[2]

On several occasions human relics have been found intermingled with bones of *Megalonyx*, particularly in the Big Bone Cavern in Tennessee.

Professor Martin of Kansas University, exhuming skeletons of *Bison occidentalis* in a Quaternary deposit, collected a flint

[1] See Wilson, Th., "La haute ancienneté de l'Homme dans l'Amérique du Nord" (*L'Anthropologie*, xii., 1901). Osborn, *The Age of Mammals*, 1910, p. 495.

[2] This fact seems also to be confirmed, on the one hand, by the stratigraphical study of the deposits, often very superficial, where these bones are found, and on the other hand, by representations of this Proboscidian (confused with the Mammoth) on two engraved objects, one of which was found in Delaware, the other in Pennsylvania. The former is a fragment of a shell of *Fulgur;* the other is a kind of stone pendant (*Lenape Stone*). It is true, however, that the authenticity of these engravings is not admitted by all archæologists.

arrowhead (Fig. 231) which lay under and in contact with the right shoulder blade of one of these skeletons. According to the palæontologist Williston,[1] this find proved that Man was contemporary with an extinct species of Bison.

Archæological Observations. Now we turn to archæological evidence. Beside innumerable stone objects of Indian manufacture, strewn all over the surface of the soil of the United States, there are some objects of ruder manufacture and more ancient appearance, which are much more like the most ancient Palæolithic implements of the Old World. At first it was thought that they were of the same antiquity. Wilson[2] was convinced of the truth of this theory, basing his belief on an important collection of objects of this nature which he had gathered together at the Smithsonian Institution. The geologist Winchell[3] recently came to the same conclusion after studying the "palæoliths" from Kansas, which he regards as belonging to four successive periods, two of which were Palæolithic. These distinctions he bases mainly on his study of the patina.

It must be admitted that discoveries of implements in the depths of geological layers would be more conclusive. Evidences of this sort do, indeed, exist, but they have been hotly disputed.

There are, in the first place, mortars, pestles and other objects, found at a certain period in the auriferous gravels of California, about which much has been written. The true origin of these objects is so doubtful, the conditions of their deposit are so obscure, and they so exactly resemble the industrial products of present-day Indians, that American anthropologists are now all but unanimous in denying their high antiquity.

Dressed stones have been found in various formations clearly of Pleistocene Age, such as the ancient alluvials of Mexico, the deposits of the ancient Lake Lahontan in Nevada,

[1] *American Geologist*, xxx., 1902, and *Congrès des Américanistes*, 1902.

[2] Wilson, Th., "La période paléolithique dans l'Amérique du Nord" (*Congrès intern. d'Arch. et d'Anthrop.*, Paris meeting, 1889).

[3] Winchell, N. H., *Ibid.*, Geneva meeting, ii., p. 365, and *Minnesota Historical Society*, xvi., 1913.

the gravels or silts of Minnesota, Indiana, New Hampshire, Ohio, New Jersey and elsewhere; but the value of all these finds has been disputed, and it seems evident that the majority of them are very open to criticism. I do no more than mention them here.[1]

The Trenton Alluvials.

But there is a deposit of the same kind, at Trenton in New Jersey, upon which we must dwell at greater length, for it has been and still is the subject of great discussion. In the year 1875, Dr Charles Abbott[2] collected, in the ancient alluvial deposits of the Delaware River, stone implements (quartzite and argillite) rudely worked in forms often similar to those of European Palæolithic flints. Soon afterwards he described this "Primitive Industry," attributing to it a very high antiquity. The Trenton gravels have been formed by the re-sorting of the moraines of the last glacial extension; they contain bones of fossil mammals; and they are therefore undoubtedly Pleistocene. The presence of dressed stones in the depths of these alluvial deposits proves the reality of the existence of an American Palæolithic Man. This result at which Abbott arrived seemed conclusive, and it was at first generally accepted.

About 1890, a strong reaction in opinion became evident. Certain ethnographers of repute in the United States, Holmes, Brinton, and MacGee, declared that the supposed implements of Palæolithic Man, including those from Trenton, were only discarded products of manufacture, identical with those found in enormous heaps around old quarries worked by the Indians. They thus denied, at one and the same time, the antiquity of the dressed stones, and both the antiquity and the authenticity of the deposit in which they were found.[3]

In 1893, two years after an expedition which I had the pleasure of making with Abbott and Wilson to the ballast pits at

[1] Wright, F., *The Ice Age in North America*, New York, 1889 (Supplement to the 3rd ed. 1891); *Man and the Glacial Period*, New York, 1912.

[2] Abbott, C. C., *The Stone Age in New Jersey*, 1877; *Primitive Industry*, 1881, etc.

[3] MacGee, W. J., "Palæolithic Man in America" (*Popular Science Monthly*, 1888). Holmes, M. H., "Are there Traces of Man in the Trenton Gravels?" (*Journal of Geology*, 1893); *Science*, 1892, 1893, *passim*.

Trenton, I considered it advisable to publish the reasons which led me to believe in the authenticity and antiquity of the implements of Palæolithic form collected by Abbott himself in the Trenton gravels. It seemed to me at that time that Abbott's case somewhat resembled that of Boucher de Perthes.[1]

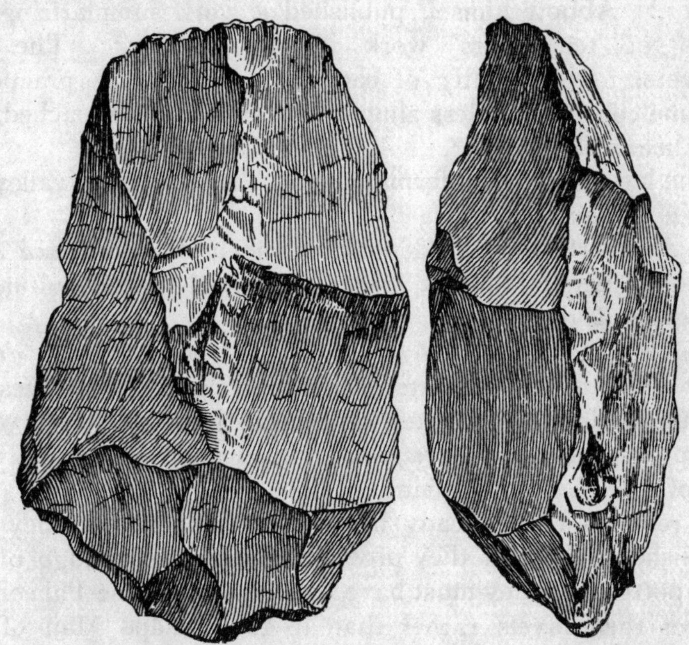

FIG. 232.—Front and Side View of an Implement of Argillite found in the Trenton Alluvials, 2 metres below the surface. Natural size. (After Wilson.)

Opposition was renewed more strongly than ever, so much so that in 1897 the geologist Chamberlin, playing the contrary part of Elie de Beaumont, went so far as to say that the existence of Palæolithic Man on American soil was not worthy of the honour of a discussion.

But Abbott's theory was supported by certain loyal partisans, first among whom must be ranked the late F. W. Putnam, Director of the Peabody Museum, Cambridge.[2]

[1] Boule, M., "L'Homme paléolithique dans l'Amérique du Nord" (*L'Anthropologie*, iv., 1893).

[2] Putnam, F. W., Numerous articles in his *Reports* as Curator of the Peabody Museum of Harvard University, 1876 to 1910.

Interested from the first in Abbott's researches, he commissioned one of his fellow-workers, Mr Volk, to make observations and excavations, which were carried out during a period extending over twenty-two years, and the results of which have been published in a large volume which appeared in 1911.[1] Abbott himself published a work summarizing the results of ten years' work in the deposit.[2] The two observers, independently of each other, arrive at practically the same conclusions, less simple than those first broached, but nevertheless explicit.

The later geological formations of the Delaware valley are of three kinds :—

1. A superficial black deposit or vegetable soil (*black soil*), containing very abundant traces of the Neolithic culture of the Lenâpe Indians.

2. This black earth covers a yellow deposit (*yellow drift*), clayey and sandy in character, containing dressed quartzites and particularly dressed argillites, characteristic of this level, which corresponds to a *pre-Lenâpian* culture, more primitive than that of the *Lenâpe* Indians. The peoples associated with these remains may already have been Indians, but they had no polished axes and they probably had no knowledge of the art of pottery. They must have been related to the Palæolithic Man of the gravels rather than to the Lenâpe Man of the surface soil.

3. Below the "yellow drift" come the true and much-discussed "Trenton Gravels," of fluvio-glacial origin and Pleistocene age. Here, and here only, are to be found not argillites alone, but also quartz and quartzites of very rude manufacture, manufacture so rude as to be in itself open to question, although the presence of these objects *in situ* can no longer be matter for doubt. Abbott gives the name *Homo delawarensis* to the fossil man who manufactured these implements found in the Trenton gravels.

The question of American Palæolithic Man cannot be cut

[1] Volk, E., "The Archæology of the Delaware Valley" (*Papers of Peabody Museum*, v., 1910), Cambridge, 1911. With bibliography.

[2] Abbott, C. C., "Ten years Diggings in Lenâpe Land," Trenton, 1912.

short by a blunt dismissal of the problem, for even if the artificial character of the quartzites in the gravels be not admitted, it is evident that the second deposit, with argillite implements and bones of the Musk-ox, dates at least from the end of the Pleistocene period. The only argument of any importance that can be brought against it, and it is only of indirect significance, is that the American caves, which are at once so rich in fossil animals and at the same time in Indian relics, have not yet yielded any trace of a true Palæolithic industry. This fact, however, may be explained in several ways. It forms but a negative argument, and a hundred reasons of this kind are not worth one positive fact like the Trenton finds.

Before touching on the discoveries of human bone remains, there must be mentioned the imprints of human steps, first seen at Carson in Nevada, and then near Lake Managua in Nicaragua. In the former locality the tracks were simply those of large Edentates; in the second the imprints would seem to have been left by feet shod in moccasins. This is all the more curious because the volcanic tuff on which these footprints were made, bears a layer of clay containing the bones of Mastodons.[1]

Imprints of Human Steps.

Examination of the bone-remains is a difficult and complicated task, in the elucidation of which the general studies of Hrdlička[2] will be of special assistance. This anthropologist has devoted himself to the re-examination of all the discoveries. None of them has been spared his criticism; in some cases it is the deposit which is questioned, sometimes the bones themselves are obviously modern; often the evidence is unsatisfactory in every respect.

Osteological Evidence.

As regards the oldest discoveries, I shall do no more than simply remind the reader of their existence, for to us they

[1] Flint, Dr, in Putnam's *Report Peabody Museum*, 1883; *Science*, 7th March 1884.

[2] Hrdlička, A., "Skeletal Remains suggesting or attributed to Early Man in North America" (*Smithsonian Institution. Bureau of American Ethnology*, Bull. 33), Washington, 1907. "Recent Discoveries attributed to Early Man in America" (*Ibid.*, Bull. 66), Washington, 1918. These works contain the history and complete bibliography of each discovery.

are lost evidence, since it is now impossible to estimate their value with full knowledge of the facts. Such, for instance, is the human skeleton from New Orleans (1884); the iliac bone from Natchez (Mississippi), which came from a deposit containing bones of Pleistocene animals, but which the distinguished English geologist, Lyell, after examining it *in situ*, attributed to an Indian burial. The same thing applies to a skeleton found by miners at a depth of 22 feet at Soda Creek in Colorado; to a few human bones exhumed at Charleston in South Carolina; to a skull found in a crevice filled with earth and pebbles at Rock Bluff in Illinois, and so on. The circumstances connected with the famous skull from Calaveras are equally unfavourable (see p. 123).

Other discoveries deserve more attention.

Trenton. Peñon.

There is first the series of bone remains collected at Trenton between 1879 and 1899, by Abbott and Volk. In 1891 I examined some of these relics at the Peabody Museum. A fragment of a right mandible (No. 33327), "found in the gravels at a depth of 16 feet," and water-worn in appearance, impressed me favourably, but a skull (No. 14635) labelled from the "Gas Work," seemed to me to be quite modern. Hrdlička indeed attributes it to an Indian.

There are two other skulls. The first, described as from Burlington County, was accidentally found in a field, and seems to have come from a superficial alluvial deposit. The second, labelled "from Riverview Cemetery," was recovered by a grave-digger at a depth of 3 feet, in a part of the cemetery still unused. The circumstances of the discovery are, to say the least of it, unsatisfactory. Now, these two skulls are equally remarkable from a morphological point of view. Not large, much depressed, with narrow face and large orbits, they indicate a human race very different from the Indians. Hrdlička has shown that they closely resemble the "Batavian" skulls described by various German authors, which have sometimes been compared with Neanderthal skulls; he does not hesitate to ascribe them to two old immigrants of Dutch origin.

In 1899, Volk himself took from the gravels a human parietal bone and a femur. If these bones are really contemporary with the gravel, as Volk affirms, they must belong to the dresser of the quartzites; they do not, however, present any noteworthy structural features.

In 1884, a piece of skull and some other fragments of a skeleton were found in a calcareous tufa at Peñon, a valley in Mexico. By some this tufa is looked upon as of Pleistocene date, while to others it appears to have been recently formed by the hot springs which rise in the neighbourhood.

The "Loess Men."

In 1902, there were discovered at Lansing, in Kansas, the skeleton of an adult man and the jawbone of a child, lying in a silt 20 feet below the surface of the soil. The locality has been examined by many expert geologists, some of whom regard the silt as a true Pleistocene *loess*; others consider it of recent formation due to the flooding of the neighbouring river. The bones are, moreover, identical with those of modern Indians in this part of the United States.

In 1894 and 1906, an earth mound at Omaha in Nebraska also yielded numerous human bones. The discoveries of the latter year, much the most important, comprise the remains of a dozen individuals. They have been examined by the originator of the excavations, Mr Gilder, and by various scientists, including Barbour the geologist, Osborn the palæontologist, and Hrdlička the anthropologist. The bones are derived from different levels, ranging from 0.80 to 2 metres in depth. According to Professor Barbour those from the upper level belong to a burial, but the others must be as old as the mud in which they were embedded. They represent "Loess men."

Hrdlička combated these conclusions. The conditions of the deposition of the skeletons, several of which retained their anatomical relations, indicated burials. At all the levels the bones showed the same coloration, consistency, and surface scratches. None present the slightest trace of fossilization. All exhibit, and at the same spots, incisions denoting certain funeral rites, such as have been found on similar specimens from the tumuli in the region. It is true that certain skulls have somewhat unusual characters: great thickness of the bones, strong orbital arches, and low and receding foreheads, features which have caused them to be compared, mistakenly however, to the Neanderthal and Pithecanthropus skulls. These characteristics certainly led to their being regarded as of very high antiquity. But Hrdlička has shown that such skulls are fairly frequent in the mounds or tumuli of the region, and that in all the remaining features of their morphology the Omaha skulls are the skulls of Indians.

Rancho la Brea.

At Rancho la Brea, near Los Angeles in California, there is a curious bed of Quaternary animals, probably the richest in the United States. This deposit is formed of beds of asphalt, pure or mixed with alluvial products. For several years, Professor Merriam, of the University of California, made in that spot rich palæontological collections of the skeletons of animals of extinct species, especially of *Smilodon* (an animal akin to our Sabre-toothed Tiger, *Machairodus*), skeletons of which are found there in thousands.

In 1914, the American newspapers made a great fuss about the discovery of a human skeleton in the asphalt deposits at Rancho la Brea. The news caused a lively sensation, for a human skeleton, contemporary with *Smilodon* and other extinct mammals, seemed to be a relic of the first importance, capable of throwing much light on the still much-disputed question of fossil Man in America. But soon after, a paper by Professor Merriam focussed the matter in its true light. Once more scientists were faced by a fact without satisfactory geological backing. Owing to the viscosity of the asphalt, layers of this material cannot really be accurately placed from a stratigraphical point of view. The human remains were found at a depth of between 2 and 3 metres, in a sort of chimney filled with asphalt, originating from a great subterranean mass of the same substance, and opening upon the surface. Such infiltrations of cavities or of spaces, in the upper layers of the detritus of the region, may have happened at various periods, and may date from very different ages.

From the palæontological point of view, the facts are just as unsatisfactory. The numerous animal remains, found at the same time as the human bones, do not belong to the now classic Pleistocene fauna of Rancho la Brea, but to the Californian fauna of the present day. From the anthropological point of view, the high antiquity of the skeleton is not confirmed, for it does not differ from skeletons of Indians from Southern California. In this instance again, therefore, there can be no question of a fossil Man of the Pleistocene period.

Discoveries in Florida.

Finally, the most recent discovery, announced in 1916, and still much discussed, was that at Vero, in Florida.

This is not the first time that the presence of fossil Man in Florida has been canvassed. A whole series of finds of human bone-remains were made, from 1852 to 1886, on the shores of Lake Monroe, and on the west coast of the peninsula, to the south of Sarasota, particularly in the neighbourhood of the little town of Osprey.

The majority of these bone-remains were embedded in a hard rock, a kind of ferruginous sandstone, rich in limonite; and chemical analysis indicates that they are deeply impregnated with mineral matter. But the geologist Vaughan has shown that, in spite of their appearance, the fossil-bearing deposits are post-Pleistocene, that the special conditions of fossilization are of no significance in this case, for they are due to the action of numerous iron springs, which consolidate recent sands and rapidly petrify all kinds of objects, particularly Indian pottery. Further, the bones, examined by Hrdlička, do not differ from those of the Indians.

The latest discovery, made by the State geologist Sellards at Vero on the east coast, has formed the subject of numerous reports by the most qualified specialists in the United States.

The subsoil of the town and neighbourhood of Vero is composed from below upwards of the following components:—

(1) Marls of marine origin, containing Pleistocene shells. These marls bear: (2) a marly sand, of river origin, with plant

remains and bones of Pleistocene terrestrial animals, notably of an Elephant (*Elephas colombi*). Above comes (3) a superficial layer, likewise alluvial, but richer in organic matter (humus) and clearly separated from the two preceding ones. In excavating an irrigation canal through all these layers, the bone-remains of two human skeletons were discovered, at a depth of 0.80 metres, in the middle layer and at the base of the upper layer. The middle layer also yielded a blade of flint and some bones which seemed to be worked. The upper superficial layer is rich in shards of pottery, and in objects worked in bone or stone. According to Dr Sellards, the contemporaneity of Man and a Pleistocene fauna in Florida is definitely established by these discoveries. The palæontologist Hay entirely agrees with this opinion, while other specialists combat it more or less energetically.

Of the geologists, Vaughan believes that the superficial layer is of recent age, and that the human remains of the middle layer have not been found *in situ*. Chamberlin at first thought the animal bone-remains of layers two and three had been re-sorted, and that they came from more ancient deposits. Later, he came to adopt the opinion of Berry, who classifies the fossil plants as Upper Pleistocene, while the animal remains seem to be more ancient. This difference is readily explained by the fact of the later survival in Florida than in the northern countries, of the fauna comprising the large Pleistocene animals. Layers (2) and (3) may, then, be much less ancient than Sellards considers them, but Man may quite well have been contemporary with their deposition. MacCurdy seems to support the same explanation, the archæological relics having no character which may be regarded as definitely archaic.

Hrdlička is an uncompromising opponent of Sellards' view. He does not believe in the antiquity of the human bones. Chemical analysis shows that they are still rich in organic matter; the conditions of the deposit support the theory of burial far better than any other theory; and in their anatomical characters, the bones do not differ from the bones of Indians. Hrdlička admits, however, that they may date from the early times of the Indian occupation of Florida.

Such is the picture, almost complete, though somewhat summary, of the facts brought forward in support of the existence of Fossil Man in North America. As we have just seen, according to certain scientists the matter is proved, from the anthropological as well as from the archæological point of view. According to others, not one of the facts adduced can be regarded as conclusive: all the osteological or archæological material must be attributed to the indigenous populations before the Conquest.

Conclusion.

It is difficult for an outsider to take a firm stand on one side or the other, for he can only exercise his judgment on facts collated by others. It seems, however, that an opinion at least intermediate, if not actually capable of reconciling the conflicting views, may be based upon the mass of positive facts acquired up to date. Certainly, the opponents of the theory of the high geological antiquity of man in America have not found it difficult to show that much of the evidence appealed to is incapable of withstanding criticism, and has no scientific sanction; but there are other evidences which, so far, can hardly be disputed, unless by pure negation or *a priori* reasoning. Thus Hrdlička starts from the principle that, according to the laws governing the general evolution of mammals, fossil man must differ from living man. Without absolutely condemning this view of the matter, it may be said that the absolute denial of the geological antiquity of human bone-remains because they resemble similar bone-remains of Indians, is going much too far; for such an assertion simply begs the question, and to give prime importance to this argument results in negation pure and simple.

It would seem that the radical difference in opinion arises partly from a misunderstanding. It is not necessary that man or any other creature, in order to be regarded as fossil, should no longer be represented in living nature. This fact, indeed, is certain, that not the least trace of a man differing structurally from modern man has been found in America, and yet this should not surprise palæontologists, who know well that America is not the land of origin of the higher Primates; but the resemblance of the old skeletons already discovered to the skeletons of Indians, does not prove that these old skeletons date from recent or Holocene times. If, in general, American Man presents a certain combination of common physical characters, this community must date far back into the past; and admitting that America was populated by migrations, mainly if not entirely from the great stem of the Yellow Races, the importance and extent of these migrations, the settling and differentiation of the populations over the whole surface of the two Americas, must have required a great lapse of time. Now, it certainly appears that the dressed stones from Trenton and other localities witness to the existence of Man in America before the dawn of modern geological times. On the other hand, among all the discoveries of human bones, some have been discovered with well-founded guarantees of authenticity and high antiquity. It becomes a question of determining this antiquity, and here the extreme camps might come nearer agreement. Among all the relics, dressed stones or bone-remains, not one seems to me to date very far back in the Pleistocene. The impression forces itself upon one that the best of them can be referred only to a closing period in this epoch, something like our Upper Palæolithic, or like our transition period between Palæolithic and Neolithic. On this hypothesis, everything might be satisfactorily explained. The mass immigrations, starting from Asia, could not take place while the ice-sheets covered the greater part of North America. They could have been possible only during an inter-glacial period, and more probably after the definite retreat of the glaciers. As soon as the ways were open, following on this event, Man was able to use them to invade step by step the American continent, where, for more than a million years, no animal of the Primate group had lived. In the present state of our knowledge, this explanation of the problem of the antiquity of Man in America seems to me the most rational. But we must not disguise the fact that this solution is very uncertain.

South America.

General Remarks.

South America, with its very ancient central massif or Brazilian plateau, with its immense river basins, with the Cordillera mountain range 9000 kilometres long, beaconed and crowned by enormous active volcanoes, forms, from the physical point of view, a very special and highly autonomous continental unit.

This autonomy is no less marked from the point of view of living things. South America constitutes one of the chief biogeographical divisions of the globe, the *Neo-tropical Region.* Its virgin forests and grassy plains show special plant associations extended over vast areas; it has also its particular forms of insects, fishes, and reptiles; it is the home of a multitude of birds of gorgeous plumage, such as the humming-bird, parrots, cock-of-the-rock, and so on. It is the almost exclusive home of the marmosets, of monkeys with prehensile tails, vampire bats, opossums, peccaries, llamas, and particularly of the edentates—ant-eaters, sloths, and armadilloes.

This special feature is a legacy from the past, a fact which we see confirmed when we retrace our steps through the geological ages. During Quaternary and Tertiary times, there were, besides enormous edentates such as *Megatherium* (Fig. 233), *Mylodon,* and *Glyptodon* (Fig. 234), so amazing to the early palæontologists but in reality only giant forms of modern sloths[1] and armadilloes, a crowd of other mammals, so different from those of the northern hemisphere that special orders had to be created for them; such are *Typotherium, Toxodon, Macrauchenia,* and still earlier, *Astrapotherium, Pyrotherium* and others. On the other hand, there has never been found in the rich Tertiary fossil-bearing beds of Patagonia, the least trace of animals which could be likened to our proboscidians, artiodactyls, perissodactyles, placentary car-

[1] It was the great French scientist, Cuvier, who recognized the true nature of *Megatherium* on examining a skeleton sent to Madrid in 1789. The King of Spain, Charles III., then ordered officials in the colony to send him one of these animals alive or at least stuffed. It goes without saying that the royal command was unfulfilled!

nivores and higher monkeys. Towards the end, during Pliocene and Pleistocene times, there were, however, mastodons, tapirs, horses, deer, and large felines, but these were not native animals; they had come from elsewhere, mainly from North America. Albert Gaudry was able to state that, after the beginning of the Tertiary era, the evolution of mammals did not proceed in these regions as in the northern hemisphere.

This is a fact of considerable importance from the point of view of the palæontological history of Man. Since South

FIG. 233.—Skeleton of *Megatherium*. Actual height 3·60 metres. Palæontological Gallery, French National Natural History Museum.

America has always lacked highly developed Primates and higher monkeys, we cannot expect to discover in this continent representatives of early Man or of his immediate ancestors. We shall see, however, that a naturalist of great merit, Ameghino, adopting exactly the opposite view, constructed a whole system of human genealogy.

South America was never buried beneath Quaternary ice-sheets. Even in the southern part of the continent they remained localized in the chain of the Andes or in the neighbouring regions. But the great climatic changes of the later geological periods have, none the less, played their part and left well-marked traces. First, there originated from the breaking down of the Cordillera range and spread out around the base of the range, great deposits laid down by glacial rivers and torrents. These gradually merge into formations which cover all the low portions of the continent with an immense coat of detritus: sands, clays, silts and so on, produced sometimes by streams, sometimes by the rush waters of floods, sometimes by the aerial transport of volcanic cinders, or by temporary encroachments of the sea, and often by the combined action of these different factors.

FIG. 234.—Skeleton of *Glyptodon*. Actual length 2·90 metres. Palæontological Gallery, French National Natural History Museum.

The Pampas Formations.

The formations to which belong the Pampas deposits, very similar to our silts or loess, are of special interest to us since they have yielded the majority of the anthropological discoveries of which I must give an outline.

All these deposits are rich in fossils. Unfortunately, because of differences in the faunas, which have in this region a special character of their own, it is difficult to establish any time correlations or synchronisms with North America, and still more difficult with Europe. Palæontologists were at first deeply impressed by the curious character and giant forms of the fossil animals discovered in the superficial deposits of South America, and, through remote analogy with Europe, they were led to regard these deposits as very ancient. This was the line followed by the Argentine scientist Ameghino.

Nowadays the tendency is rather to regard them as more recent, and this is consistent with purely geological data. Many curious fossil beasts from the subsoil of the Pampas may have lived up to a period relatively not far distant from our own times. In the case of some of them this seems even to have been proved, as we shall see directly.

The Pampas formations were at first regarded as a unit mass. Then an attempt was made to divide them, based on the one hand on their physical characters, and on the other, on their palæontological content. The division was carried to excess by Ameghino, who distinguished a dozen stages. But certain good geologists, like Roth, Steinmann, Burckhardt, Bailey Willis, far from following this example, only make a small number of divisions in the thickness of the Pampas formations.

The early observers, d'Orbigny and Darwin, regarded these deposits as of recent formation. Ameghino dated them from much more remote times, regarding the oldest as belonging to the Miocene. At the present day, geologists and palæontologists seem to be agreed in regarding them as a complex, ranging from the Pliocene to the present period. Thus the superficial deposits of the Pampas may be divided into four stages, the characters of which are indicated below.[1]

HOLOCENE or RECENT		*Post-Pampean.*
PLEISTOCENE	*Pampean*	Upper, *Bonarean.*
		Lower, *Ensenadean.*
PLIOCENE	*Pre-Pampean*	or *Hermosean.*

Below these are to be found in Patagonia more ancient formations belonging to Miocene and Oligocene times.

The *Hermosean*, readily seen at Monte Hermoso, near the Bay of Bahia Blanca, is sometimes referred to the Pampean as *Lower Pampean* or *Pre-Pampean.* It is formed of sandy or clayey layers of a ginger-bread-brown colour and of volcanic ashes. Ameghino regarded it as Miocene; most geologists attribute it to the Pliocene.

[1] The bibliography is large and too special to be given here. I simply recommend to anthropologists the clear summary to be found in a charming little book by Outes and Bruch, *Los Aborigenes de la República Argentina*, Buenos Aires, 1910.

The true Pampas or *Pampean* deposit is superimposed upon the Hermosean, and almost everywhere forms the covering of the Argentine plain. It is essentially of sub-aerial origin, is warm brown in colour, and represents, with its calcareous concretions (*tosca*), a formation similar to our loess; but occasionally there are to be observed fresh-water deposits, beds of volcanic ash, and, towards the shore, intercalated beds of marine sediments. It may be fairly easily divided into two stages.

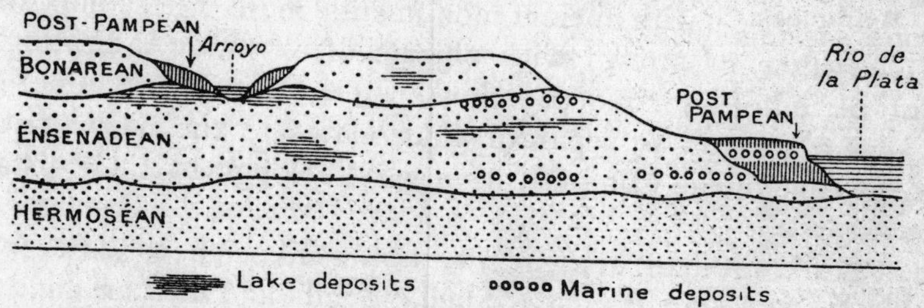

FIG. 235.—Diagrammatic Section of the Pampas Formations between Buenos Aires and La Plata. (After Outes and Bruch.)

The lower stage, or *Ensenadean*, is formed of plastic clays, compact, light brown in colour, containing calcareous concretions with concentric layers or branched. It is from 15 to 20 metres deep in the region of La Plata.

The upper or *Bonarean* stage is formed by a less compact unstratified clay, a true air-borne loess, of yellow colour, containing concretions of fantastic shapes ("fairy stones" or "*poupées*"). This portion of the Pampas formations is most easily observed, as it crops up everywhere from under the humus. It, likewise, sometimes attains a depth of 20 metres.

Above the Bonarean at certain points are to be seen certain deposits which Ameghino regarded as still Pleistocene, but which may be correlated with the European Holocene, for they really represent modern times. They are sedimentary, grey in colour, formed in depressions of the soil, chiefly on the banks of water-courses. Some deposits show signs of slight encroachments of the sea; and there are also coastal dunes.

Palæontology shows an imperceptible transition between the different stratigraphic series, from the Pliocene fauna of the

Hermosean to the exclusively modern fauna of the Post-Pampean.

After these general statements, we may now enter upon our main subject; and continuing the method I have hitherto followed, I will first examine the archæological discoveries and then the discoveries of human bone-remains.

Archæological Evidences.

The whole of South America is rich in archæological relics. From the Antilles to Cape Horn, it is covered with ancient monuments, more or less imposing in character, or strewn with objects of all kinds indicating a long period of human occupation. Here, as in North America, it is very difficult to separate what relates to the native tribes at the time of the Conquest from what belongs to their distant ancestors.

Certain human settlements, however, seem to date from fairly far back in the past. Shell mounds and kitchen middens are numerous, both on the shores of the Pacific, in Chili[1] and Peru, and in Guiana, Brazil, the Argentine Republic, and Patagonia as far as Tierra del Fuego. In this region, where the present day natives are still in process of forming them the shell mounds have been examined by Lovisato.[2] They are sometimes of considerable size (more than a kilometre in length); the shells they contain are stronger than those of the same species of living molluscs; and it would seem that since these mounds were formed, the level of the sea has altered. All these circumstances lead us to attribute to them a relatively high antiquity, a matter of considerable interest because of their position in the extreme south; but there is nothing to prove, as has been claimed, that they date from beyond the modern geological period.

This also applies to the *paraderos*, the sites of villages and burial mounds of the ancient Patagonians.[3] The archæological material belonging to them comprises numerous stone objects, but though some specimens resemble Palæolithic work, th

[1] See Appendix, p. 486.

[2] Quoted by Keane, *Ethnology*, p. 96.

[3] Verneau, R., *Les anciens Patagons*, Monaco, 1903. Outes, F., "La edad de piedra en Patagonia" (*Anales del Museo de Buenos Aires*, 1905). With a summary French, and a bibliography.

general character of the knives and scrapers, arrow-heads, mortars, polished stones, pottery, etc., clearly corresponds to the general Neolithic culture of the Indians before the Conquest. It is permissible, however, to adduce, as evidence in favour of a certain degree of antiquity, changes of climate which have supervened since these centres of habitation were abandoned, in a country now arid and desolate.

Similar layers in Brazil, known as *sambaquis*, may also be very ancient, without, however, dating from the Pleistocene.

These are surface deposits. Numerous finds have been made even in the depths of the Pampas formations. In the short description which I must give of them, I shall group them under four heads: (1) Stone objects; (2) Cinders and baked clay, said to be relics of hearths; (3) Bones, split, engraved, worked, burnt, and so on; (4) Other archæological facts of such a nature as to establish the contemporaneity of Man and of lost species of animals.

Stone Relics.

Speaking generally, stone objects are rare in the depths of geological layers more ancient than the Post-Pampean or recent deposits. And this is a significant fact, when we contrast it with the abundance of human bones exhumed from the same deposits. Outes, in his fine work on "The Stone Age in Patagonia," tells us that superficial layers containing stone implements of Palæolithic manufacture are fairly numerous in Patagonia. Ameghino is said to have taken two dressed pieces of quartz from a rather deep bed in the gravels of the brook *Observación* (Fig. 236), a fact which justifies one in attributing to the Pleistocene (Upper Pampean) the generality of this industry, with its oval stones, dressed on both surfaces and greatly resembling those from Trenton in the United States. The objection may be made that the Pleistocene Age of the gravels has not yet been proved, and that similar object sare numerous in Post-Pampean deposits.

In addition to this discovery, which is by far the most important, there have been recorded a great number of isolated finds of quartzites and flints, apparently showing signs of intentional dressing, and sometimes even of real implements, in the Upper Pampean, the Lower Pampean, and even in the

Hermosean. All this proves little; either we are dealing merely with formless stones, split by natural agencies, or with simple flakes, or even with objects identical with those of the Indians. Nowhere has there been found *in situ* a collection of stone objects belonging to a settlement or factory.

Ameghino described an industry of flaked or split stone, which he attributed to his *Homo pampæus* of the Lower Pampean, and which he declared to be more primitive than the European eoliths. In the Hermosean, considered by him, as we have said, to be of Miocene age, he found an industry of

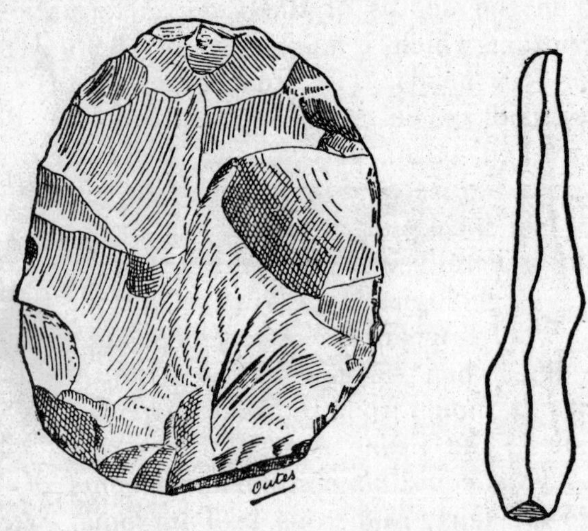

FIG. 236.—Dressed Quartz from the brook Observación. Four-fifths natural size. (After Outes.)

"broken stone, the simplest that could possibly be imagined." All these stones are not eoliths, as I had at first thought. Examinations made on the spot by Hrdlička and Willis showed that the bed in which these pebbles rested was in recent sand-dunes, and that the pebbles themselves represent only the waste products of manufacture of an Indian industry barely a few centuries old—the authors put it at "a century old!"

I must add that Ameghino[1] has recorded facts of the

[1] Ameghino, F., "Une nouvelle industrie lithique" (*Anales del Museo de Buenos Aires*, xx., 1910); "Vestigios industriales in el Eocene, etc." (*Congreso cientifico internacional americano*, Buenos Aires, 1910).

same kind from the much more ancient deposits of Patagonia, dating, according to him, from the Oligocene and even from the Eocene! They would seem to have been rudimentary implements manufactured and used by the small apes of these remote periods, the supposed ancestors of the human kind!

Cinders and Baked Clay.

There have long been known in the Pampas formations, cinders and hard red rocks, resembling brick or baked earth. They are usually in the form of little pebbles, often of large masses, sometimes forming actual beds or interstratified layers.

They had hitherto been regarded either as materials of volcanic origin or as products of natural fires, when in 1875, Ameghino,[1] affirming that the pieces of baked earth were sometimes associated with burnt or striated bones and stone implements, considered them as relics of ancient human hearths, that is to say, as having "an anthropic origin." Since that time, Ameghino has often returned to this subject; he has described a great number of localities where ashes and baked earth are to be found, at all levels in the Pampean and even in the Hermosean formations.[2]

Ameghino's view has been keenly discussed. Appeal has been made for the co-operation of mineralogists of both continents in the solution of the problem. Here I cannot summarize this prolonged polemic,[3] but state only what seem to me to be the results of the different enquiries.

The ordinary Pampas formations are very rich in minerals of volcanic origin, and this is not surprising, since these deposits have originated through the breaking down of the chain of the Andes. The cinders or scoriæ are clearly of

[1] Ameghino, F., "Nouveaux débris de l'Homme et de son industrie" (*Journal de Zoologie*, iv., 1875).

[2] Ameghino, F., *La antigüedad del Hombre en el Plata*, 2 vols. Paris, Buenos Aires, 1880-1881; *Contribución al conocimiento de los mamiferos fósiles de la República Argentina*, Buenos Aires, 1889. And in particular, "Productos piricos de origen antrópico" (*Anales del Museo de Buenos Aires*, xix., 1909); "Enumération chronologique et critique des notices sur les terres cuites et les scories anthropiques" (*Ibid.*, 1910).

[3] See especially: Outes, F., Ducloux, E. H., and Bücking, H., "Estudio de las supuestas e corias y tierras cocidas" (*Revista del Museo de La Plata*, xv., 1908). Bailey Willis, Wright, E., Fenner, N., and Hrdlička *in* Hrdlička's *Early Man in South America*, 1912.

volcanic origin, but they have undergone more or less extensive transport. The burned earth is of exactly the same chemical composition as the Pampean silts; apart from certain examples, which exhibit a case of fragments of loess becoming adherent or forming concretions, these baked clays must have been submitted to the prolonged action of great heat (800° to 1000° C.). According to some authorities, they may be volcanic tuffs; according to others, they are produced by prairie or forest fires, plant and tree impressions being still noticeable on the baked earth. But do fires of this kind produce such results? Certain authorities deny it; other observers are strong in affirmation. In any case, the question to be answered is whether these fires were spontaneous, caused by lightning, or whether they were originated by Man. The latter theory has few supporters among scientists who have studied the question on the spot.

Apart from the resemblances of those specimens of ashes and baked earth which I have seen, to the re-sorted volcanic products from the Central Massif of France, I, for my part, am struck by the abundance of these substances at all levels of the great Pampas formations (Lehmann-Nitsche says that "he found them at every step"), by the great extent of the layers in which they are found, and by the stratigraphical regularity of certain layers containing ashes or baked clay—all facts which scarcely agree with the theory of their human origin. To be sure, we have still to find satisfactory explanations of these facts, but it would seem that the theory of their "anthropic origin" must now and henceforth be dismissed.

Bones of Fossil Animals, Used, Worked, or Cut.

The whole long series of bones, burnt, split, striated, engraved, polished, or dressed, which have been collected from the different levels of the Pampas formations, even from the Hermosean, and have been figured or described at length by Ameghino,[1] do not seem to have any greater value as proof, and may in general be accounted natural objects.[2]

[1] Ameghino, *La antigüedad del Hombre.*

[2] See Lehmann-Nitsche, R., "Nouvelles recherches sur la formation pampéenne et l'Homme fossile de la République Argentine" (*Revista del Museo de La Plata*, xiv., 1907).

A certain number of observations have been advanced in support of the contemporaneity of Man with vanished species, About forty years have passed since Ameghino and Roth spoke of the carapaces of Glyptodons buried in an inclined or inverted position, which could only be explained by human intervention. Quaternary hunters, finding in the Pampas neither cave nor shelter of any sort, would make use of the carapace of a dead Glyptodon. They would empty it, place it in a horizontal or inclined position, then, hollowing out the earth beneath, they would obtain a shelter, uncomfortable but nevertheless valuable in that distant period. Ameghino saw that the soil was heaped up and hardened underneath some of these carapaces. Near by, he repeatedly found bones of different animals, intentionally broken, as well as charcoal, ashes, and flints apparently dressed. Finally, as we shall presently see, a human skeleton was found under one of the carapaces.

On several occasions there have been collected, in more or less close association with the bones of large extinct animals, such as *Glyptodon*, *Mylodon*, and *Toxodon*, stones which might have been used to slay them, or teeth of these fossils shaped as implements. The majority of these finds, chiefly due to Ameghino, are of doubtful value. The last of them caused a great sensation, and is still the subject of lively discussion.

In 1914, Carlos Ameghino,[1] brother of the late palæontologist, while exploring, near Miramar, a layer of an age intermediate between the Hermosean and Lower Pampean, and regarded by him as Miocene, found a femur of *Toxodon* which retained, actually in the body of the bone at the level of the great trochanter, a quartzite point which must have transfixed it. This point, of the modern form known as "willow-leaf," might have been the head of a lance used to attack the animal from behind. Shortly after, and not far from the spot of his first find, C. Ameghino and Dr Keidel found a portion of the vertebral column of *Toxodon*, with

[1] Ameghino, Carlos, "El femur de Miramar" (*Anales del Museo de Buenos Aires*, xxvi., 1915). Articles in *Physis*, ii. and iv., etc.

two dressed quartzite arrow-heads embedded between the vertebræ. These discoveries seemed at first sight to be perplexing. In Europe, we were debarred from criticism, through lack of having seen the specimens and their beds. But they were discussed in America. While the geologists in La Plata Museum affirm that the objects were found *in situ* in the "Miocene" layer, and that they were indeed contemporary with this layer, Colonel Romero,[1] though a warm admirer of Ameghino, declares that they came from the upper beds, which formed the site of a *paradero* or ancient Indian settlement, and that they were found to-day in the Tertiary bed only as a consequence of disturbances and re-sortings which that bed had suffered. The archæological data support this conclusion, for the same Tertiary bed yielded dressed and polished stones, *bolas* and *boleadores*, identical with those used as missiles by the Indians. An excellent ethnographer, Boman,[2] recording these facts, wrote: "The chief difficulty lies in this: without exception, all the objects exhumed from the Chapalmalean bed at Miramar are absolutely identical with similar objects found everywhere on the surface and in the upper beds of La Plata and Patagonia. Could it be possible, then, that Man lived in the Pampas, from the Miocene to the Spanish Conquest, without changing his habits and improving his primitive industry in any way whatsoever?"

On the other hand, the significance of the archæological fact, the presence of an arrow-head in a Toxodon vertebra, is not beyond all criticism. It may, perhaps, be a case of trickery, for Colonel Romero is of opinion that the arrow-head has been embedded in the bone after it had already become fossilized. And, even supposing it were true that the arrow and the vertebra were indeed contemporaneous, one would still wonder if the fact could not be better explained by post-dating the disappearance of the extinct species, than by attributing so extraordinarily high an antiquity to Man in America?

[1] Romero, A., "El *Homo pampæus*" (*Anales de la Soc. cientifica Argentina*, lxxxvi., 1918).

[2] Boman, E., "Encore l'Homme tertiaire dans l'Amérique du Sud" (*Journal de la Soc. des Américanistes de Paris*, vol. xi., 1914-19).

Neomylodon.

The theory I have just mentioned can be supported by another discovery to which I have already referred, made in the extreme south of Patagonia. About 1899, various explorers, Ramon Lista, Moreno, Nordjenskiöld and Hauthal, found at Ultima Speranza, in a cave named *Cueva Eberhard*, large shreds of skin full of ossicules and covered with the hair of an animal (Fig. 237) to which various names have been given (*Glossotherium* and *Grypotherium*), but which really seems to be the descendant of *Mylodon* of former times, that is to say a *Neomylodon*.[1]

FIG. 237.—Piece of Skin of *Neomylodon*, with hair and dermic ossicules. Natural size. Palæontology Gallery, National Natural History Museum, Paris.

Even its excrement has been collected. All these remains are so well preserved that they can only have belonged to animals quite recently dead. They were associated with human bone-remains in a layer of "manure" more than a metre thick, which led to the belief that *Neomylodon* was domesticated. Furthermore, the Indians have preserved in their legends the tradition of a great shaggy animal armed with strong claws, and contemporary hunters allege that they have seen and followed it in its nocturnal wanderings. If so the monster may be still alive! One thing appears to be certain, that its bones bear traces of the hand of Man: "It would seem that the animals had been killed by blows upon the head, dealt with large stones. They would then be cut up and eaten, and this would explain the very broken condition of their bones."

[1] Hauthal, R., Roth, S., and Lehmann-Nitsche, "El mamifero misterioso de la Patagonia" (*Revista del Museo de la Plata*, ix., 1899). Lehmann-Nitsche, R., "Zur Vorgeschichte der Entdeckung von *Grypotherium*" (*Naturwissenschaftliche Abhandlungen*, Heft 29, 1901). With bibliography.

The same cave yielded burnt bones and pieces of skin of a horse belonging to an extinct species, *Onohippidium*. Finally, Outes declares that in the neighbourhood of the river Salado and of the brook Tapalqué, the bones of a sort of Glyptodon (*Dœdicurus*) and of the large Tiger, *Smilodon*, are found in a state of remarkable preservation.

If facts of this kind were to be multiplied, they would lead to a general and remarkable post-dating of the superficial deposits of South America, and notably of the Pampas formations; and this would help to explain many of the facts which seem to be contradictory, and to reconcile to a certain extent the most divergent opinions.

Human Bones. Lagoa-Santa.

The palæo-anthropological discoveries in South America have been made, some in caves in Brazil, and others in the Pampas formations of the Argentine Republic.[1] We shall study them in order.[2]

Lund, a Danish naturalist, devoted forty-eight years of his life to the study of the fossil fauna of Brazil. He claims to have explored more than 800 caves in the province of Minas Geraes. Between 1835 and 1844 he exhumed human bones from six of these caves. One of the caves, situated on the shore of Lake Sumidouro, near Lagoa-Santa, yielded the remains of about thirty individuals, young and old, lying mingled together with the bones of animals of living and extinct species. Lund at first regarded these discoveries as affording little proof of the contemporaneity of Man with the extinct species; the cave is subject to periodical flooding by the waters of the lake which might have resulted in the inter-

[1] Apart from these territories, I may record two more finds. First, that of a lower jawbone, found by Dr Montané under a stalagmite deposit in the cave of Sancti Spiritu in Cuba. This jawbone seems to lack a chin. Ameghino has named it *Homo cubensis*.

Again there are the human bones taken from gravels near Cuzco in Peru, by a scientific expedition from Yale College, Newhaven. American scientists at first regarded these bones as of Pleistocene age. They then very loyally recognized the fact that they were simply relics of a burial.

[2] Hrdlička, A., "Early Man in South America" (*Smithsonian Institution. Bureau of American Ethnology*. No. 52, Washington, 1912). In this work will be found the history and complete bibliography of all the discoveries, edited and critically reviewed with the collaboration of distinguished specialists, the archæologist Holmes, the geologist Bailey Willis, and the mineralogists Wright and Fenner.

mingling of deposits, and furthermore really ancient layers, seemingly undisturbed, contained no human bones. He thought that the cave had served as a burial-place. Later, he became less hesitating. In 1844, he had become almost convinced: "The occupation of South America by Man *probably* dates back to geological times." An examination of the skulls enabled him to add: "The race which then occupied this part of the world was already that which Europeans found here at the time of the discovery [of the continent]."[1]

With the exception of a skull left in the Historical and Geographical Institute of Rio de Janeiro, the greater part of the human remains collected by Lund were sent to Copenhagen, where they may still be seen. These relics have been examined by many anthropologists: Lacerda and Peixoto, Blake, Reinhardt, de Quatrefages, Kollmann, Soren Hansen,[2] and others. De Quatrefages[3] regarded the Sumidouro skull as the type of a particular race, the *Lagoa-Santa race*, clearly characterized by the shape of the head, which was both long and high. The present-day Botocudos originate from this type, which also enters into the constitution of many of the South American peoples. Later, Ten Kate recorded its presence in the Indians of Lower California, and its existence was likewise affirmed among the "sambaquis" [prehistoric tombs] of Brazil, and in the old cemeteries of Tierra del Fuego and of Patagonia (Verneau). In 1908, Rivet[4] again met with it in Equador, on the Pacific slope. This race has, then, played a great part in the early peopling of South America.

Rivet thus summarizes the distinguishing features of the Lagoa-Santa type: small, dolichocephalic skull, very highly vaulted; face small, forehead broad, nose medium, orbits medium, palatal vault very large. Seen in full face, the skull presents a characteristic pyramidal appearance, due to

[1] Lund, P. W., "Lettre à Rafn, 28th March 1884" (*Mém. de la Société royale des Antiquaires du Nord*, 1845). Reproduced in part in *Matériaux*, xvii., 1882-1883.

[2] Hansen, Soren, "Lagoa-Santa Racen" (*Museo Lundii*, i., Copenhagen, 1888).

[3] Quatrefages, A. de, *Introduction à l'étude des races humaines*, Paris, 1887.

[4] Rivet, Dr, "La race de Lagoa-Santa chez les populations précolombiennes de l'Équateur" (*Bull. et Mém. de la Soc. d'Anthropologie de Paris*, 1908).

the widely separated zygomatic arches (Fig. 238). Following de Quatrefages, Rivet lays stress on the points of resemblance between the Lagoa-Santa skulls and the Papuan skulls from Melanesia, resemblances which seem to him to indicate close relationship. He admits the possibility of Melanesian migrations to the Californian coast.

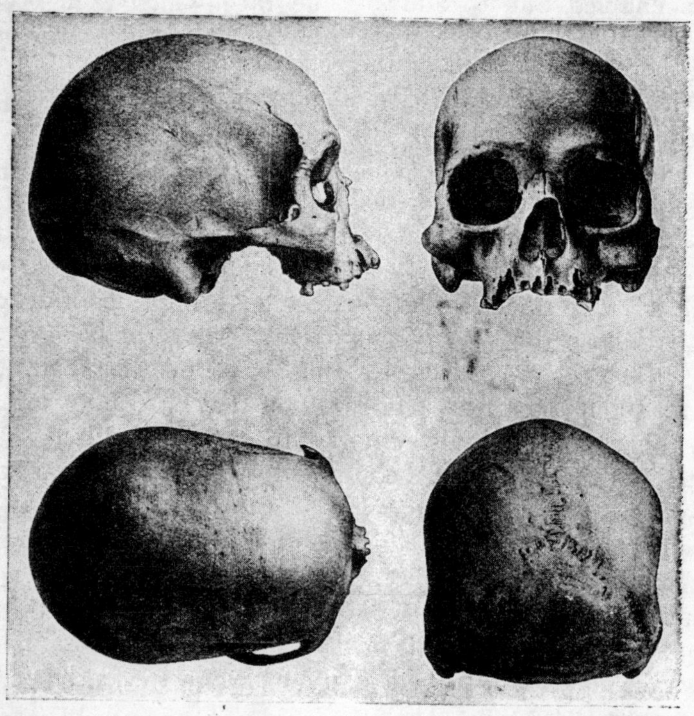

FIG. 238.—One of the Human Skulls from the Sumidouro Cave, seen in profile, in full face, from above, and from behind. One-fourth natural size. (After S. Hansen.)

According to Hrdlička, the presence of thirty individuals in a single bed seems to prove that it was a case of burial. The skulls possess all the fundamental characters of the "American race." The relationship with the Papuans, if it exists, can only be extremely distant.

To sum up, in the present state of matters, and pending new investigations and discoveries in the Brazilian Caves which will bring forth clear and precise facts, it seems evident that a certain antiquity must be granted to the Lagoa-Santa race, whose importance as an ethnical stock is beyond doubt. Does this antiquity date back to Pleistocene times? That is possible, but it has not been scientifically proved.

Discoveries made in the Argentine Republic by different investigators are numerous, and still more numerous are the works published about these discoveries, particularly by Ameghino, Roth, Lehmann-Nitsche, Outes, Mochi, Hrdlička, and others.

Argentine Republic.

Before describing the discoveries, I must introduce to the reader, in a few words, the scientist who has played the principal part in the study of Argentine palæontology, and whose discoveries and theories made a great sensation throughout the scientific world.

Florentino Ameghino, who died in 1911, was Director of the Natural History Museum of Buenos Aires and a great worker. He early made himself known by publications on palæontology and prehistory. After having devoted a large work to the descriptions of the animals of the Pampas formations, he turned his attention to the study of more ancient fossils, which his brother, the explorer Carlos Ameghino, excavated for him from the Tertiary deposits of Patagonia. In this field he has made important palæontological discoveries on which there is here no space to dwell.

In the domain of prehistory, as we have already seen, Ameghino was no less active. From the commencement of his research work, about 1875, he devoted himself to the establishment of the existence of Fossil Man in South America. In 1881, he published an important work in two volumes entitled *La antigüedad del Hombre en el Plata*, in which he put forward a mass of facts and observations. He has since written a great number of papers on all kinds of discoveries, some of which we have already described and others we shall now discuss.[1]

[1] Ameghino has himself summarized his various works in: *Geologia, Paleogeografia, Paleontologia, Antropologia de la Republica Argentina* (special number of the newspaper *La Nacion*, 25th March 1910). Ambrosetti, J.-B., has published a biography and complete list of Ameghino's works (*Anales del Museo de Buenos Aires*, xxii., 1912).

In the end Ameghino became convinced that South America was the land of origin of all the groups of mammals. Since Man himself could not appear simultaneously over the whole surface of the globe, said he, "he must have had a beginning and a starting-point. Now, as in other regions of the earth, no human bones of Tertiary Age have been found, we must

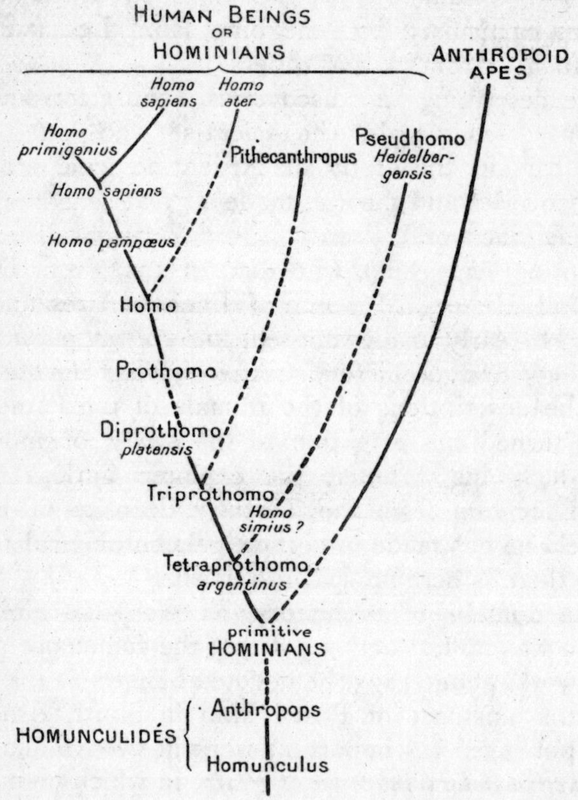

FIG. 239.—Human Genealogy according to Ameghino.

conclude that Man's place of origin and centre of dispersal is the southern portion of South America, where his fossil remains dating from the Tertiary are found in abundance." This theory took definite shape in his mind, and he sought to bring to its support new observations of every kind, whether stratigraphical, palæontological, or archæological. Ameghino ended by constructing a genealogical tree of human beings, starting from his *Homunculids*, minute monkeys from the Lower Tertiary in Patagonia (see p. 80), and ending in the genus *Homo*, by way of several intermediate genera, *Tetraprothomo*, *Triprothomo*, *Diprothomo* and *Prothomo* (Fig. 239). Having conceived these transitional forms in his mind, Ameghino then set himself actually to discover them in the Pampas formations.

Let us now review the discoveries of human bone-remains, proceeding from the most recent to the most ancient deposits.

Discoveries in the Upper Pampean.

The deposits of the Upper Pampean or *Bonarean* which, as we have said, spread over almost the whole surface, yielded numerous discoveries of human bones.

In 1864, a naturalist-traveller, Séguin, found on the banks of the *Rio Carcaraña*, to the north of Rosario, in the province of Santa-Fé, the remains of four skeletons which, together with a collection of fossil bones of animals from the same source, he sold to the Paris Museum. Gervais[1] described and figured nine separate human teeth and some dressed stones, identical, however, with those of Indians. This find is open to criticism. Moreno, Burmeister, and Hrdlička deny its antiquity, while Ameghino, Roth, and Lehmann-Nitsche affirm the contemporaneity of the human bones with the extinct species of the Upper Pampean. Séguin's discovery is the first of this kind which was made in the Argentine.

The succeeding discoveries, due to Ameghino, were made, in 1870 and 1873, in a terrace of the little valley of the *Arroyo de Frias*, near Mercedes. They consist of human bones in a state of great deterioration, found in company with the remains of *Glyptodon*, and with pierced and incised bones of animals, dressed stones, charcoal, etc.

In 1876, Santiago Roth first observed a human skeleton at *Saladero* near Pergamino, in circumstances which suggest a burial. Five years later, in 1881, Carl Vogt[2] announced that Roth had just taken from deposits at Fontezuelas or Pontimelo, near the Rio Arrecifes in the Province of Buenos Aires, an almost entire skeleton, of small stature, which lay

[1] *Journal de Zoologie*, ii., 1873.
[2] Vogt, C., *Bull. de la Soc. d'Anthrop. de Paris*, 1881.

under a well-preserved Glyptodon carapace, or at least close by. The skull belongs to the Copenhagen Museum. S. Hansen attributes it to the Lagoa-Santa race, and declares that we cannot regard as absolutely proved the contemporaneity of the Man and of the Glyptodon, as Roth related it.[1] This opinion, shared by Hrdlička, is not that of Lehmann-Nitsche, who believes in the antiquity of the Fontezuelas skeleton.

In 1889, Ameghino recorded the discovery, in the same region and in the same deposit, of a skull known as the *Arrecifes* skull. The circumstances of its occurrence are insufficiently established; the skull, somewhat impregnated with mineral matter, is of the same type as the preceding.[2]

In the same year, Ameghino described the so-called *Samborombón* skeleton, found by a naturalist-traveller from the Museum of Buenos Aires in a lake deposit interbedded in the Pampean formations. At a higher level, the remains of a large fossil edentate, *Scelidotherium*, were collected. The human skeleton, now in the Museum of Valencia, Spain, presents the double peculiarity of having eighteen dorso-lumbar vertebræ and a perforated sternum. Kobelt has named it *Homo pliocenicus*.

The skeleton from *Chocori*, described by Lehmann-Nitsche in 1907, was found nineteen years previously by an assistant from the La Plata Museum, near the *Arroyo Chocori* on the Atlantic coast. The only point, and that not a definite one, favouring a certain degree of antiquity, is the presence of some calcareous incrustations.

Several series of human bones, collected from various places in the neighbourhood of Ovejero, in the province of Santiago del Estero, belong, according to Ameghino, to two distinct races, of which one is a pygmy raçe (1.30 metres [4 ft. 3.2 ins.] high) akin to the Negritos of Africa and Asia. From an examination made on the spot by Hrdlička and Willis, in the course of which the scientists themselves found another skeleton, it appears that these are burials of Indians whose descendants

[1] This account, both the original text and a French translation, is to be found in Lehmann-Nitsche's *Nouvelles recherches*. . . .

[2] A bibliography of this, and also of the following discoveries in the Pampas formations, will be found in the works of Lehmann-Nitsche and Hrdlička.

are still living in the same spots. The bone-remains do not represent two races, and do not in any way resemble the Negritos. The deposit which contained them is a wind-borne formation of recent date.

If we return to the Atlantic shore of the province of Buenos Aires, we come upon a new series of finds, which have become celebrated owing to Ameghino's interpretations of them.

There is, first, the skeleton found in 1910 near the settlement of Villanueva in a *barranca* or steep bank of the *Arroyo Siasgo*, a tributary of the Rio Salado. According to Ameghino's account, the skull of this skeleton, very small, very long, very low in front, and very high in the hinder portion, presents in these respects very primitive and even simian characters. Its possessor must have held his head bent downwards, whence the name *Homo caputinclinatus* given by Ameghino to this new species, in which the Samborambón skull falls also to be included. Unfortunately, Mochi and Hrdlička find this specimen to be only the skull of a child artificially deformed, and Willis declares it to be a case of burial in a recent deposit of wind-borne origin.

Much farther to the south, on the Atlantic coast, to the west of Miramar and of the *Arroyo Chocori*, which I have already mentioned, there flows the *Arroyo del Moro*. On its desolate shores, in 1909, a sailor exhumed two human skeletons. Ameghino described them as belonging to a species distinct from the various species he had already created, and he designated it *Homo sinemento*. The El Moro Man seems to present a curious mixture of characters that are primitive and of characters in which its development surpasses that of *Homo sapiens*. Of small stature, and having a chinless lower jaw, he seems to have possessed a straight-jawed (orthognathic) dentition, without wisdom teeth. This species must be extinct. Now, as a matter of fact, anthropologists actually recognize these as being the skeletons of Indian women, with a chin, weak but still present. Willis regards the deposit in which the skeletons were buried as distinct from the true Pampean and of recent origin.

On the whole, even according to Ameghino, all the human

species from the Upper Pampean seem to belong to the genus *Homo*, but the latter seems already to be represented by several species.

Discoveries in the Lower Pampean.

The Lower Pampean or *Ensenadean* is less rich in anthropological evidences, due evidently to the fact that its outcrops are much less extensive.

There must first be recorded the skeleton exhumed by Roth in 1887, near *Baradero*, a locality on the Parana, about 1301 metres above Buenos Aires. This poorly preserved skeleton is in the Polytechnic School at Zurich. R. Martin finds that it does not differ from the skeletons of modern man in South America. The fact that its various parts have retained their anatomical relations favours the theory of a burial.

Prothomo.

Returning again to the Atlantic shore, we have to record the new discoveries which enabled Ameghino to establish still another new species of fossil man, *Homo pampæus*. There is first a skull found accidentally in 1888 near the brook *Tigra*, south of Miramar. Next there are three skulls, accompanied by other bones, which were lying at no great depth near Necochea, in a deposit regarded by Ameghino as belonging to the Lower Pampean (and therefore Pliocene). *Homo pampæus*, characterized by a very low receding forehead, a character in which he differs greatly from *Homo sapiens*, might perhaps be a true *Prothomo*.

Examination of these relics by expert anthropologists, Lehmann-Nitsche, Mochi, and Hrdlička, revealed nothing of the sort. The skulls, artificially deformed after the fashion of recent Patagonian and Peruvian skulls, in other respects resemble those of Lagoa-Santa. They still contain much organic matter; their position in the layer rather indicates burials, and, according to Willis, these burials were carried out in a deposit of recent age, not Pampean at all. *Homo pampæus* has never existed but in Ameghino's imagination.

Diprothomo.

This scientist's imagination carried him still further. In the course of work carried out in 1896 in the port of Buenos Aires, during the making of a dry dock, the workmen found several human skulls.

One of them, consisting of a portion of the vault only, was sent after long delay to Ameghino. The deposit from which it had been taken seemed to belong to the extreme base of the Pampean, and would thus, according to Ameghino, date from the Lower Pliocene. As to the bone fragment, it appeared to the Argentine palæontologist's eyes so different from the types already known, that he saw in it the representative, not only of a new species, but even of his genus *Diprothomo*, till then merely a hypothetical genus, and he named his new fossil *Diprothomo platensis*.[1]

Here again Ameghino made a serious mistake. Owing to his having examined this fragment of skull with a wrong orientation, he found in it strange resemblances to that of one of the lower monkeys, *Arctopithecus*. When it is correctly posed, the brain-box has all the characters of a modern man. Numerous anthropologists[2] had no difficulty in proving this fact, and Schwalbe showed that the frontal bone of *Diprothomo* can be exactly superimposed over that of an Alsatian!

The exact age of the relic is more difficult to establish, for it passed successively from the hands of a navvy to the hands of a foreman, who handed it to his master, and only after three years did it come into Ameghino's possession.

Tetraprothomo.

I have now but to say a word regarding the still more extraordinary finds said to date from the Hermosean—the Miocene according to Ameghino. The La Plata Museum has long possessed an atlas bone, resembling a human atlas, of small size and thickset form. This bone is said to have been taken from the cliff at Monte Hermoso. In 1908, Ameghino[3] described it at length, and made it the type specimen of his *Tetraprothomo argentinus*. He was induced to do so by the sight of a small femur from the same locality, which he likewise attributed to the genus *Tetraprothomo*, found at last.

[1] Ameghino, F., "Le *Diprothomo platensis*" (*Anales del Museo de Buenos Aires*, xix., 1909.)

[2] See particularly: Mochi, A., "Nota preventiva sul *Diprothomo*" (*Revistá del Museo de La Plata*, xvii., 1910); "Appunti sulla Palæontologia argentina" (*Arch. per l'Antrop. e l'Etnol.*, xi., 1910).

[3] Ameghino, F., "Notas preliminares sobre el *Tetraprothomo argentinus*, un precursor del Hombre" (*Anales del Museo de Buenos Aires*, xvi., 1908).

It was but a fresh dream! Lehmann-Nitsche immediately recognized that the atlas bone was more human than simian, and attributed it to a primitive man of Tertiary Age, *Homo neogeus.* Then Hrdlička followed, and clearly proved, by means of numerous comparisons, that all the peculiarities shown by the atlas bone from Monte Hermoso clearly differentiate it from the atlas bone of apes, and do not surpass the limits of variation to be observed in the atlas bones of modern Indians.

As to the femur, it could be attributed neither to a Man nor to a Primate; it was a femur of a Carnivore, and probably of a Feline.

It is indeed pitiful to see a naturalist of Ameghino's gifts end his scientific career, so meritorious in many respects, with a series of works which exhibit such a disordered imagination and an altogether fantastic interpretation of morphological facts.

Conclusions.

What conclusions may be drawn from the mass of facts which I have felt obliged to describe at some length? One reply is easy, that Ameghino's theories are radically mistaken, and that there has never been found in South America the least relic of a fossil human being differing from modern men. Can we likewise deny the existence in the Pleistocene period of fossil men, the direct ancestors of these modern men? Yes, is the decided reply of many scientists, notably Hrdlička and his fellow-workers in the United States. According to them, nothing, absolutely nothing, gives us any ground at the present time for affirming the existence of Fossil Man in South America. There is no more proof of it here than in North America.

This opinion seems to me to be too absolute. In the contradictory state of the arguments, it is better to be less categorical. Yet, on the one hand, how can one but be struck by the contrast between the abundance of human skeletons in the Pampas formations and the poverty of the archæological finds made *in situ* in the same deposits? This is exactly contrary to the case in Europe, where true fossil human bone-remains are most rare in comparison with the innumerable archæological relics. And how can one but be unfavourably impressed by the statement that, apart from a few broken or splintered pebbles, and a few split or striated bones without any conclusive value, the industry of the alleged fossil men is identical with that of modern Indians? Such a fact seems incompatible with our knowledge of those other regions of the globe which have been most thoroughly studied, and in which industrial evolution seems to follow a general law throughout the ages.

But, on the other hand, among so many finds there are some the criticism of which is not so easy, and for which reliable scientists can vouch. How can we deliberately reject them, just because they do not seem to conform to *a priori* notions?

What will be the outcome of all Ameghino's "discoveries"? Certainly much less than some of his warm admirers believe; probably more than his ruthless detractors admit. It is greatly to be desired that young South American naturalists should bring a new and untrammelled spirit, free from all prejudice, to bear on the solution of those intensely interesting problems which their country contains.

At the present moment my own conviction is that in South America, as in North America, Man would really seem to be much more ancient than many anthropologists believe, and that the peopling of the New World must be dated from at least the dawn of modern geological times.

CHAPTER XI

GENERAL CONCLUSIONS

WE have just completed an inventory of the main contributions to knowledge made by human palæontology and prehistoric archæology. It is now our duty to summarize these facts, in an endeavour to construct a unified, if provisional story, so that we may see how far they illumine the great questions of the origin and evolution of Mankind.

* * *

Naturalists, from Aristotle onwards, recognized at a very early stage that the human body showed great resemblances to the bodies of other mammals, and especially to those of the apes; they have not hesitated to classify Man along with the latter, while affirming that he is superior to all his fellows in the same zoological class, that he is premier amongst the Primates, that is to say the very "First of the First."

Some scientists, it is true, regarding such company as humiliating to Man's dignity, have desired, on the ground of the attributes of his intelligence and his religious nature, to place Man not only above but even quite apart from all other living beings, and create for the protection of his "threatened dignity" the "nebulous sphere" of a *Human Kingdom.* But it was easy to prove that nothing could be more irrational, and with Darwin one may say, that "if Man had not been his own classifier, he would never have thought of founding a separate order for his own reception."

With the progress of zoology and comparative anatomy, the morphological relationships between Man and the other Primates have become more accurately known. Buffon first taught us that, in his physical structure, Man differs less from the anthropoid apes than these differ from the lower monkeys.

This view, clearly expressed by Huxley a century later, and renewed and developed by Broca in his excellent work on *L'Ordre des Primates*, has since been confirmed by numerous works, bearing not only on the skeleton, but also on the soft parts of the Primates, their muscular and nervous systems, their sense organs, viscera, dentition, genital organs, spermatozoids, hair arrangement, papillary lines of the extremities, and so on.

It has become possible to estimate numerically the relative affinities of their various groups, and to establish a scale based on the number of the human characters presented by each group. According to Keith, the Chimpanzee presents the greatest number of points of resemblance to Man. The Gorilla closely follows the Chimpanzee; then comes the Orang and after it the Gibbon, but at greater distance. The other apes end the procession.

Embryological studies, contributing their share, accentuate these resemblances and show that many differences shown by adult Men and apes are much less marked or may even be absent in the embryo.[1] Thus we are led to admit descent from common ancestors.

Other phenomena can be similarly explained only by admitting more or less direct genealogical relationships. There are, first, certain *anomalies*, that is, certain structural arrangements which are occasionally found in Man, but which are, in the normal course of things, present in related animals. At first regarded simply as curiosities, they now appear, in the light of the theory of evolution, to be marks of retrogression, or, to put it otherwise, atavistic characters, that is to say, abnormal relapses to an ancient state of things which was normal in the case of the common ancestors. These anomalies are countless, and concern all the organic systems; they have furnished anatomists with matter for many important works.

Such are the true "rudimentary organs," structural arrangements which, normal and well-developed in other

[1] To take only one example, such is the inter-maxillary bone, the existence of which in Man has for long been misunderstood, and which might pass for a distinctive character; but embryology shows it to be present, with its simian features, in the human embryo less than two and a half months old.

mammals where they fulfil a more or less important function, are reduced in Man to the point of becoming physiologically useless. The significance and importance of these rudimentary organs have been clearly brought out by Darwin. They furnish the strongest arguments that comparative anatomy, on its own authority, can bring forward in support of the transformist theory in general, and of the animal descent of Man in particular.

Physiology likewise contributes evidence in its favour. There is now a science of comparative bio-chemistry, according to which each group of beings possesses a specific chemical constitution accompanying its specific morphology, and distinguishing it, like the latter, from neighbouring groups. The very striking experiments made in the last few years by the method of serum precipitation by numerous physiologists, have enabled an accurate estimation to be made, in a marvellously delicate manner, of the degrees of consanguinity between the different Primates. The relationship of Man especially with the anthropoid apes, and with the Chimpanzee in particular, has been confirmed. With other monkeys of the Old World, such as the Macaque, his relationship is much less close and it is still more distant with the Flat-nosed (Platyrrhine) Monkeys of the New World.

Comparative pathology also testifies in the same sense: these creatures are most closely akin to us, from the morphological point of view, which take our infectious illnesses with the greatest readiness.

So much for the physical or bodily aspect. There is yet the moral or spiritual side, the extreme importance of which cannot be doubted, and regarding which some explanation must be given.

Since psychology lost its scholastic character and became scientific, that is to say physiological, the great barrier it had sought to erect between Man and "the beasts" has indeed been demolished. It can no longer be maintained that the mental faculties are essentially different in the various kinds of living beings. They exhibit only differences of degree, and their mechanism is everywhere the same. We now speak of "comparative psychology," "animal psychology," and even of "cellular psychology" in a way which would have startled the majority of the professors of philosophy who prepared the youth of my time for his Arts degree. Most interesting books on the origin of the intelligence, on the psychic life of animals, and on mental evolution have appeared. We can observe the progressive development of the psychical phenomena in each of the great zoological groups, and this development takes place parallel with that of the nervous centres, the real psychic machinery. We can see a continuous chain of phenomena, from the first manifestations of a vague consciousness to the most complicated mental operations.[1]

Although the human brain is much larger than the brain of the highest anthropoid ape, although the human intelligence is much superior to that of the ape, yet all the manifestations of the former are to be found, merely in less degree, in the latter. And these differences become still less when our comparison is made between the most intelligent of modern apes and those living Men who most approach the natural state. Agassiz himself declared that he could not tell "wherein the mental faculties of an infant differ from those of a young Chimpanzee."

Human reason is not then a special creation; it made no sudden appearance, but came gradually into being. Here again, in the kingdom of the spirit, as well as in the kingdom of the body, the close ancestral relationship between Man and the other Primates crops up.

It has not gone so far as an articulate language, though the rudiments of such are to be found amongst many animals, particularly amongst birds, which are capable of associating certain sounds, that is to say certain vocal signs, with certain acts or objects. And in the case of our neighbours, the monkeys, the constitution of their brain, larynx, genio-glossal

[1] For long, "human reason" and "animal instinct" were placed in irreconcilable opposition. But all psychological naturalists, along with Darwin, E. Perrier, Romanes, Bouvier and others, now recognize that we must of necessity place intelligent operations at the origin of instincts, and that the latter are only "hereditary habits," a sort of automatic reasoning, examples of which are not lacking even in Man. This, after all, does not seem to me to be irreconcilable with Bergson's theory of the duality and the independence of intelligence and of instinct.

and genio-hyoid muscles is such that it may be safely affirmed that very little more is necessary, from the organic point of view, to secure the exercise of a function which must have gradually developed in Man along with his intelligence, after a mechanism which physiologists and linguists are now beginning to understand and to reconstruct. It was animal intelligence that prepared the way for the birth of language, and it is language that has made possible the wonderful development of the human intelligence.

To sum up, the facts relating to modern Primates which have been acquired by the different branches of biology, may thus be expressed :—

1. Man is a Primate, the highest of the Primates. He is much more akin to the anthropoid apes than the latter are to other monkeys.

2. The individual development of Man shows that his various systems and organs pass through transitory phases corresponding to the final stage of lower animal forms (ontogeny repeats the stages of phylogeny).

3. The *anomalies* of his various anatomical systems are very often only reappearances of morphological features of these lower types, and many *organs* called *rudimentary* can only be explained by the theory of evolution; they represent relics of ancestral states.

4. Man is much superior to the highest apes in the size and organization of his brain. As a result, the highest physiological product of this brain, that is intelligence, is likewise much higher in Man. But the difference is only one of degree and not of kind. We may say, paraphrasing and completing the sentence with which Darwin ends his admirable work on *The Descent of Man*, that Man still retains, in his spirit as well as in his body, the indelible seal of his lowly origin.

Embryologists have sought to go still further; they have made an attempt to reconstruct the different stages of evolution in Man, on the basis of the parallel between ontogeny and phylogeny. Haeckel made himself notorious by the daring with which he launched upon and followed this course, which led to the production of his *Anthropogenia*. Praiseworthy as his attempt is, it nevertheless seems foolhardy, especially in the eyes of palæontologists, who know, from the genealogical history of other zoological groups, how easy it is to err in attempting such reconstructions in the absence of sufficient material evidences.[1]

Further it must be noted that opinions expressed on the subject have themselves been very diverse. First, on the question of our knowledge as to whether Man's origin be single or multiple: as to whether the genus *Homo* comprises only one species originating from a single ancestral form, or whether it comprises several species each one of which must have had its particular line of descent.[2] And, secondly, on the degree of relationship which exists between the human branch, whatever may be its degree of ramification, and the other branches of the great stem of the Primates.[3]

Differences of opinion even among equally expert scientists are very considerable. They evidently show that, in respect to the establishment of human genealogy, as in respect to everything relating to the evolution of organized beings in general, the last word must lie with palæontology when that science is in a position to give a clear pronouncement. The finest anatomical works, the most subtle, the most ingenious theories on the structure of living creatures, cannot have the conclusive value of relics extracted from the rock where they

[1] To prove how dangerous inductions on such a subject may be, compare the list of characters which Abel Hovelacque ("Notre ancêtre," *Revue d'Anthrop.*, vi., 1877, pp. 62-69) attributes on *theoretical* grounds to fossil Men, with the list of characters which have since been recognized as actually theirs, particularly in the case of Neanderthal Man. The following are a few of the characters given by Hovelacque: small cranial capacity; presence of a parieto-occipital crest; division of the malar bone; early fusion of the nasal bones; slightly developed nasal spine; pterions X-shaped or curved; great alveolar subnasal prognathism; strong canines; incurved cubitus bone; platycnemic tibia; forearm very long in relation to upper arm. . . . This is all simply so much *pithecomorphism*.

[2] The most notable supporter of this theory to-day is G. Sergi, "Le origine umane," 1913. "L'evoluzione e le origine umane," 1914, etc.

[3] Is it worth while mentioning that the idea of a direct genealogical relationship between Man and modern apes has long since been abandoned, if indeed such an idea was ever entertained, by all true naturalists? It is now found only among writers totally ignorant of science, or in the sermons of a few country curates. But of necessity, a common origin must be attributed to apes and Man. Differences of opinion are confined only to modes of regarding the branching of the common stem.

were deposited and embedded in their actual chronological order. Such evidences form a new kind of archive, such anatomical specimens, preserved for thousands or hundreds of thousands of years, represent in concrete, tangible, and measurable forms the transitory stages, the different links of a chain uniting a series of organized types, the extreme terms of which appear at first sight to be, and indeed are, so very different.

The discovery of the remains and the study of ancestral fossil forms constitute the precise end of palæontology, and particularly, in the present case, of human palæontology. It cannot be too often repeated that the origin of Man is a problem the solution of which can only be looked for from palæontology. The prime importance of the study of fossils and the advantages to be derived from it are made evident by the fine results which have crowned the science of animal palæontology.

* * *

It is true, as we have seen, that anthropology owes a great deal to zoology, but it must be confessed that the latter science shows Man's place in Nature in a false light, since it presents him as a creature almost isolated among the Primates, in the same fashion as, though in less degree than, the horse among Perissodactyles, the elephant among Proboscidians, the camel among Ruminants, and so on. Since palæontology has broken down the isolation of these animals by relating them to others through intermediate forms, and by discovering their genealogy from generalized forms, there was reason to hope that it would also break down, or at least likewise diminish the isolation of Man, and that it would also enable us to discover the main evolutionary stages in human genealogy. Have these hopes been realized?

Towards the end of last century, the naturalist-philosophers who studied the mystery of human origins derived little help from the study of fossils. Relics of the kind were too rare and too incomplete: only a few remains of apes closely akin to living species, a larger number of human remains which revealed no difference of importance from modern Man, and only one human brain-box which, indeed, showed a few simian characters, but the antiquity of which was questioned, as well as its very nature, for by some it was regarded as normal and by others as pathological.

Since then our palæontological knowledge has increased. But even now does it reveal to us the principal evolutionary stages of the human branch? To reply to this question, we must first recall the conclusions reached in the chapter in this book dealing with fossil monkeys.

First we have seen the order of the Primates take its rise, like all the other orders of mammals, at the commencement of the Tertiary era. The zoological group to which Man belongs is not, then, the slow and splendid crowning of a unique, and indeed imaginary, series in the whole range of the mammals. In the Lower Eocene its early representatives break away from a crowd of related creatures, which, however, themselves show tendencies towards other orders of mammals. It is not, as was long believed, because they are, taken as a whole, the most intelligent of the mammals that the Primates necessarily appeared latest. Their early differentiation, which took place in a particular direction, dates as far back into the past as the early differentiations, in different directions, of the Pachyderms, Ruminants, Perissodactyles, Carnivores, and so on. And this fact is of the greatest interest from the point of view of a philosophy of nature.

We have next seen the representatives of the different modern groups succeed each other in chronological order, in an order conforming to their places in the animal kingdom: first the lemurs, then the tailed monkeys, followed by the anthropoids. Unfortunately, the scarcity and the fragmentary state of the palæontological relics have not allowed, so far, of a comparative study sufficiently detailed to determine the ties of relationship which link together the various fossil apes. On this point only vague indications have been gained. It would seem, however, that the different branches of the tree of the Primates may have had very ancient origins, that they may have begun to differentiate early, at least from Oligocene times. *Propliopithecus* from Fayum may be taken as a generalized type of anthropoid, related to, if not in very truth, that synthetic form which theory demands should be placed at the base of the branch of the great apes, and from which present-day Gibbons seem to descend almost directly, through the intermediary of the Miocene *Pliopithecus.*

While tailed monkeys in their various kinds increase till they spread over vast areas of the Old World, the anthropoid branch in its turn split into several branchlets, at first so close to one another as to be difficult to distinguish, judging by the few fragments we possess, and then becoming more and more distinct and increasing so as to produce, on the one hand, forms which are manifestly the ancestors of modern types, and, on the other, special forms, sometimes greatly differentiated and very large, which have died out without leaving any descendants. The relative frequency of their remains in the fossil-bearing deposits of the Siwalik Hills leads us to believe that Southern Asia was, if not the chief, at least one of the chief centres of habitation, of multiplication, and of differentiation of those ancient anthropoids, some of which show affinities with modern Orangs, others with the Gorilla-Chimpanzee group, whilst others again represent independent and extinct forms. Among the latter, the curious *Sivapithecus* is noteworthy for certain peculiar features which suggest a resemblance to the human group. In reality, up to the present we have either not found, or have not been able to distinguish in this series, which is unfortunately too short and consists of fossils too fragmentary in character, one which could be regarded as having belonged to a pre-human form. *Pithecanthropus* itself, on which some have been inclined to confer this dignity, because its skull really possesses characters exactly intermediate between those of the skull of a large anthropoid ape, like the Chimpanzee, and of a primitive Man, must, it seems to us, be considered rather as a large specialized form which belongs to a twig of the anthropoid branch independent of the true human branch.

All that we can safely affirm, therefore, is that, thanks to *Pithecanthropus*, the history of the fossil monkeys, however incomplete and disjointed it may be, greatly lessens the morphological gap separating modern monkeys from modern Man.

On the other hand, the history of fossil Man helps in the same way to show a closer relationship between the human and the simian branches. Although it is but in its first youth, human palæontology has already made noteworthy progress. We now know at least two types of fossil man which, on account of their osteological characters, clearly take their place at a lower level than modern types, and which exhibit a combination of structural features separating them less markedly from the apes than from the generality of modern men.[1]

There is first the Heidelberg Man, *Homo heidelbergensis*, who dates from the dawn of Quaternary times. His lower jaw, the only relic known to us, exhibits an extraordinary mixture of human and simian characters. Had only the teeth of this jaw been found, they would certainly have been attributed to a man not differing, in any important character, from certain races of *Homo sapiens*. If, by some accident, the jaw had been found without its teeth, there would have been no hesitation in making it the type of a new genus of anthropoid ape. This example of the miscarriage of Cuvier's famous law of the correlation of characters is particularly instructive, since it relates to a palæontological relic representing, in an almost ideal way, a form intermediate between the structure of an ape and of a human being. Unfortunately, these remarks refer only to the merest fragment of the skeleton. When the day comes for the discovery of a complete skull or limb-bones, palæontologists will be called upon to make investigations and verifications of the greatest interest, perhaps with results entirely unforeseen.

At first it was thought that *Eoanthropus dawsoni*, from Piltdown, represented a second very primitive type, not on

[1] After what I have said, in the course of this work, concerning *Tetraprothomo*, *Diprothomo*, and so on, I need not return to the reasons which lead me to reject the fossil men of S. America and Ameghino's ideas on the subject. Certain specimens from the Pampas formations may indeed possess a certain antiquity, the word being used in an archæological rather than in a geological sense, but their morphology contributes nothing to the discussion of the problem in which we are interested.

account of his skull, which scarcely differs from that of a modern man, but because of his lower jaw. But the simian characters of this bone are so different from those of the Heidelberg jaw and so like those of a Chimpanzee's jaw, that zoologists and palæontologists come more and more to share the opinion that this jaw really belonged to a Chimpanzee.

Further, and most important of all, there is Neanderthal Man, *Homo neanderthalensis*, descended perhaps from Heidelberg Man, perhaps from an unknown and still more archaic form. We now know a good deal about his structure. We know that, in the organization of his skeleton and of his brain, he combined not only the majority of the rare simian characters present in certain representatives of modern man, but also several features of inferiority unknown in the latter.

Human beings, the Hominians, have always been regarded as forming a family or an order, or even a class or separate kingdom, placed much higher in the scale than the other Primates. Up to the present time, this group of high degree corresponded to the genus *Homo* alone, and this genus itself comprised, in the opinion of many anthropologists, only the single species *Homo sapiens.*

Palæontological discoveries have lessened the distance separating Man from the animals most closely related to him, and in so doing have lessened the isolation in which some have always been inclined to place him. This isolation, which was looked upon as not the least of his attributes, seemed to augment his greatness and nobility in the eyes of those who regarded him from a religious or metaphysical point of view.

At the present day, apart from the genus *Pithecanthropus*, classed by many scientists among the human beings, there has been created by Bonarelli, and perhaps rightly so, the genus *Palæanthropus* for the Heidelberg jaw, while the Neanderthal type likewise, according to several naturalists, represents a special genus. In any case, it is difficult to deny nowadays that there may have been several species of the genus *Homo* and that *Homo heidelbergensis* and *H. neanderthalensis* may not be very distinct from the main mass of *Homo sapiens*, fossil or living.

This conclusion should not be surprising to naturalists who, believing in the theory of evolution, cannot refuse to apply it to all living beings, to Man as well as to his neighbours the apes and other mammals. But it is not theory that we are dealing with here. These are facts of conclusive value. Palæontology contributes new *concrete* data to the natural history of the human zoological group, and these data clearly show that the evolution of this group proceeded in the same way as the evolution of other groups of mammals.

The origin of mankind must be relegated to a much more distant geological past than is usually supposed. General palæontology hints as much, for the problem of the origins of the different forms of life is much oftener relegated than solved. Human palæontology teaches us, moreover, that, as early as Mid Pleistocene times at least, there existed other human types than those of Heidelberg and Neanderthal, and that these types already in striking degree resembled *Homo sapiens*. It is probable that these fossil types represent the direct ancestors of modern Man; from times so remote they have formed an individual branch, long distinct from the branch of which *Homo neanderthalensis* represents the now withered end.

The fact that Neanderthal Man existed at the same time as the ancestors of certain types of *Homo sapiens*, and the additional fact that he seems to have become extinct without leaving any posterity, agree with what is tending to become a palæontological law, namely that the development of creatures is not accomplished as simply as was believed in the early days of the science; that unilateral series appear to us as more and more rare; that, if they exist, it is extremely difficult to discover them or trace them for any distance.

Each grouping of beings related to one another in the generality of their structural characters, whether it be a family, a generic, or a specific group, may be compared to a tree or a bush more or less branched, of which each branch, branchlet, or twig represents either a genus, a species, or a race. The development of each of these offshoots has been more or less vigorous, its duration more or less long. Modern forms are only the developments, the latest blossomings of certain terminal twigs, the majority of which are dead and fossilized.

The human group is no exception. At an early stage it must have divided into several branches, which must have borne branchlets, and these in their turn twigs. In terms of the polygenist theory, it might be said that several of these branchlets or twigs have survived up to the present period; according to the monogenist theory, it is claimed that the mass of *Homo sapiens*, with its various races, forms but one single branchlet. Yet even a few years ago we did not know, and it was palæontology that taught us, that side by side with these branchlets which are still vigorous and full of sap, the human branch formerly gave rise to branchlets which are now withered, and of which we are just beginning to discover the fossilized blooms in the depths of the geological layers.

So the original stock of the genus *Homo* must plunge its roots deep into a much more ancient past than *a priori* was supposed. Sometimes I have been reproached, from certain quarters, for being a determined opponent of Tertiary Man; but such an accusation entirely contradicts what I have so often written. I am convinced that a being, already in possession of the main physical if not even the psychical attributes of Man, must have existed somewhere during the Pliocene and perhaps during the Miocene period. On the other hand, I do not regard as conclusive any of the concrete discoveries brought forward as evidence of his existence. No discovery of bones made in an alleged Tertiary deposit has yet been able to withstand criticism, a point on which, it must be acknowledged, almost everyone is agreed. I combated the eolith theory, not because it seemed to me improbable, but because long geological experience had shown me that it is often impossible to distinguish stones splintered, dressed, and retouched by purely physical forces from certain products of deliberate but rudimentary manufacture. To carry conviction in such a case, the evidence ought to be absolutely demonstrative. Now I hold that, up to the present day, there exists no solid proof of the existence of a human being in our regions before the dawn of Quaternary times. To-morrow, perhaps, may bring forth irrefutable evidence. Let us bide our time.

* * *

So far I have singled out for special consideration only certain intermediate structural forms. Can we, in the present state of our knowledge, form a reasonable and accurate opinion regarding the palæontological origin of the human group? Or to put it otherwise: What genealogical relationships can we establish between the human branch and the other branches and trunk of the tree of the Primates? Replies to this new question are not lacking, but, from all that has previously been said, it is not difficult to foresee that they must be hesitating and incomplete.

Of late years, several naturalists have drawn up "Genealogical trees" of the Primates, living and fossil, according to their morphological and chronological relationships. I would mention in particular the attempts of Gregory, Keith, Pilgrim, Sera, and Bonarelli.[1] A comparison of their diagrams is calculated to increase, if possible, our caution, for between the various diagrams there are such great, sometimes even fundamental differences. The human group is shown as having relationships so different that the wisest plan is to conclude that this group is still "in the air," and that we do not know exactly the place where the human branch should be inserted among the branches or twigs round about it. The more authors, the more theories.

Together with Gaudry, Dubois, Schlossen, Schwalbe, Osborn, Klaatsch, Friedenthal and others, Gregory would place Men in close relationship with the Anthropoids, both together forming a common branch, which had long been distinct from the neighbouring branch of the tailed monkeys

[1] Gregory, W. K., "Studies on the Evolution of the Primates" (*Bull. Amer. Museum Nat. History*, xxxv., 1916). Keith, A., "The Antiquity of Man," London, 1915. Pilgrim, E., "New Siwalik Primates and their Bearing on the Questions of the Evolution of Man and the Anthropoidea" (*Records of Geolog. Survey of India*, xlv., 1915). Sera, G. L., "La testimonianza dei fossili di Antropomorfi per la questione dell'origine dell'Uomo" (*Atti della Soc. ital. di Scienze naturali*, lvi., 1917). Bonarelli, G., "Alcuni problemi d'Antropologia sistematica" (*An. de la Soc. cientif. Argentina*, lxxxv., 1918).

(Fig. 240, H). We know that Darwin and Haeckel regarded the human group as forming an autonomous branch, early detached from the mother branch of the Catarrhinians or Dog-faced Monkeys (Fig. 240, H′). Carl Vogt, Ameghino, and Sera prefer to attach it to the older branch of the Platyrrhinians or Flat-nosed Monkeys (H″). Cope again would place it still lower, at the level of the most ancient of the Primates,

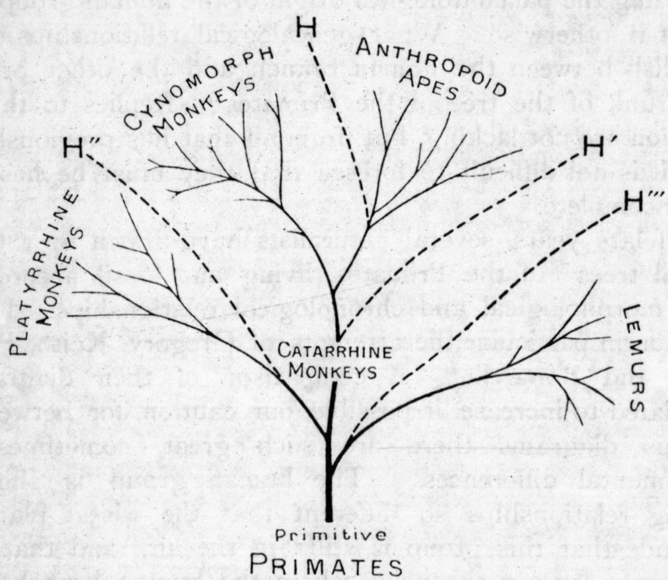

FIG. 240.—Diagram showing the different theories of the genealogical relationships of Mankind with the other groups of the Primates. For explanation see text.

the Lemurs (H‴), and the anatomist, Wood-Jones,[1] does not hesitate to regard the curious monkey, Tarsius, as the living animal most akin to Man.

Each of these theories has been supported by good arguments, for the arguments are based on more or less common generalized structural features, which give a marked unity to the whole Primate group. According as the stress is laid on one or other special character, insistence on a particular comparison becomes necessary.

[1] Wood-Jones, F., *The Problem of Man's Ancestry*, London, 1918. Miller, G. S., "Conflicting Views on the Problem of Man's Ancestry" (*Amer. Journ. Phys. Anthrop.*, 1920).

Morphology and physiology, including physiological psychology, obviously support the first theory, and are strongly reinforced by recent facts in biochemistry. But there is room for discrimination.

The views of Klaatsch, the German anthropologist, on the alleged close relationships between the different types of Quaternary Man and certain genera of modern anthropoids, views which have led him to conceive a group he has called the "Orang-Men of Aurignac," along with another group, the "Gorilla-Men of Neanderthal," are not based on any reasoned argument. They are only a variation, under a kind of scientific guise, of the way in which the general public sums up its ideas on the question of our origin when it asks if "Man is descended from Monkeys."

The much more scientific views of Osborn, Gregory, and others, while attaching the human group to the anthropoid group so as to form a single whole, clearly separate the human branch from the neighbouring branch of the great apes, and only attribute to them a very distant common origin (dating from the Miocene) from which they must have diverged. This is somewhat the same as Keith's view. It is, indeed, the view most easily vindicated, and we have seen that, to the weight of anatomical arguments in its favour, may be added the biochemical argument, the value of which apparently is undeniable.[1]

This view is, however, open to serious objections, and I would be inclined to reconcile it with the second theory, that of a more ancient origin, starting from the great Catarrhinian stock. For this purpose it would be enough to admit that the anthropoid form leading up to Man very early became separated from the neighbouring forms leading up to modern anthropoid apes; that its point of insertion on the ancestral

[1] We should like to know if phenomena of convergence, on a par with those of structural convergence, may be admitted in biochemical characters; whether a similar morphological evolution in two different groups may not be accompanied, from the physiological point of view, by a similar evolution of biochemical phenomena; in other words, whether the morphological resemblances resulting from convergence under the influence of the same factors do not involve corresponding resemblances in the molecular constitution of the elemental substances. The future of biochemistry, so full of promise, will certainly shed light on this point.

parent-branch of the Catarrhine Monkeys was independent of the points of insertion of those other forms which led by evolution to the fossil or modern types of anthropoid apes.

Zoologists who are specialists in the comparative anatomy of the Primates, have long since observed that many human characters cannot be explained by direct descent from one single stock of large apes. "It would be necessary to merge in one," said Carl Vogt, long ago,[1] "the anthropoid characters of the *three* anthropoid apes and *even of several other monkeys*, to obtain the combination from which Man could have descended." This is the view which palæontology seems to confirm.

A study of Neanderthal Man, especially of the skeleton of his limbs, shows that, in many osteological characters, it is easier to associate this fossil Man with the lower monkeys than with the various forms of living anthropoids. Without creating too serious difficulties, then, we may place the insertion point of the human branch upon the branch of the dog-faced Catarrhinians, at a lower level than the starting-point of the anthropoid branch.

It would perhaps be wiser, without going so far as the Lemurs, as Cope proposes, to descend still lower, even to the common stem of the Monkeys. Palæontology, indeed, teaches us that the various types of modern monkeys are very ancient, and that the independence of each of the groups massed about these types has been acquired at a very early period.[2] The same must have been the case with the human group, the ancestral forms of which must have possessed, from the Catarrhinian stage and perhaps even from the Platyrrhinian stage, certain features of organization different from those of neighbouring types; and the progressive development of these forms must have attained to a distinct anthropoid stage,

[1] *Congrès internat. d'Anthrop. et d'Archéol. préhistor.*, 2nd meeting, Paris, 1867, p. 442.

[2] Do we not find, even in the Lower Eocene, that quite small-sized Primates, *Anaptomorphus* for example, exhibit the main characters of the modern Tarsius, remarkable for its development of brain-pan, its shortened face, its combination of lemurian and simian features, and even for many human characters? (See Wood-Jones, *loc. cit.*)

the actual forerunner of the prehuman and human stages. The recent discoveries at Fayum to a certain extent confirm this view, for *Propliopithecus* seems to present a mixture of characters both very archaic and yet specialized in the direction of a higher evolution.

Palæontology and zoological palæogeography also teach us that if our very remote ancestors must have passed through a Platyrrhinian stage, at least as regards their dental formula, it would be impossible to maintain now, like Ameghino, that the human branch only represents an exceptional development of certain elements of the Platyrrhinian branch.[1] From Upper Eocene or Oligocene times, the latter have been confined to South America; there they developed and diverged, but their progressive evolution seems to have stopped at an early stage.

These first conclusions are not purely imaginary conceptions; they are scientific theories, having the advantage of being based on palæontological observations, or of being in agreement with these observations.

We must not deceive ourselves, however. We are still far from knowing accurately the main stages of the human line, from its commencement with the original lowly forms. So far it is quite impossible to establish a progressive series, based on concrete evidences, as has already been done for many mammals, such as the horse, the elephant, the bear, and others.

It is very probable that our most distant ancestors early became distinguished from the mass of other Primates by certain characters, which we may describe as fundamental, characters which already exhibited a tendency towards human superiority and by which these distant ancestors are recognizable. But it is only by slow degrees that the relationship can be established and traced with certainty. The solution of the

[1] M. Sera, in a paper entitled *I caratteri della facia e il polifiletismo dei Primati* (1918), takes up Ameghino's theory, the main ideas of which he adopts, and he expounds his personal views regarding the group of the Primates, basing them mainly on the comparative study of certain facial characters. He divides the Primates into six fundamental groups, each of these branches of the great primary trunk comprising both simian forms and a human form. There would thus be as many human types as large simian types. This exaggerated multiple phylogeny forms a new system which runs directly counter to accepted notions.

problem of our origin and especially the accurate determination of the different elements in our descent, demands fresh discoveries of fossils, and indeed of many fossils.

We must confess then, however damaging the confession may be to our *amour propre*, that we are still too ignorant to give a direct answer to Huxley's "supreme question," or to solve in full the perplexing problem of our origin. And this ignorance, I cannot refrain from repeating, is due to the great blanks in our palæontological evidence, these blanks which Darwin deplored, which made my old teacher Gaudry say that palæontology was at once splendid and poverty-stricken, and which can be filled in only with painful slowness. The discoveries of these later years, however, give us the right to place great faith in a future perhaps not far distant.

It must be recognized, indeed, that science has made real progress since the last discoveries of human palæontology were made. Alongside of all that we are yet ignorant of, we must place all that we have already learned, all that we have securely laid hold of.

The human type is no longer isolated. We know that there have been several species, and probably several genera of the human kind, and that these ancient human beings were structurally much lower in type than living men; that they exhibited numerous characters in which they were less distantly related than we are to the other Primates, particularly to the highest representatives of these, the Anthropoid Apes. In our museums we possess actual remains of these structurally intermediate forms; but before these discoveries were made, such intermediate forms were only imaginary or theoretic.

We know that there was a human *branch*, and that this branch was divided into many more branchlets than had been supposed; the human branch now assumes in our eyes the same character as the other branches of the Primates or of other groups of mammals. Its evolution, as we are now beginning to perceive, in every way resembles the evolution of these groups. The science of palæontology is a *unity*, whether it concerns itself with Men or with animals.

So much then we know to-day, with certain knowledge. It is, indeed, little in comparison with what still remains to be learned; but it is a great deal in comparison with our former ignorance, or our very imperfect knowledge of not so long ago.

* * *

The unsatisfactory nature of palæontological evidence still makes it difficult to decide exactly in what countries the human kind originated. The first idea to occur quite naturally to the minds of anthropologists, was that the present habitat of the races which we call "primitive" must be regarded as their place of origin; and hence arose an extraordinary number of "centres of creation" or "centres of appearance." This was the convenient solution, naturally suggested by such adherents of the polygenistic theory as Agassiz. But de Quatrefages rightly pointed out that this *initial* cosmopolitanism was opposed to the general facts of zoological distribution. It is also contrary to our present understanding of the conditions and causes of the migrations of mammals in general, as palæontology has revealed them.[1]

Next there have been suggested in turn the northern countries, the central massif of Asia, subtropical regions, Africa, South America, the Antarctic continent, and Australia. These propositions are, for the most part, mere conjectures. It is likely that this new problem is also much more difficult and more complex than we suppose; that the immediate ancestors of the human kind, or the first of the human beings, were often displaced in the course of their long evolution corresponding to geological ages the vicissitudes of which, along with the other mammals, they must have kept pace with and endured. The main fact which palæontology seems to have firmly established is that, starting from very primitive stages, the Lemurian and Platyrrhinian stages, the evolution

[1] Giuffrida-Ruggeri (*Rivista italiana di Sociologia*, xix., 1915, and *Rivista ital. di Paleontologia*, xxiv., 1918) has revived and modified this first theory. In his opinion there is good reason for admitting the existence of as many "cradles" as there are elementary species, Whites, Yellows, and Blacks, without necessarily imagining as many geographical regions far distant from each other. He admits, however, at the foundation of all, one single "phylum" of ancestral representatives.

of that group which potentially comprised the human branch did not take place either in North America, whence all Primates seem to have disappeared since the Upper Eocene, nor in South America, where the Platyrrhinian branch has dominated exclusively. It is therefore in the Ancient Continent that we must seek our "cradle." Mankind is a product of the Old World.

Does the present state of our knowledge enable us to make a more definite statement? The part played by Asia, and, especially by Southern Asia, must have been considerable. The Siwalik fossils show that in that region, about the Upper Miocene and Lower Pliocene periods, there was a most extraordinary flux of life, especially among the higher Primates. In view of the number and the diversity of form of the great fossil apes already described, the impression arises that, at this time, Asia was the laboratory where the differentiation of the ancestors of Mankind must have been in process of elaboration.

The American palæontologist Matthew,[1] in a very suggestive memoir, states the reasons which lead him to believe that the centre of dispersal of Mankind should be placed farther to the North, towards the great central Asiatic plateau. He lays stress on a statement, frequently made previously, that the living human races regarded as the most primitive are found in regions farthest from this centre: the Australians, Andamans, Veddahs, African Negritos and Negroes, Bushmen, the Fuegians in the extreme south of South America, and the Esquimaux in the far north. But his theory is confronted by a difficulty connected with the probable coolness, if not indeed actual rigour, of the climate of Central Asia at this period. Yet the hypothesis, often suggested, that early Men were first of all adapted to a tropical climate can only be partially true. It is more reasonable to consider the loss of hair in the human race as due to the wearing of clothes than to attribute it to simple climatic changes.

Matthew's theory agrees with that which de Quatrefages

[1] Matthew, W. D., "Climate and Evolution" (*Annals of the New York Academy of Science*, xxiv., 1915).

long since put forward and ably supported, arguing both from the study of modern human populations and the correlated distribution of the three great linguistic groups, the monosyllabic, the agglutinative, and the inflected languages.[1] But the speculations of de Quatrefages bear on anthropological facts very near to us in time, and on circumstances probably very different from the original conditions. So that, even if we admit that Asia must have played a great part in the diffusion and distribution of very ancient human groups, we are much less justified in declaring that Asia was the scene of the transformation from the pre-human anthropoid stage into the human stage. There is no justification for excluding, in this respect, the continent of Africa, still so full of mystery, and perchance even some continent now buried beneath the ocean waves.

* * *

Prehistoric anthropology is still too poor in osteological evidences to be able to reconstruct the genealogies and migrations of the human beings who have peopled the different continents, starting from the centre of development and dispersal of the first human beings. The rare skeletons or fragments of skeletons of our most distant ancestors which we possess have been exhumed from the soil of Western Europe. Now this region, I cannot too often repeat, is but an advanced point of Eurasia, a sort of cul-de-sac, into which successive waves of many human tides have broken; it is in no respect a continental centre where, any more for Man than for the other mammals, we may watch the scene of an unbroken evolution.

The great antiquity of the Heidelberg jaw obliges us to admit a still more unbelievable antiquity for the first representatives of Mankind, and this view agrees with those geological observations which lead to the belief that certain layers in India containing dressed flints perhaps date from the Tertiary era. It is impossible to relate Heidelberg Man to one of the great human types, from which it entirely differs, judging at least from the jaw-bone, the only relic of it we

[1] *Introduction à l'étude des races humaines*, p. 131.

possess. For it, it is necessary to make a special species, the ins and outs of which so far we do not know, but which is still very simian.

Countless centuries separate Heidelberg Man from Neanderthal Man, for geology and palæontology teach us that the archæological periods called by prehistorians Pre-Chellean, Chellean, and Acheulean, correspond to a long series of physical and biological events, and, in consequence, to an immense lapse of time—much greater than that of all the more recent archæological periods added together.

The distribution of amygdaloid implements extends not only in the direction of Europe; it occurs also over the whole of Africa, as well as in Australia and America, and this presupposes a past of vast duration, provided we admit the single origin of these primitive industries rather than the spontaneous and independent origin in every continent, a possible but improbable occurrence. It may be that this origin was rather African than Asiatic, for it would really seem as if our oldest European Palæolithic industry had come from Africa, by way of land-bridges which then united that continent with Europe.

However that may be, we know nothing or almost nothing of the bodily structure of the men who lived in our country during the long periods of time which separate the deposit of the Mauer sands from the cave deposits whence the skeletons of Moustierian Man were exhumed.

On the contrary, we have good grounds for affirming that Neanderthal Man, the successor in Western Europe of these men, is still an isolated type compared with the various representatives of modern Mankind. Perhaps he is descended from Heidelberg Man, perhaps he represents a new type, entirely different from those of the warm fauna, a type which may have come to us from the North with the glaciers and the cold fauna which accompanied him. In any case, he represents a peculiar species, the terminal bloom of a twig, now withered and dead, of the human branch.

We begin to suspect that there existed other contemporary

twigs. The Grimaldi Negroids are of a geological age equal to that of the majority of our Moustierian skeletons, and these Negroids, inhabitants of warmer countries, already form part of the great group of modern forms. The origin of the latter must be lost in a still more distant past, as the Piltdown discovery seems to show, and perhaps as do also certain other finds, of which we have purposely not taken much account because their origins are not sufficiently authenticated.

Only when we come to the Upper Pleistocene, the Reindeer Age, relatively much less distant in time, can we affirm, positively and indisputably, that there existed on our soil human forms higher in every respect and now readily comparable to living types in various parts of the globe. And it is interesting to observe that even if these Men, grouped around the type known as "Cro-Magnon," are already Whites, they occasionally exhibit many points of resemblance, on the one hand, with the Yellow, and on the other with the Black races.

The differentiation of human types was, then, already far advanced before the end of Quaternary times. Evidences collected by prehistoric anthropology enable us, very imperfectly it is true, to follow the progress of this differentiation under manifold influences, first of a physical and geographical and later of a political kind, and to see it result in the extraordinary medley of ethnical groups which correspond to prehistoric, proto-historic, and modern times.

Sparse and fragmentary as they are, our European evidences may be looked upon as rich compared with those of the other great regions of the earth's surface. And no attempt at synthesis, even of a rudimentary kind, can be attempted until Asia and Europe reveal some of the secrets they now withhold.

We know almost nothing of Asiatic human palæontology, apart from a few archæological facts, which are, nevertheless, of great interest. But, from the point of view which is of particular interest to us, great efforts still remain to be made on this continent. If, by some magic art, we could divest it of its great covering of superficial deposits, what discoveries of prime importance we should be called upon to record. According to the latest work of geologists, the great continental formation of "yellow earth," or loess, is not simply the result of recent wind action. Like the Pampas formation of South America, it represents a complex formation of layers of different origins, the earliest of which, dating from beyond Quaternary times, contain abundant remains of various mammalian faunas.[1] There is every reason to hope that these faunas comprise human or prehuman beings which, one day or other, science will be able to study.

From Africa, apart from the very doubtful skeleton from Odolway and the very curious Broken Hill skull with its recent appearance, the Boskop fragments are the only osteological human evidences we can attribute to a geological period previous to the present. But is it not interesting to observe that here already we have an African type?

The same remark applies to the skulls from Talgai and from Wadjak, which, in all their characters, may be recognized as those of pre-Australians.

Finally, the most ancient skulls from the two Americas, whatever their degrees of antiquity, already possess the main features of the skulls of Americans or rather of Amerindians.

From these relics, so few in number, so scattered, so widely separated from each other, it would seem that a fact of the utmost interest may be gathered, namely that the differentiation and the geographical distribution of the principal human types everywhere date very far back into the past. From this new point of view, Mankind appears more immensely old than ever; its development is bound up in a great series of events which we cannot yet follow nor even picture in complete sequence, either from a geological or from an anthropological point of view.

* * *

Our knowledge of the physical evolution of Mankind is still very rudimentary; and here care must be taken to distinguish physical evolution from moral and intellectual evolution. The

[1] Anderson, J. G., "Preliminary Description of a Bone-deposit at Chow-Kou-Tien in Fang-Shan-Hsien, Chili Province" (*Geografiska Annaler*, 1919).

study of the latter is rather more advanced, owing to the vast accumulation of facts revealed by prehistoric archæology in every country. In this case also, however, only a partial or regional synthesis can be attempted. As J. de Morgan[1] has said, "The unfortunate thing is that the majority [of countries] where the first civilizations developed present conditions which make research there particularly difficult."

In Western Europe, which is specially favoured in this respect, it is possible to trace a general and fairly accurate picture, at least so far as the foreground is concerned, that is to say so far as the prehistoric ages which lie nearest to us. Thus, away beyond the times corresponding to the ancient history of historians, which is in reality only ultra-modern history to the prehistorian and even more so to the palæontologist, we are fairly well acquainted with the habits of Neolithic Man and of the Men of the Reindeer Age.

Their intellectual and moral development and their culture can be readily compared with the state of certain peoples still living, or but lately living, in more or less pronounced conditions of savagery. Parallelisms can be drawn even in a great number of details relative to their psychic and moral life as well as to their material life. A whole series of ethnographical facts clearly reveals the same mentality and the same level of intellectual development. Mankind in the Reindeer Age of our country was already a superior Mankind, essentially resembling modern Mankind, and endowed with the same intelligence, the same inventive genius, and the same sentiments.

Our Moustierian Men, much more ancient, lived in a state more primitive in every respect. Nevertheless it is possible to compare them, if not from the physical aspect, at least from the moral point of view, with some specially retrograde peoples in certain portions of the globe, which lead a life extraordinarily like that which *Homo neanderthalensis* must have led. Already, in spite of the structural inferiority of his brain, the latter is a Man, a *Homo*, and in nowise a pre-Man, for accompanying his skeleton, lie in confusion the stone implements

[1] *Les premières civilisations*, p. 39. See also, by the same author, *L'Humanité préhistorique*, Paris, 1921.

which he knew how to make, along with charcoal and cinders from the hearth-fires he knew how to light and feed. Already his methods are those of certain modern savages. And were naturalists, abandoning their general methods, to give prominence to intellectual characters in classifying the creatures they study, there would be no occasion to separate *Homo neanderthalensis* from Modern Man in a specific sense, although, as we have seen, we cannot refuse him this distinction on account of his physical characters.

We know very little of the ethnography of the peoples who for so long a time, in the Acheulean and Chellean periods, occupied our own land and a great part of the surface of the globe. We can, however, affirm that they also were true Men, in the full sense of the word, in habit as well as in physique; for, with selected materials, they knew how to make implements, indeed very fine implements; already an æsthetic feeling accompanied their spirit of invention. They knew how to make fire, that distinctive human accomplishment which lay at the foundation of all future progress, which contained the potentiality of all civilization, and the discovery of which "constitutes the most characteristic act of genius of which Mankind can boast."[1]

Indeed, the invention of primitive implements and the production of fire result from intellectual qualities as marvellous as the greatest modern inventions they made possible of achievement. And in this respect we cannot but admit Rémy de Gourmont's law of the intellectual constant, and, up to a certain point, the doctrine of "psychic unity" held by certain philosophic anthropologists.

As for Heidelberg Man, *Homo heidelbergensis*, he was possibly only a pre-Man, a forerunner. It may, indeed, be possible, though we are not justified in saying so, that he spoke an articulate language, that he knew how to light a fire and to dress flints, that he already represented Bergson's *Homo faber*.[2]

[1] Rémy de Gourmont, *Promenades philosophiques*, 2nd series, 1908, p. 11.

[2] "Were we able to divest ourselves of all pride; were we, in order to define our species, to adhere strictly to what history and prehistory show to be the constant characteristic of Man and of intelligence, perhaps we would not say *Homo sapiens*, but *Homo faber*" (Bergson, H., *L'évolution créatrice*, 22nd edition, 1920, p. 151).

To fix the main point in the evolution of Mankind, it would be most important to know the exact moment when the pre-human anthropoid, attaining at a single bound the dignity of the human status, for which his physical and mental evolution had prepared him, on the one hand learned how to kindle and maintain a fire, and on the other, passed from the habit of using rude stones to the manufacture of an implement.

Regarding the invention of fire, we have only very vague ideas. The oldest archæological cave deposits, beds in which only Acheulean and even Chellean industries are represented, already contain pieces of charred wood and cinders. But we are not sure that certain still more ancient deposits, likewise attributed by prehistorians to their Chellean or pre-Chellean, do not afford similar evidences. The invention of the first implements constitutes a problem of which the scientific explanation may still perhaps be long delayed, for reasons I have explained in connection with the fascinating "eolith" theory.

* * *

Such, to my mind, are the principal conclusions to be drawn meantime from our knowledge of human palæontology. The day will come when much more abundant material, collected from nearly all parts, will make the history of early Man less obscure and less discontinuous. At the present moment, it is possible to state that the human group is more branched and more diversified than had been supposed, and that the evolution of this group obeyed the biological laws regulating all creatures, from the lowliest to the highest.

Already certain writers have been able to disentangle, from the mass of historical facts, certain causes of a general character, independent of and dominating the more artificial factors. Dimly they have perceived the link between human history and natural history, the close relations between the laws of empire and those of organized beings. As Edgar Quinet in particular remarked: "There are points in common between the revolutions of the globe and the revolutions of the human kind, as if they both belonged to one single plan which unfolds from age to age. . . . All the elementary laws of Palæontology are to be found and can be verified over great fields in the history of human society. Words change, but the principle remains the same in fossil nature and in the world of Mankind."[1]

In showing that Mankind has steadily progressed during the few thousands of years which it records, history has taught us that this progressive evolution is not effected in a continuous manner, that is, that it is not brought about by the whole of Mankind acting simultaneously or on the same spots, but by different and successive groups acting in different countries. From history we have likewise learnt that the purely ethnographical groups have, like the more natural groups, only a limited duration of life, that they are born, grow, and die, and that often, if not always, their final decline follows very soon after the climax of their development has been reached. Such disappearances, however, do not entail the stoppage of the progressive movement of Mankind as a whole. The movement is renewed by other groups, which have hitherto remained in the background, and which, taking advantage of the results acquired by their predecessors, develop more or less rapidly, and in their turn add riches to the common intellectual treasure, before succumbing like the rest. The general perfection of the human race, in historic times, may be represented as the sum total of the partial advances made through the repetition of this kind of cycle.

This succession of relays in time and in space, which, once again, confirms the great solidarity of Mankind (a familiar idea in history), is not peculiar to history alone. Human palæontology teaches us that it acted in the same manner in prehistoric and even in geological times. It is only a special aspect of the mechanism of evolution which general palæontology reveals in its reconstruction of the genealogy of any group of animals.

Obvious progress has indeed been made from the Lower to the Upper Palæolithic, thence to the Neolithic Age, from the Neolithic to the Ages of Metals, and from the latter to

[1] Edgar Quinet, *La Creation*, vol. ii., p. 227.

historical times. But this progress is the collective work of peoples very different from each other and inhabiting various countries.

The continuity of the gradual perfecting of Mankind, since the use of the first flints and the use of the first fire, cannot, therefore, be represented by a straight ascending line, but by a succession of broken lines, the different segments of which are joined together like the branchings of certain plants. This continuity is thus only the apparent result of partial and discontinuous progress, which projects itself into the terminal branch, the highest of all. This branch, to continue the metaphor, is formed of a succession of different components, each of which has served as a basis of progress at a given moment, but not one of which can claim to have been the exclusive factor. At all periods and in every group of creatures, there are individuals which have advanced and others which have been left behind. The latter yield under pressure from the former, biding the time when these new masters of the moment shall undergo the same fate. How striking are the examples of this truth which the great events of prehistory reveal to us: the quick succession from the Reindeer Age to the Moustierian, from the Neolithic to the Reindeer Age, and so on.

In the long series of changes and transformations, the final result of which is to raise higher and higher the terminal twig of our branched system, many components of the system have disappeared never to return. There has been the extinction of certain groups, such as that of Neanderthal Man. Their place has been taken by other groups, whose development, until then latent, suddenly sprang into fresh life and vigour.

This is true, not only of the successive representatives of the genus *Homo*, but also of those forerunners of his who link Mankind genealogically to the great trunk of the Primates. The ramifications of the latter took place in the same way, by divisions into branches, branchlets, and twigs; some of these simply vegetated before dying, while others gave rise to finer products and then withered in their turn, to be replaced by new shoots still higher in the scale, whose vigour, nevertheless, lasted or will last only for a time.

Man, in spite of his superior attributes, thus takes his place in the scheme of general organization, and forms no exception among living creatures. In stepping backwards from history to prehistory and from prehistory to geological times, we see that always he is subject to the laws which regulate the evolution of every living thing. And it is because of this that human palæontology and general palæontology are alone found to be capable of guiding us to a full understanding of the true place of Man in nature. His true superiority, of a purely intellectual order, gradually acquired in the course of a slow and laborious evolution, now enables him to raise a corner of the veil which conceals from him both the lowliness of his origin and the glory of his ascent.

APPENDIX

The Problem of Tertiary Man.[1]

Recently Mr Reid Moir has collected in many places among the Pliocene deposits of the Ipswich region numerous flints which he considers were dressed by the hand of man.[2] In particular, the new specimens which he obtained in the Red Crag layers at Foxhall seem to have clinched the opinions of several naturalists and historians, who up to that time had remained doubtful or undecided.

The most important of these conversions is that of my learned collaborator, M. Breuil, to whom Mr Moir showed, in the upper portion of the Red Crag, a level where "flints are scattered, sparse, sharp cornered and not large, as in the case of a level presenting a true industry," where "the flaking is well defined, accompanied by trimming and retouching"; where he noted that "the cutting was produced by successive parallel strokes," and where "burnt flints are not lacking."[3]

While I am quite prepared to admit the great interest of these observations, and quite recognize the fact that they bring forward new arguments in favour of the theory of human handiwork, I cannot share the opinion of certain of my fellow-workers, more enthusiastic than critical, that they definitely settle the question. On the one hand, indeed, the intrinsic characters of the new finds, convincing though they appear, seem to me to be not incapable of an interpretation other than that of intentional dressing; on the other hand, I am still under the influence of the unfavourable impression regarding the human hypothesis produced in 1911 by my studies of the

[1] See p. 137.

[2] Among Mr Reid Moir's numerous papers, see especially *Pre-palæolithic Man*, Ipswich, 1919; "Further Discoveries of Humanly Fashioned Flints in and beneath the Red Crag of Suffolk" (*Proc. Prehist. Soc. of East Anglia*, 1920-21).

[3] Breuil, M., "Les industries pliocènes de la région d'Ipswich" (*Revue anthropologique*, July-August, 1922, p. 228).

geology of the layers *in situ.* Many of the objections explained in the preceding pages are still valid, and constrain me to maintain, at least until new evidence is forthcoming, the caution which is imposed in such a matter on those who have learnt prudence through long acquaintance with science. As Pascal said, "It is necessary to know how to doubt when doubt is necessary."

The Weimar Fossils.[1]

In 1916, two years after the discovery at Ehringsdorf of the jaw described by Schwalbe, a few bones of the skeleton of a child aged about ten years, in particular a lower jaw and some isolated teeth, were

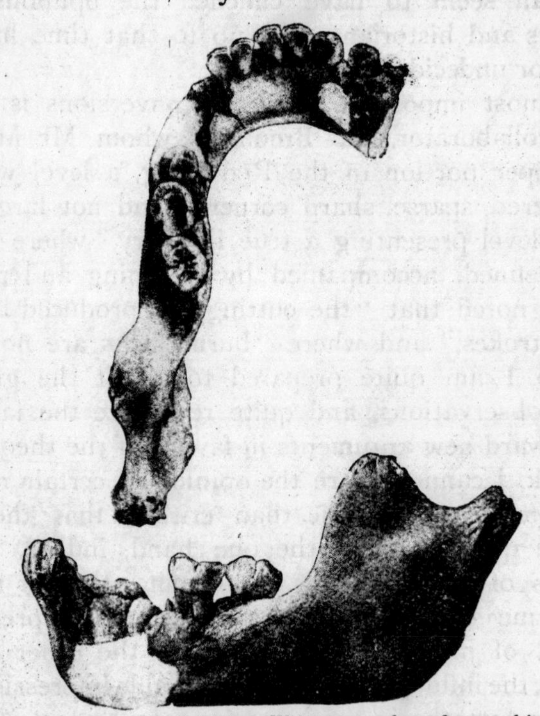

FIG. 241.—Lower Jaw of a Child from Weimar, seen from above and in profile. Two-thirds natural size. (After Virchow.)

collected at the same level. These new relics were studied by Hans Virchow,[2] according to whom the lower travertine here contains, along with dressed flints consisting of points

[1] See p. 147.

[2] *Die menschlichen Skeletreste aus dem Kämpfe'schen Bruch im Travertin von Ehringsdorf bei Weimar*, Jena, 1920.

and scrapers of a Moustierian type, bones of Merck's Rhinoceros, of the Brown Bear, of a Horse, Red Deer, and Elk, shells of molluscs and imprints of plants, all denoting a climate similar to that of Thuringia at the present day.

This child's jaw (Fig. 241) is of the same type as the adult jaw described by Schwalbe, except that certain characters are less pronounced. The alveolar arch is relatively larger, the dental prognathism less marked. It is remarkable for the same absence of chin, the same thickness of the lower border, the same horizontal arrangement of the digrastic muscle insertions (see p. 153). In this case, one of the ascending rami, in good preservation, resembles the ascending rami of jaws of Neanderthal Man. The teeth are intact; the canine does not rise above the level of the neighbouring teeth; in all their structural details the molars are essentially human, and the last true molar is not reduced in size.

Is Eoanthropus an Artificial and Composite Creature?[1]

After long hesitation I reached the conclusion, which I expressed in the first edition of this book that the remains of *Eoanthropus* included those of two creatures, a Man and a Chimpanzee. This conclusion was based on anatomical facts, but I recognized, nevertheless, that the conditions of the bed pointed to a contrary conclusion. It is certainly difficult to imagine the presence, in the same spot in the depths of an ancient alluvial formation, of remains belonging to two species of large Primates, and to explain that, by a freak of chance, these remains present the same physical characters, belong to beings of the same stature, and supplement and complete similar portions of the skeleton.

The discovery of 1915, found at a distance of two miles from the original site, seems to reinforce this argument. It consists of (1) a fragment of a frontal bone and a fragment of an occipital, similar to the corresponding bones in the individual first discovered; (2) a first molar, which also is similar to the first molar found in position in the jaw. This association, which exactly repeats that of the first discovery, is regarded as "providential" by the American palæontologist,

[1] See p. 171.

Osborn, and has brought about his conversion to Smith Woodward's views.[1]

On the other hand, a new reconstruction of the Piltdown skull seems to have enabled Elliot Smith and Hunter[2] to note that in its occipital region, which is more vertical in direction than had been supposed, this skull presents characters resembling those of the skulls of young anthropoid apes. The new reconstruction would seem to be more in keeping with the structure of the jaw, and this should eliminate or at least greatly lessen the "anatomical paradox" which has been so much discussed.

In face of these new facts, I am not inclined to be so positive as formerly. I recognize that the balance inclines somewhat more to the side of Smith Woodward's theory, and I am glad of this, for I esteem both the knowledge and personal attributes of this scientist. But I must add that my doubts have not been completely laid to rest, and that the arguments for and against do not seem to me decisive; it is still permissible to suppose that the Piltdown skull and lower jaw may have belonged to two different beings.

Neanderthal Man: the Child's Skull from La Quina.[3]

The fine discovery, recently made at La Quina, of the well-preserved skull of a child (see p. 192) justifies the statement that the majority of the very special facial characters of Neanderthal Man which I have described (p. 202), seem to be present in Neanderthal individuals even from youth (Fig. 242). According to H. Martin[4] the skull of the child from La Quina, aged about eight years, presents an ovoid appearance and a cephalic index of 77 (sub-dolichocephalic), whilst modern children

[1] Osborn, H. F., "The Dawn Man of Piltdown" (*Natural History*, Nov.-Dec., 1921, p. 581). The fact does not seem to me to possess the importance attributed to it by Osborn, for the opinion of the authors who attribute the Piltdown jaw to a Chimpanzee is chiefly based on the structure of the jaw. Now, no such specimen was found on the second occasion in association with fragments of the human skull; a tooth only was found, and a tooth particularly difficult to identify in that it was the first lower molar, which of all the true molars in Man is most difficult to distinguish from those of anthropoid apes.

[2] *Nature*, 3rd June 1922, p. 726, and personal communications from Prof. Elliot Smith to the author.

[3] See pp. 192 and 205.

[4] Martin, H., "Un crâne d'enfant néanderthalien du gisement de la Quina" (*L'Anthropologie*, xxxi., 1921, p. 331).

are generally brachycephalic. The general outline is depressed. It has a receding forehead, without frontal bosses; the orbital arches are already continuous and projecting. The occipital region is beginning to protrude in chignon-like form. The mastoid apophyses are rudimentary, almost negligible. The face is Neanderthal in type, both on account of its muzzle-like appearance due to the absence of canine fossæ, as well as on account of its immense rounded orbits and broad nose.

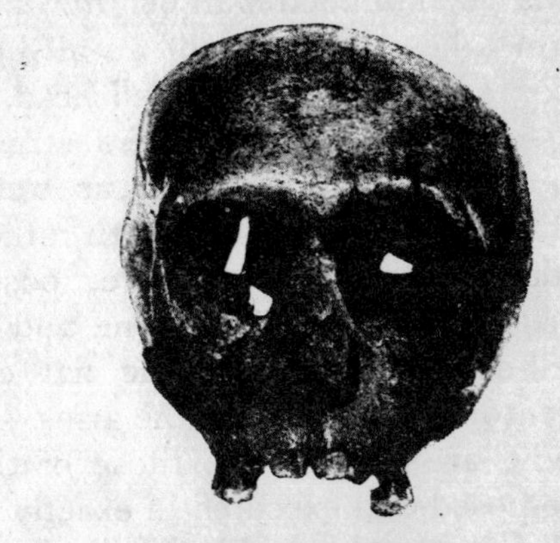

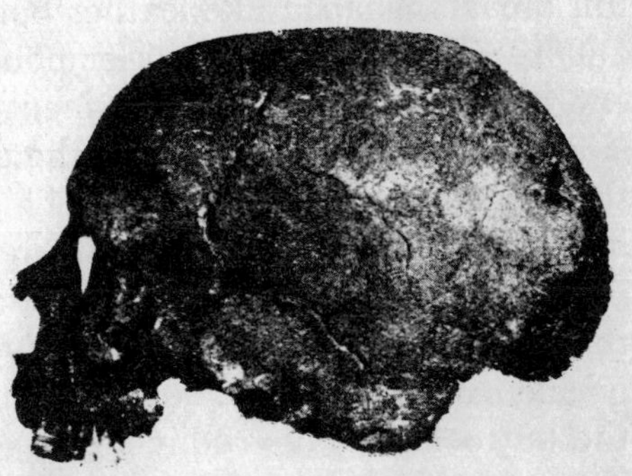

FIG. 242.—Child's Skull from La Quina, seen full face and in profile. One-third natural size. (Photo, H. Martin.)

This skull therefore differs greatly from the skulls of children of *Homo sapiens*. It shows that the cranial morphology of *Homo neanderthalensis* is a very ancient morphology, dating from a very distant past, and that it is really specific in character.

Neanderthal Man: Brain Capacity.[1]

The intracranial cast of the Man from La Ferrassie which I have just had made, is at least as large as that of the specimen from La Chapelle-aux-Saints. This new addition to the little series raises the average figure to 1450 cubic centimetres.

Aurignacian Human Figures.[2]

In August 1922 a discovery of a new human figure was made by M. and Mme. de Saint-Périer[3] in the Grotte des Rideaux, at Lespugue in the Haute-Garonne. It consists of a statuette in mammoth ivory, well

[1] See p. 231. [2] See p. 303.

[3] Saint-Périer, Dr René de, "Statuette de femme stéatopyge découverte a Lespugue (Haute-Garonne)," *L'Anthropologie*, xxxii., 1922.

preserved, almost complete and of such beauty as to establish it as a queen amongst the Aurignacian "Venuses" (Fig. 243). It is 147 millimetres in height. In general outline its graceful form takes the shape of an all but regular lozenge, the upper apex occupied by the small head, the lower by the attenuated legs. It presents the general structural characters of the works of art already described, as well as their peculiar features. Thus the head, covered with rather short hair, which partially hides a face without features, possesses the graceful outline of certain heads from Mentone and Brassempouy. The slender, well-defined neck and the flat chest are also found in the Mentone specimens. The arms, standing out slightly from the body, are slender like those of the Willendorf statuette, and are posed and executed in exactly the same fashion. The long, full breasts, shaped like leather bottles, hang over the abdomen and recall one of the Brassempouy ivories. The buttocks are enormous, not specially prominent behind but greatly enlarged sideways. The lower part of the abdomen is damaged, so that the sexual parts are absent. Thighs and legs are tapered as in the Willendorf and Laussel specimens; the feet are barely indicated by slight forward projections. But what distinguishes this statuette from all others is the presence of a sort of garment, a waist-cloth, covering the back of the thighs below the buttocks. This rudimentary garment seems to be made of vertical strips or straps starting from a transverse cord and ending below in a fringe.

M. and Mme. Saint-Périer have been so good as to offer this superb specimen of Palæolithic art to the Palæontological Gallery of the French National Natural History Museum.

Azilian Human Remains at Mugem.[1]

The study of the human skeletons from the kitchen-midden of Mugem has been taken up again in recent years by Mendes Corrêa.[2] In these skeletons, to the characters of which, as indicated on p. 334, may be added the facts that the cranium is high

[1] See p. 334.

[2] Mendes Corrêa, A. A., "A propos des caractères inférieurs de quelques crânes préhistoriques du Portugal" (*Archivio de Anatomia e Anthropologia*, vol. iii., 1917); "Origins of the Portuguese" (*American Journal of Physical Anthropology*, vol. ii., 1919).

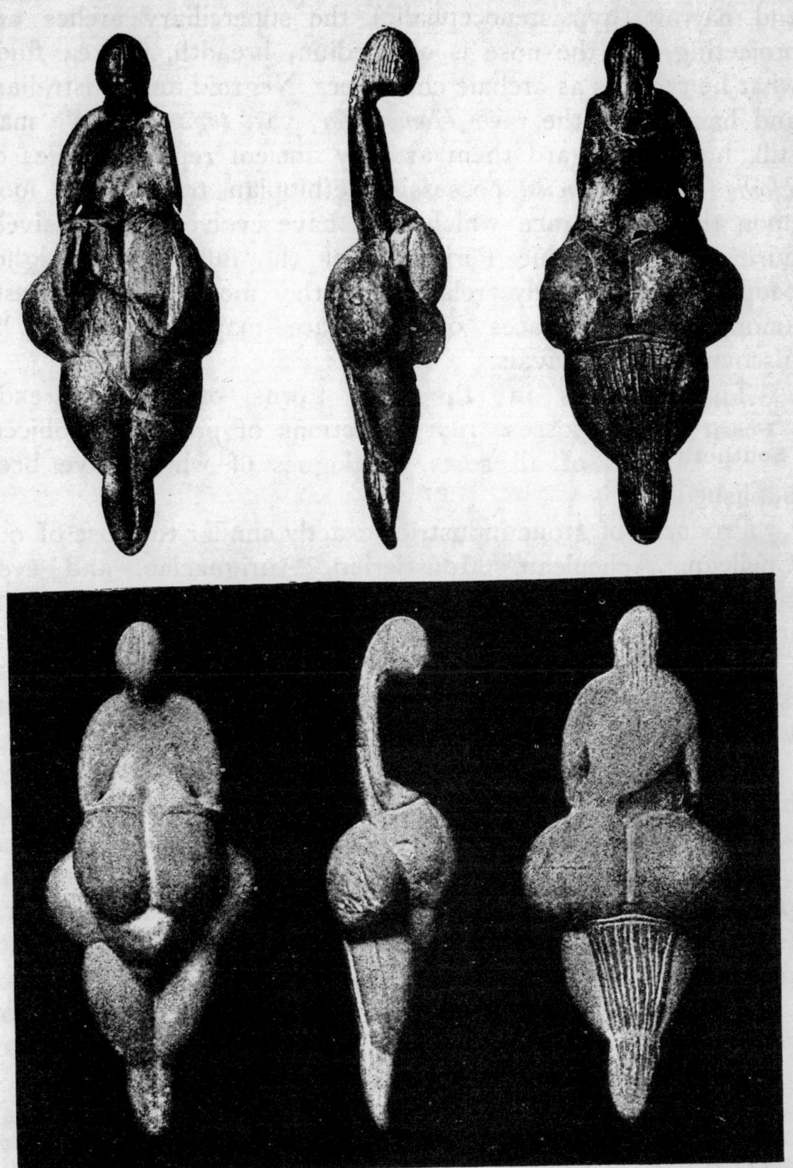

FIG. 243. The "Venus" from Lespugue, front view, profile, and back view. Above, the statuette as found. Below, the statuette restored (plaster cast). Half natural size.

and narrow (hypsistenocephalic), the superciliary arches are projecting and the nose is of medium breadth, Corrêa finds what he regards as archaic characters, Negroid and Australian, and has named the race *Homo afer*, var. *taganus*. We may still, however, regard them as very ancient representatives of *Homo mediterraneus*, possessing ethiopian traits; and look upon them as a race which may have evolved progressively during the Neolithic Period under the influence of higher peoples more closely related to the modern Portuguese, amongst whom traces of this race may occasionally be discovered as survivals.

Fossil Man in Southern Asia.[1]

The museums in the chief towns of India already possess rich collections of prehistoric objects of all sorts, catalogues of which have been published.[2]

Products of stone industries exactly similar to those of our Chellean, Acheulean, Moustierian, Aurignacian, and even Solutrean periods are widespread on the surface soil, especially in South-East India. The neighbourhoods of Madras and of Cuddapah, which are particularly rich in this respect, constitute a region where the technique was gradually perfected, and which became a great centre of dispersal for Palæolithic culture.[3]

The caves in the Karnul region, still imperfectly explored, have yielded implements resembling those of our Magdalenian, although they seem to date from a more ancient period. There can still be distinguished in India a late Palæolithic phase—represented by pygmy flints, Tardenoisian in form (see p. 330), from Banda and the Vindhyan Hills—which passes without break into the Neolithic period.

Discoveries of cave paintings and engravings are now relatively numerous, and of a most interesting nature. Red

[1] See p. 357.

[2] Foote, R. Bruce, *Catalogue raisonné of the Foote Collection in the Madras Museum*, 1914. *Notes on the Age and Distribution of the Foote Collection of Prehistoric and Protohistoric Antiquities in the Madras Museum*, 1916. M. A. Rea's *Catalogue of the Prehistoric Antiquities from Adichannallur and Perumbair*, 1915. Brown, J. Coggin, *Catalogue raisonné of the Prehistoric Antiquities in the Indian Museum*, Calcutta, 1917.

[3] Mitra, Panchanan, "Prehistoric Cultures and Races of India" (*Calcutta University Journal of the Department of Letters*, vol. i., 1920; "Prehistoric Arts and Crafts" (*Ibid.*, vol. iii., 1920).

paintings on rocks situated near Singanpur, in the Raigahr district, represent hunting scenes and dances with masked figures, resembling those at Cogul in Spain. One of them depicts Kangaroos, animals now restricted to Australia. Wall engravings at Ghatsila, in the Singbhurn district, are remarkable for their Australian character. According to Mitra, these facts point to the existence of an ancient Indo-Australian culture extending from the Upper Palæolithic to the Neolithic.

FIG. 244.—Painting from a Cave in the Mirzapur district, representing a Rhinoceros Hunt. (After Mitra.)

Cockburn discovered numerous paintings in the caves in the Mirzapur district. Here hunting scenes show men, armed with harpoons having stone shaftheads, attacking Rhinoceroses (Fig. 244). They may be attributed to the end of the Palæolithic period, represented in this area by Capsian and Azilian industries. In the Bellary district, more than twenty groups of animal drawings and hunting scenes with men armed with javelins and shields are known, and these are probably Neolithic.

Human Remains from Japan and the Philippines.[1]

Skeletons recently exhumed from accumulations in Japanese kitchen-middens exhibiting a Neolithic industry, according to Matsumoto,[2] present (with certain peculiarities pointing to at least three types) a common stock of characters which relate them to the skeletons of the Ainos, but which, more than these, recall certain skeletons of men from the recent Palæolithic and European Neolithic.

D. Sanchez y Sanchez has just described a human skull found in the town of Manila in the Philippines at a depth of 2.50 to 3 metres in the subsoil, which is formed by the alluvial deposits of the river Pásig. He finds certain Negrito characters in it, and attributes it to a pre-negrito race, naming it *Homo manillensis.* Unfortunately, it is impossible to date this bone relic.

The Wadjak (Javan) Skulls.[3]

Dr Eugène Dubois recently made an important communication to the Academy of Amsterdam regarding two human skulls found at Wadjak in Java in a deposit of rubble and alluvium.[4] This discovery dates as far back as 1890; it was therefore prior to the discovery of *Pithecanthropus* which Dubois made at Trinil a short time afterwards.

The skulls, found in a fragmentary condition, are sufficiently fossilized to be considered of Pleistocene age, in spite of the absence of other proofs of their geological antiquity.

The first, Wadjak I., is that of a female; the second, Wadjak II., is that of a man. They are both dolichocephalic, very different from the Malay type, and though of more robust appearance, have all the characters of Australian and Tasmanian skulls: keeled cranial vault, scaphocephalic; receding forehead; superciliary ridges prominent; low orbits, etc. (Fig. 245).

[1] See p. 361.

[2] Matsumoto, H., "Notes on the Stone Age People of Japan" (*American Anthropologist*, vol. xxiii., No. 1, 1921).

[3] See p. 361.

[4] Dubois, Eug., "The Proto-Australian Fossil Man of Wadjak, Java" (*Proceedings of the Academy of Amsterdam*, vol. xxiii., No. 7, 1921.) English translation of a paper first published in Dutch in the *Verslagen* of the Amsterdam Academy.

The skull, Wadjak I., is exceptionally large for a woman (length 200 mm.; breadth 145 mm.); its cranial capacity may be reckoned at 1550 c.c., a figure larger

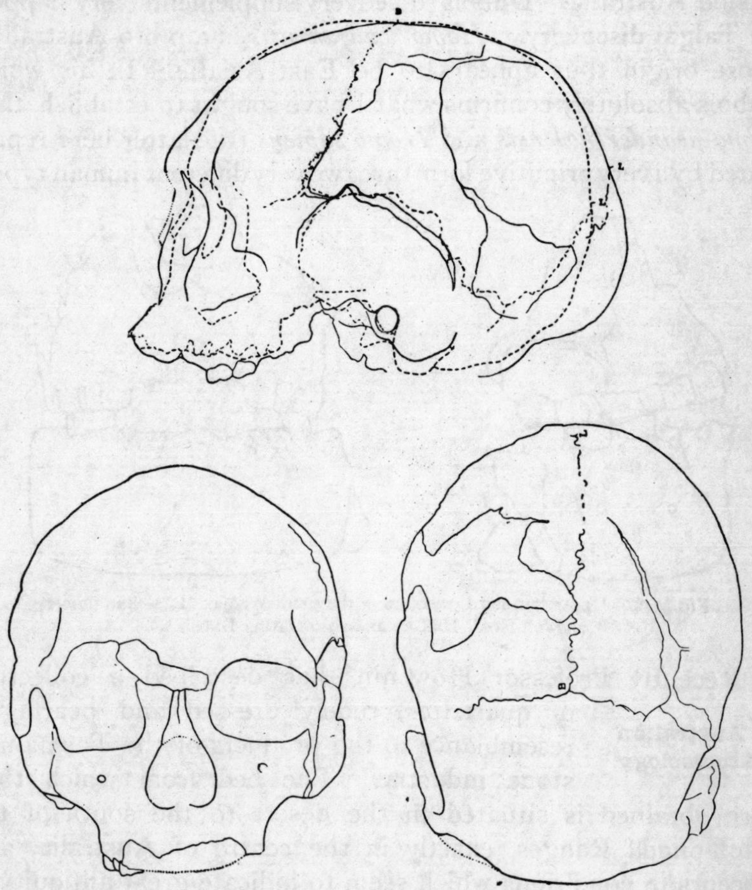

FIG. 245.—Skull, Wadjak I., seen in profile, full face, and from above. One-third natural size. In the drawing of the profile (above) there has been superimposed the outline of a typical Javanese skull to show the differences. (After Dubois.)

than the maximum capacity for Australians and Tasmanians, and also larger than the average for European women. The palate is immense; the upper dental arch overlaps the lower arch.

The jaws, fashioned after the Australian type, are massive

(Fig. 246), as strong, according to Dubois, as the jaw of Heidelberg Man.

For the first time skulls of such character have been recorded outside Australia. Dubois' discovery supplements very happily the Talgai discovery. *Homo wadjakensis* is a proto-Australian whose origin thus appears to be East-Asiatic. In his work, Dubois absolutely confirms what I have sought to establish, that *Homo neanderthalensis* and *Homo sapiens* (the latter here represented by a very primitive form) are two very different human types.

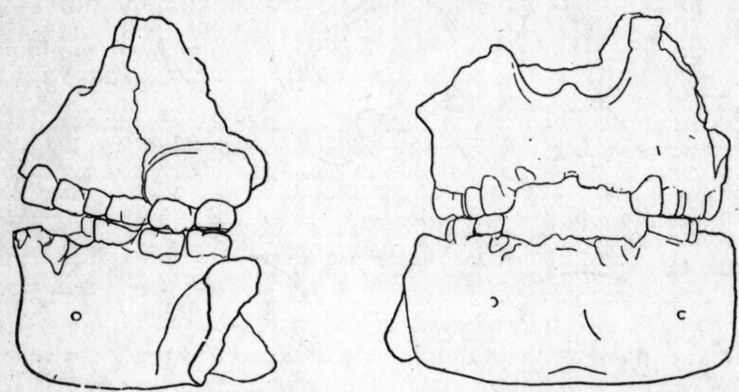

FIG. 246.—Upper Jaw and Lower Jaw of the Skull Wadjak II., seen in profile and full face. Half natural size. (After Dubois.)

Australian Archæology.[1]

Recently Professor Howchin[2] has described a collection of quartzites, rudely dressed and bearing a resemblance to the products of the Tasmanian stone industry. The bed from which they were obtained is situated in the desert to the south of the MacDonnell Ranges, exactly in the centre of Australia, and it occurs in conditions which seem to indicate great antiquity.

Stone Age in Egypt.[3]

Dressed flints, principally showing Moustierian characters, are to be found *in situ* in the ancient and conglomerated alluvials, which are certainly of Pleistocene age, in the neighbourhood of Thebes in Upper Egypt. On this subject the reader should

[1] See p. 366.

[2] Howchin, W., "On the Occurrence of Aboriginal Stone Implements of Unusual Types in the Tableland Regions of Central Australia," (*Trans. and Proc. Roy. Soc. of South Australia*, vol. xlv., 1921).

[3] See p. 374.

consult C. G. Seligman's "The Older Palæolithic Age in Egypt" (*Journ. Roy. Anthrop. Institute* of Great Britain and Ireland, vol. li., 1921).

Stone Relics in Algeria.[1]

In recent years, in the southern portion of the Constantine area, M. Reygasse[2] has explored certain isolated and homogeneous stations where the typical Moustierian industry is accompanied by tanged or hafted flints, such as are ordinarily attributed to the Neolithic Age. One of these beds, situated in the heart of a geological formation, at the wady Djebbana, contains also bones of animals, the examination of which has not yet been completed. M. Reygasse is of opinion that another industry, with a Solutrean facies, originating directly from the Acheulean industry, may be contemporaneous with this Moustierian.

The Sahara.[3]

For a recent discussion of the meteorological changes which have taken place in the Sahara, see R. Chudeau's "L'hydrographie ancienne du Sahara, ses conséquences biogéographiques" (*Revue scientifique*, 23rd April 1921); and as regards the "Oueds fossils," see E. F. Gautier's account in *Annales de Géographie*, 15th January 1921.

Southern Africa.[4]

Reginald A. Smith has reported observations, similar to those from Robinson, Orange River Colony, which he has made at Victoria West and at Vosburg, where an ancient conglomerated alluvium contains stones dressed in almond shape.

Human Remains in Southern Africa: Broken Hill Skull.[5]

The latest discovery in Southern Africa was recently brought to the knowledge of the public in November 1921 by *The Times* and *The Illustrated London News*. The first purely scientific papers on the subject are by Dr Smith Woodward.[6]

In Northern Rhodesia there is a hill, known as *Broken*

[1] See p. 376.

[2] Reygasse, Maurice, *Nouvelles études de palethnologie maghrébine*, Constantine, 1921.

[3] See p. 379. [4] See p. 384. [5] See p. 395.

[6] Woodward, A. S., "A New Cave Man from Rhodesia, South Africa" (*Nature*, 17th November 1921). "A Guide to Fossil Remains of Man in the British Museum," 3rd edition.

Hill, where the minerals lead and zinc are mined. It is tunnelled by a long cave long known in the district for its decorative stalactites, and for the abundance of animal bone-remains, more or less fossilized and mineralized, which it contains.

The work of mining, carried on in the open air, has already partially demolished the hill, thus exposing the great subterranean recess. Towards the end of the summer of 1921, at the further end and lowest depths of this cave, a skull and other human remains were found in association with stone and bone implements and broken bones of animals which had evidently been used as food. The implements hardly differ from those of modern Bushmen. The broken bones all belong to species still living in Rhodesia. Their appearance is very fresh, although the surface is incrusted with a fine layer of silicate of zinc (hemimorphite).

The human remains consist of a skull, unfortunately lacking the lower jaw, a portion of the upper jaw of a somewhat smaller skull, a sacrum, and portions of femora and of a tibia. They present exactly the same physical characters as the animal bones found along with them; they have not lost their organic matter and are not fossilized.

The skull is totally different from that of any other type of modern African whatsoever. On the contrary, it resembles in an extraordinary degree the skulls of the Neanderthal Man of our old European Palæolithic. It has an even more brutish or simian aspect (Fig. 247).

Its length is 210 mm., its greatest breadth 145 mm. It is therefore elongated in form, dolichocephalic. The height of the cranial vault is 131 mm. These various figures agree remarkably closely with those of the skull of the Neanderthal Man from La Chapelle-aux-Saints. But the brain must have been much smaller. The cranial capacity of the skull from Broken Hill was only about 1280 c.c.

The orbital arches are as prominent as in Neanderthal Man; the forehead is even more receding. The face thus greatly resembles the well-known French fossil: it has the same large flat maxillary bones, without canine fossæ, and consequently the same *muzzle-like* appearance, still further accentuated by the size of the sub-nasal space. The large nasal aperture merges insensibly in the face, as in the Gorilla.

The palate is immense; the dentition is quite human; the canines are normal; the wisdom teeth are reduced in size. All the teeth are affected with caries; this pathological feature has never so far been observed in truly fossil European skulls. The lower jaw must have been very massive.

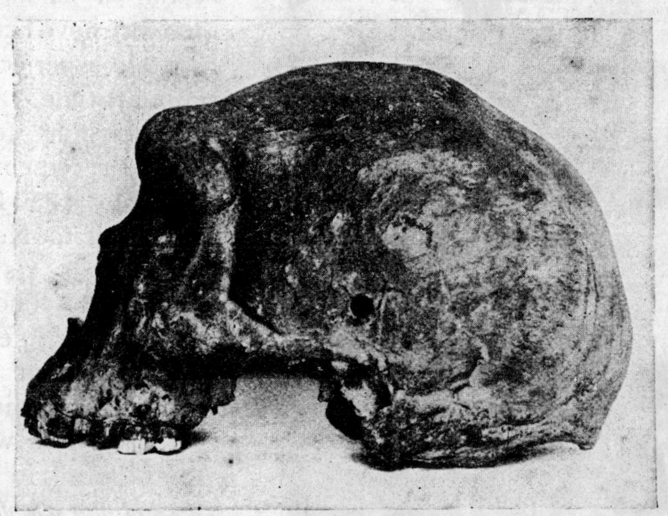

FIG. 247.—Broken Hill Skull, seen in profile. (After Smith Woodward.)

According to Elliot Smith, who has studied an endocranial cast, the brain of this Man is of quite a low type; its prefrontal and lower temporal regions are reduced in size, but it shows an unexpected prominence of the auditory region.

A glance at Fig. 248 suffices to show that the Broken Hill Man is much more like *Homo neanderthalensis* than any other race or variety, living or fossil, of *Homo sapiens*. Smith Woodward recognizes these points of resemblance, but he has given a special name to the skull he has studied, *Homo rhodesiensis*, because, while in Neanderthal Man the foramen magnum occupies a more backward position at the base of the skull than in modern Man—which is in keeping with the general attitude of the body, still not absolutely upright—in

Homo rhodesiensis the foramen magnum occupies a more forward, more central position, so that the skull must have rested with perfect ease and in perfect equilibrium on the vertebral column. The other bones found with the skull likewise bear out this fact. *Homo rhodesiensis*, though possessing so coarse and bestial a skull, must therefore have been able to attain a perfectly upright attitude.

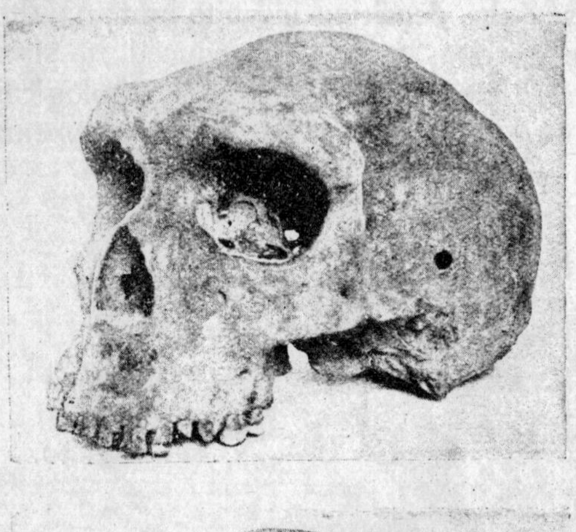

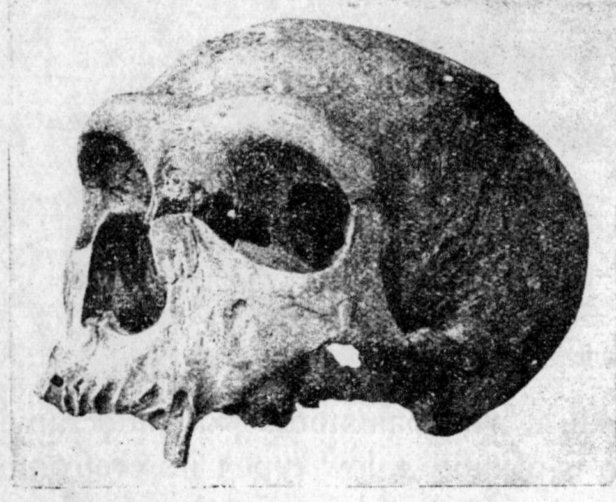

FIG. 248.—Skull from Broken Hill (above) and from La Chapelle-aux-Saints (below), three-quarters view, at almost the same angle to enable comparisons to be made.

According to Dubois,[1] the resemblance borne by the Broken Hill skull to Neanderthal skulls is only superficial. There are numerous differences. The points of resemblance, according to him, are rather with the Australians, in which case *Homo rhodesiensis* would represent a type of *Homo sapiens* akin to the Australian type, but even more primitive than it. Even more than the Australoid skull from Wadjak (see p. 478), the Broken Hill skull deserves the title of "proto-Australian."

In an attempt to draw any conclusion from the structural details, the age of the relics obtained from the Broken Hill

[1] Dubois, E., "On the cranial form of *Homo neanderthalensis* and of *Pithecanthropus erectus* determined by mechanical factors," (*Proceedings of the Academy of Amsterdam*, vol. xxiv., Nos. 6 and 7).

Mine must be taken into account. The depth at which the discovery was made, and on which so much stress has been laid in the newspapers, has not in the present case any great value, as the human skull was taken from a cave into the depths of which it might have found its way in various diverse manners and at any period.

The skull was contained in a superficial deposit covering the floor of the cave, which had yielded "tons" of animal bones. Now these bones, accompanied by rude implements of stone and bone, resemble, as we have said, species still living in the country. The occupation of the cave by these animals and the filling up of the opening by bone-bearing deposits could not, it seems, date back to the Pleistocene period. The excellent state of preservation of the skull, and the absence of fossilization also argue in the same sense.

How can these facts be reconciled: on the one hand, the likeness between *Homo rhodesiensis*, *Homo neanderthalensis*, and modern Australian Man; on the other hand, the presence in Africa, almost in the centre of the Dark Continent, 900 kilometres from the nearest shore, of a human being of relatively recent age (in the geological sense of the word, be it understood), so different from all the races and varieties of African Negroes? The following conclusion may be drawn, at least provisionally.

Neanderthal Man, Rhodesian Man, and the modern Australian race present a common stock of primitive characters. In spite of differences which distinguish them, it may be admitted that the three forms have a common origin; they must have spread and lived for a long period over vast extents of territory. In our country, Neanderthal Man seems to disappear somewhat abruptly after the Glacial Period, but perhaps it was not a total extinction. He may have continued to live in other regions. Indeed, it seems that *Homo rhodesiensis* reveals to us the persistence in Africa of a human type, long since become fossil in France. This type seems to have preserved in his skull and in his face the primitive features of brutishness, but, in the course of ages, he seems finally to have attained a perfectly upright attitude. We are thus led to think that he must have survived for a long time in the Dark Continent, as the last representative of a very ancient human form, now obsolete, in the midst of the modern black races, several of whom are themselves archaic in type and on the point of extermination.

The physical and pathological characters of the skull from Broken Hill seem to indicate that the owner of the skull did not die very long ago. And perhaps there may be still, in some unexplored corner of Africa, living examples of the last representatives of *Homo neanderthalensis* or his variety *rhodesiensis.* To discover them would not be more extraordinary than the discovery of the Okapi, the large and strange ruminant whose ancestors we have long known from bones obtained from the Miocene deposits of Europe.

Stone Relics in Chili.[1]

Dr Uhle, in the first volume of the *Publicaciones del Museo de Etnologia y Anthropologia*, of Chili, has recently described two sites of the kind referred to (p. 418), one near the port of Taltal, the other near Constitución. The lower beds contain only implements of Palæolithic character, notably amygdaloids; arrowheads of Neolithic fashion are found only in the superficial beds.

[1] See p. 418.

INDEX

D

K

L

N

Q

S

T